DESERT PLANNING

DESERT PLANNING

International lessons

Edited by
GIDEON GOLANY

The Architectural Press : London
Nichols Publishing Co : New York

Dedicated to my dear ones
Simcha
Shaul
Dan, and
Ram Ben-David

First published in 1982 by
The Architectural Press Ltd: London

Published in the USA by Nichols Publishing Company,
PO Box 96, New York, NY 10024

Library of Congress Cataloging in Publication Data
Main entry under title:

Desert planning.

Includes index.
1. Architecture—Arid regions—Addresses, essays, lectures. I. Golany, Gideon.
NA2542.A73D47 1983 711'.0915'4 82–14086

ISBN 0 85139 153 2 (British edition)
ISBN 0–89397–119–7 (United States edition)

Set in 9pt Palatino and printed in Great Britain by Mackays of Chatham Ltd

Contents

Preface

Desert Planning is international in scope and focuses on the urban and regional development of arid regions. More than one third of the total land mass of the world may be classified as arid or semiarid: areas with too little rainfall to support dry farming, very low humidity, greater evaporation than precipitation, great temperature fluctuations between night and day, and (in some) relatively intense solar radiation.

Contemporary civilizations that have primarily established their urban centres in humid and semihumid regions have become increasingly interested in the development of arid zones, as is illustrated by the disproportionate population growth of the American Southwest and the Pilbara region of Australia. Moreover, it is evident that established arid zones are growing rapidly. In the United States, for example, the two cities of Phoenix and Tucson have the fastest growing populations in the country. Egypt, Saudi Arabia, and Iran are but a few of the countries of the Middle East whose populations are expanding greatly in arid-zone cities, old and new.

Early civilizations, such as Mesopotamia, the Indus Valley, Egypt, and Persia, were the first to formulate impressive urban and rural centres under the extreme climatic stress of arid zones. It is unfortunate that few planners around the world have fully utilized the immense body of knowledge which may be gained by studying the experience of the ancients. Indeed, the actual planning and implementation of designs for houses, neighbourhoods and urban centres in arid regions reveal that most architects and planners have merely superimposed theories, methods, and solutions which they practised in the development of humid areas.

Of the few schools that have programmes in this area, most are located either in humid areas and have dealt primarily with problems of climatic stress in humid zones (where most growth has traditionally taken place) or in developing countries of arid areas where experience and expertise are necessarily limited.

The successes of ancient civilizations and limited successes of modern cities and individual countries must be shared. There has been a noticeable gap in the literature on urban and regional planning in arid zones. It is this editor's intent to bring together a large number of leading interdisciplinary scholars who have been involved directly in arid-zone planning and development on both theoretical and practical levels. The comprehensive approach to this subject makes this book unique, and it should appeal to planners, community builders, developers, policy makers, social scientists, economists, and others who are oriented toward problem solving.

The contributors come from a variety of countries, disciplines, and life experiences. Some are theoreticians; some planners, architects, and private developers are practitioners; still others are decision makers, public and semipublic officials. Their topics within the framework of arid region development are as varied as their backgrounds. They encompass such fields as urban planning and design, architectural form, environmental quality, groundwater supply and desalination, social development, the responsiveness of town planners to culture, and many other aspects of the past, present, and future uses of arid zones for population growth. Each chapter defines problems, introduces alternative solutions, and, when possible, focuses on an existing implementation. The conclusion of each chapter includes lessons to be learned from the experience along with recommendations or future orientations to be followed.

Preparation of this work began in 1975, and the contributors were selected from among 250 scholars with whom I have corresponded. It has required an immense amount of resources and coordination because of its international scope. All the chapters were written especially for this book and were designed to meet its goal and objectives. Many of them resulted from thorough research conducted by the authors. Other chapters reflect life experiences and elaborative local observation by authors who probably are the most qualified to comment about their countries.

I would like to express my deep appreciation and thanks to all those who have worked with me throughout the preparation of the volume or have given me their advice on the many technical details involved. My special thanks are extended here to my editorial assistants, who have worked closely with me on the project: Mr Ted Sheckels, now on the faculty of the Department of English at Beaver College, Glenside, Pa, and Mrs Deborah Jean Kuritz, a former English student and instructor in the Department of English at The Pennsylvania State University. Nancy Lord Daniels, my editorial assistant, has worked intensively on the final preparation of the volume.

I am grateful to my colleague, Professor Raniero Corbelletti, Head of the Department of Architecture, for his support and encouragement.

I certainly hope that this book will be of assistance to students, faculty, planners, and developers who are involved in urban and regional planning in arid zones.

The Editor

The Pennsylvania State University
University Park, Pennsylvania

About the contributors

Yitzhak Abt is Director of the Center for International Agricultural Cooperation of the Israeli Ministry of Agriculture. Before he accepted this position in 1972, he coordinated comprehensive and rural development programmes for nations in Asia, Africa and Latin America. Among his numerous publications are planning reports for several governments and a chapter entitled 'Types of Agricultural Projects Used in Developing Countries' in *Rural Development in a Changing World* (1971).

John F. Brotchie is an Assistant Chief of the Division of Building Research in the Commonwealth Scientific and Industrial Research Organization in Victoria, Australia. He graduated in civil engineering from the University of Melbourne in 1950 and received a Doctorate in Engineering from the University of California at Berkeley in 1961. He spent some time as a visiting associate professor at the Massachusetts Institute of Technology. His research includes urban, regional and transport studies including the development of planning techniques and models and measures of benefits and costs. He has authored more than 100 publications in these and other fields.

Miles Danby is Professor of Architecture and Director of the Project Office at the University of Newcastle-upon-Tyne. He holds degrees from the Architectural Association School of Architecture, London and the University of Newcastle-upon-Tyne. He has held academic and private posts in Ghana and the Sudan and has also served as a consultant in Bangladesh, Gibraltar, India, Kenya, Saudi Arabia and Zambia.

Kennith E. Foster is Associate Professor and Associate Director of the Office of Arid Lands Studies at the University of Arizona, where he received his PhD in Watershed Management. He has been involved in extensive research in resource development and management.

Gideon Golany is Professor of Urban and Regional Planning in the Department of Architecture at the Pennsylvania State University and former chairman of its graduate programme. He holds the MA, MSc, and PhD degrees and a DipCP from the Institute of Social Studies, The Hague. He has taught at The Technion, the Virginia Polytechnic Institute and State University, and Cornell University and has served as a consultant and planner. He is listed in *Who's Who in America* and in *Dictionary of International Biography* (among others). Besides numerous monographs, articles and papers, he has written or edited 13 other books including *Housing in Arid Lands* (1980), *Arid Zone Settlement Planning* (1979), *International Urban Growth Policies: New Town Contributions* (1978), *Urban Planning for Arid Zones: The American Experience* (1978), *New Town Planning: Principles and Practice* (1976) and *Innovations for Future Cities* (1976).

Herbert C. Green is Senior Lecturer in the Department of Architecture and former Dean of the Faculty of Architecture at the University of Western Australia. His academic achievements include a MS from Iowa State University and diplomas from Adelaide University. He has also maintained a private practice since 1955. He received the C. L. Alexander Prize in Town Planning and the Donald F. Thiele Prize from Iowa State University. His most recent publications reflect his interest in towns in arid zones.

E. G. Hallsworth is director of an international study of the socioeconomic constraints on soil conservation in the developing countries. Previously, he was chairman of the Land Resources Laboratories of the Commonwealth Scientific and Industrial Research Organization, Australia. He has received PhD and DSc degrees and is a chartered chemist. At the University of Nottingham he served as Professor of Agricultural Chemistry, Head of the Department of Agricultural Sciences and Dean of the Faculty of Agriculture and Horticulture. He has served as President of the International Society of Soil Science. His publications include *Where shall we build our new cities—Definitions and Constraints* (1978), *A Forward Look at Land Use Policy for Australia, and Land and Water Resources of Australia.*

William Jacques Kidd is Professor of Architecture at the Queens University, Belfast. He received his B Arch with honours from the University of Liverpool. His numerous academic appointments include Acting Head of the Department of Architecture in Khartoum, Dean of the Faculty of Architecture at Ahmadu Bello University in Zaria, Nigeria and Visiting Professor in the Department of Architecture, University of Nairobi, Kenya.

Peter Mahony is an associate partner of Conklin Rossant, an architectural firm in New York City. He received his BA in Architecture, an MFA in Urban Planning from Princeton University and a B Arch from Cambridge University in England. In his present position, he has directed and coordinated projects in Delaware, Colorado, Wyoming, New York, Oklahoma, Maryland, Washington D.C., New Jersey, and the new capital city of Tanzania, Dodoma. He has also served as an urban design consultant for the City of New York and for the City Planner of Philadelphia.

J. C. Moughtin is Director of the Institute of Planning Studies and Professor of Planning at the University of Nottingham. He received a PhD from Queens University Belfast and went on to serve as Professor and Head of the Department of Town and Country Planning at that university. He has also served as a Visiting Professor at Johns Hopkins University Center for Metropolitan Planning and Research, visiting architect at the University of Khartoum, and as a planning and architectural consultant. In addition to preparing journal articles and television programmes, he authored *The Architecture and Settlement of the Hausa People*.

Jonathan Eric Reyman is Professor of Anthropology at Illinois State University. He received his PhD from Southern Illinois University. In the course of his research at Southern Illinois University, he received National Science Foundation grants, fellowships from the university and grants and research appointments from Illinois State and Southern Illinois universities. His publications include 'Astronomy, Architecture, and Adaptation at Pueblo Bonito' which appeared in *Science*.

Anatoly N. Rimsha, expert in town planning for the United Nations, Honoured Architect of the Russian Federation, and Member of the Union of Soviet Architects, present Professor of Civil Engineering Institute, Moscow. Professor Rimsha holds a PhD (Arch) and has over 25 years of practical experience in the field of construction and architecture. Six scientific research institutes of the Academy of Sciences of the USSR and more than ten office buildings have been constructed as his projects in Moscow and other Soviet cities. He is a designer of the general reconstruction plan for the city of Kabul and, in particular, its centre and two residential communities as well as of the general plan for reconstruction of Volgograd and other Soviet cities and towns. A consulting expert on building engineering in hot climates, he is author of nine publications, manuals included, on architecture and town planning.

Balwant Singh Saini is Professor of Architecture, University of Queensland, Australia. He received his PhD from the University of Melbourne. He has served as a consultant to several UN agencies and as a member of editorial panels of *Architecture in Australia, Man-Environment Systems, Design* in New Delhi, and *Ekistics* in Athens. A Fellow of both the Royal Institute of British Architects and of the Royal Australian Institute of Architects, he has studied the application of natural and behavioural sciences and systems approach to environmental planning and design. This interest is reflected in two books, *Architecture in Tropical Australia* (1970) and *Building in Hot Dry Climates* (1980).

Lütfi Zeren is Professor and holds the Chair of Physical Environmental Control and is President of the Building Research Center at the Istanbul Technical University where he also served as Dean of the Faculty of Architecture, 1970–2. His M Arch and all subsequent degrees were received from ITU. He studied and did research as a visiting fellow at Princeton University's Architectural Laboratory in 1957–8, and the following year he taught at the School of Architecture, University of Texas. He has been a President of the Turkish Institute of Architects and has practised as a professional architect. He has won several awards in architectural competitions and has published several articles in professional journals and books, primarily on climate and the climatic zones in Turkey.

PART ONE

URBAN PLANNING AND DESIGN

1 Arid-zone development policy and strategy

GIDEON GOLANY

Arid zones are not uniform. There are at least four primary groups:

1 Cold and dry polar zones, such as those in northern Canada and at the poles
2 Cold, continental dry zones, as in central Asia and central Australia
3 Cold, upland dry zones, blocked by high mountains, such as in central Anatolia
4 Hot, dry global zones bounding both sides of the tropics, as in North Africa and the Middle East

Common to all are dryness, diurnal fluctuation in temperature, and the resulting stressful climate.

In this chapter, I am primarily concerned with the last type—hot, dry zones. I define such arid zones as those areas where the average rate of evaporation is higher than the average annual rate of precipitation, where the annual rate of precipitation does not support dry farming, where temperatures between day and night vary greatly, and where high degrees of solar energy radiation are experienced, especially during the summer. The little rain which does fall in the arid zone is sparse, torrential, turbulent, and brief.

INTRODUCTION

Half of the world's states have part or all of their territory within arid zones. The desert contains 15 per cent of the world's population and more than one-third of the world's land mass.[1] Using a different measurement, 22 per cent of all potentially arable lands, or roughly 700 million hectares (ha), are in arid or semiarid zones (table **1.1**). Arid and semiarid areas contain most of the world's oil reserves. By the year 2000, the world population is expected to double; and arid zones will continue to grow and contribute their share to the global growth.

Like many nonarid regions, the arid zones are not homogeneous in their resources. The most dominant of their features is the stressful conditions of their climates. It is also true that most arid land has much empty cultivatable land. Such land requires large investments in fertilizers, in watering, and in special treatments, however, to make agriculture productive, not to mention that water is already scarce. In addition, generally, the wages of imported agricultural labour in arid zones, especially in the newly developed zones, are quite high. Among all the arid-zone resources, the freest and most available and reliable is solar energy, which technologically we are just beginning to be able to use.

Because agriculture in the hot, arid zone requires too great an investment to warrant the attendant risk of crop failure, economic alternatives to agriculture are not only consistent with, but mandatory to the establishment of a secure economic basis for settlement of these hot, dry areas. Despite this, man has usually tied the need for settlement of arid, and especially semiarid, zones to agricultural development. However, the contemporary urbanization process in arid zones reveals that the supportive economy of arid land settlements should have little to do with agriculture at all.[2]

Delicate environment

Despite the arid area's promise and because of the scarcity of water resources and the low humidity, life, especially flora and fauna, in arid land is balanced at the threshold of survival. The equilibrium achieved permits no disturbance if life is to continue.

The study of ancient civilizations in arid zones reveals a close relationship between the ecological equilibrium and the quality of the social order.[3] In the past, the survival of civilization in arid zones depended primarily on man's understanding the delicate environment and establishing an equilibrium between society and that environment. Disruption of this balance, such as through land salinization, led to the collapse of those societies, such as in Mesopotamia and in the Indus Valley where overirrigation eroded the agricultural base. Any development interferes, to a certain extent, with the ecosystem; however, salination is a major problem resulting from the intensive irrigation so necessary in these climates. Also, overirrigation can foster the development of malaria and bilharziasis.

Because of the delicate environment, only industries which can meet the established ecological planning goals and standards should be accepted. Land, water, or air polluting industries must be prevented from being constructed in the region.

Potential of the arid zone

In hot, dry areas there can be seven areas of advantage for potential development: agriculture, recreation, wind energy,

Table 1.1 Potentially arable lands in the arid and semiarid zones

Region	Millions of hectares	Percentage
Africa	177.9	25.5
Asia	188.3	26.9
Australia	84.9	12.2
Europe	2.0	.3
North America	111.1	15.9
South America	53.1	7.6
USSR	80.9	11.6
Total	698.2	100.00

Source: Thorne, Wynne. 'Agricultural Production in Irrigated Areas. In *Arid Lands in Transition*. Ed. Harold E. Dregne. Washington, DC: American Association for the Advancement of Science, 1970. P. 33. Adapted from The President's Science Advisory Committee. *The World Food Problems*. 2 Vol. Washington, DC, 1970, 2: 432–433.

Table 1.2 Arid zone resource potentials and their usage

Resource potential	Usage	Associated issues
Agriculture	Production of food — with dripping irrigation — in greenhouses	High cost of water Necessity for highly regulated water supply system Possibility of high wages for imported labour Low priced land with high fertilization requirements Requirement of advanced technology Good market during off-seasons Provision of more than one crop per year Development of food factories
	Production of livestock feed and graze	
	Development of industrial agriculture	Support of a healthy economic base Stabilization of employment Strengthening of regional cooperation Stimulation of development of infrastructure
	Establishment of fisheries — with brackish water — with seawater — with recycled or mixed water	Possibility for artificial water bodies with improvement of climate
Recreation	Centres for tourists — for archeological sites — for unique geological parks Areas for hunting Health centres for asthmatics, arthritics, etc. Recreational facilities for vacationers	Need for improvement of transportation, social services, facilities, hotels, restaurants and other amenities Investment for infrastructure Potential for seasonal economy
Solar radiation	Development of dehydrated food industry Production of energy — solar energy to electricity — solar ponds to steam for low pressure turbines	Necessity of technological advancement and skill Reduction of daily household electrical consumption
Wind energy	Energy for household consumption	Instability of supply Support of early independency Reduction of imported energy
Air navigation	Crossing routes and centres for national and international navigation Centres for military and civilian training Stations for astronomic observation	Visibility of pollution and increased negative effects of pollution
Mining	Centres for exploitation of natural resources	Possibility of environmental deterioration High cost of infrastructure
New settlements	Centres of urban development	Support of infrastructure of the villages in the region Necessity of research on the suitable settlements and configuration of houses Necessity of construction of lengthy and expensive infrastructure networks Reduction of nomadism Population issues of interregional shifts
	Villages for agrarian—industrial labour and families	Improvement and stabilization of economy
	Villages for recreational facilities	Seasonal fluctuation of population Improvement of services
	Villages for mining labour and families	Disruption of ecology and environment Insecurity of economic base Transience of population and culture

air navigation, mineral deposits, and land for various new settlements (table **1.2**).

The year-around cloud-free sky and the corresponding high solar radiation provide potential for a variety of agricultural innovations, such as food dehydration, grain drying, and evaporation for desalination of water. Grain drying, for example, by air heated by solar energy has been tried at six US Department of Agriculture experimental stations and is becoming a big industry in the United States.[4] With technological improvements in irrigation, desalination, and the use of brackish water, the hot, arid areas offer new potential for food production. In particular, the continuous warmth will enable crops (which have traditionally been grown only during the warm summer season of humid zones) to be grown throughout the year.

Wind energy is an endlessly abundant resource. The intensity of solar radiation throughout the day and the extreme diurnal temperature produce active aerodynamics which support the use of wind energy at the local level at least. This is true in the plains as well as between valleys and slopes. Also, the absence of flora will support the air movement more. Although air movement cannot be constant at all times, it can support the generation of energy in a variety of forms to supply at least housing needs. These forms were used throughout history and are known as: Persian sail windmill (circa 600), 'Dutch' windmill (circa 1100), Mediterranean sail windmill (circa 1400), American farm windmill (1840), Flettner Rotor (1925), Savonious Rotor (1929), Jacobs Wind Plant (1930), Darrieus Rotor (1931), Smith-Putnam Wind Turbine (1941), Venturi Shrouded Wind Generator (1950), Enfield-Andreau Wind Turbine (1954), Hütter Wind Turbine (1961), Windworks 12 Footer (1973), Princeton Sailwing (1974), SST Wind Turbine (1974), Noah Wind Rotor (1974), and NASA 100kw Wind Turbine (1975). Arid zones have much potential wind movement which can be utilized.

The cloud-free, rarely stormy skies of the hot, dry climates could provide an excellent location for air navigation centres, both military and civilian.

Finally, arid zones contain potential space for future settlements and large-scale development. The availability and low price of land, as well as the healthful climates, when contrasted to the existing congestion of urban settlements in nonarid zones, make the arid zone highly suitable for future tourism, recreational and health centres, as well as permanent settlements.

During the development of agricultural, energy or mineral resources and the growth of air navigation and recreation, some form of settlement would be necessary. The creation of urban centres would not necessarily involve any one of the other three specific land uses. Thus, all seven types of potential growth in hot, arid zones lead, sooner or later, to the planning and construction of urban or rural settlements. Unfortunately, a void exists in the development of a method for settlement planning in the stressful climate of hot, arid zones.

POLICY FOR ARID-ZONE DEVELOPMENT

Policy, the generalized guiding principles designed to influence a course of action, is the framework necessary to unify diversified parties under one umbrella toward common goals. Policy, however, is also not one, but several interrelated principles and coordinated systems—not only the framework itself, but also a process determining the means to reach defined ends.[5] A policy must be general as well as functional. Because arid-zone development requires immense resources and effective planning and coordination, it is doubtful that the arid zone can be developed to meet its potential without a national policy. Moreover, because of its delicate ecological equilibrium, the necessity for a coordinated policy becomes essential. The land's mostly virgin and empty nature and public ownership make such a policy feasible. This is primarily true in the less technologically oriented nations.

Although arid-zone development may seem to be a national-local or regional endeavour, the policies for arid land development stem from four universal problems which can be solved if the potentials of arid-zone development, as outlined above, are achieved. These four well-known universal needs are caused by population growth, increased food consumption, the shortage of energy resources, and the increasing consumption of space for the expansion of settlements, especially the urban type.

The world's population is growing faster now than in any other period of history, and a large portion of this growth is in regions considered to be arid or semiarid. Moreover, much of those populated arid or semiarid zones are in developing countries or countries with low technological progress where the *per capita* gross national product is low.

In the United States the region which is growing the most is the Southwest, and the fastest growing cities are Tucson and Phoenix, Arizona. The arid-zone developing countries such as Iran, Egypt, and Mexico have also been signalling serious population growth in the recent decade. Arid land has the definite potential for absorbing population growth.

The periodic famines throughout the world, especially in countries less technologically developed but with rapidly growing populations, are, unfortunately, well known. Food shortage crises associated with rapid population growth make the arid and semiarid zone an attractive alternative space for mass food production as technology perfects the effective use of greenhouses and drip irrigation using desalinated and brackish water for out-of-season crops and for year-round cultivation. Such advances continue to reduce the risk related to arid-zone traditional agriculture discussed earlier.

World consumption of natural resources, especially minerals, is rapidly increasing. A reasonable portion of these resources, such as oil, which has not yet been depleted exists in arid and semiarid lands. World energy crises have accelerated the search for alternative energy sources. Solar radiation is still one of the promising resources, for we cannot deplete the sun's energy as we can that of fossil fuels. Furthermore, use of solar radiation is already technologically and economically feasible, at least on a small scale, for house heating and cooling and even cooking. The humidity of nonarid areas detracts from the effectiveness of solar radiation. In addition to the low humidity, the skies of arid zones are essentially cloudless throughout the year.

Space for future settlement expansion, especially urban expansion, is more and more a pressing need. The existing urban expansion, mostly associated with negative social and economic effects, has been taking over much of the good agricultural land and consequently reduces our basic resources for food production. This process threatens the less technologically advanced countries which lack the management resources to ease the strain.

The development of arid zones seems unavoidable. In fact, urban planners, developers, and decision makers are already involved in such development on a large scale in Iran, the Middle East, the Southwestern United States, Mexico, South America, and Australia. Our knowledge of settlement planning for arid zones, unfortunately, is minimal or non-existent.

Traditional policies

Arid-zone developers have the past and present experience (especially in the developing countries) to look at. In these countries, most existing conditions and experiences are inherited from past generations, and some may go back to ancient time. It is impossible to outline one or even a few forces without considering all comprehensively. The conditions result from socioeconomic, political, and other forces rather than from climatic conditions only. For the sake of our discussion, we can simplify the question by outlining six elements of developmental policies existing in general and in the developing countries of arid zones in particular.

1 Coexistence of traditional agriculture using very low technology with public subsidies or experimental agriculture (especially in semiarid zones) under very risky conditions. Under such policy, development has focused on agriculture of many types and on livestock.

2 Either coexistence with the nomadic population with subsidies provided to them, or sometimes, attempts (usually not

successful) to resettle the nomads without substantial comprehensive planning.

3 An autarkic economy, which is, by definition, not export-oriented; thus, the acceptance of low income *per capita* (supplemented with public subsidies) and a low standard of living as a way of life.

4 Exploitation (and eventual depletion) of natural resources for economic benefit (of the public or private sectors) without any socioeconomic comprehensive plan after such exploitation ends.

5 With the exception of a few countries, little or no research, possibly resulting from a lack of expertise, management, financing, or other resources.

6 Sporadic development or haphazard planning coming mostly after catastrophic events such as drought; no comprehensive multi-faceted planning and development. As a result of these single-shot policies, stagnation and low economic growth have been unfortunately dominant.

New trends

Many facets of new national and international economic, social, and political developments since the end of World War II have brought about dynamic changes which have influenced the arid zone. First, there is the research, planning, and development effort initiated by the United Nations and its various agencies in the arid zone, especially in the developing countries of Asia, Africa, and South America. Second, many of the arid-zone countries who were colonized (such as in the Middle East and North Africa) have become independent. The new political situation has led to the formulation of national planning policies and a native effort to implement them and has generated considerable construction.

Third, the intensification of transportation development (especially air transportation) and the increased seasonal recreation of the middle class in developed countries have accelerated tourism in some developing arid-zone countries (especially those with ancient archeological sites, such as Egypt, Israel, and Iran, and those which serve as seasonal recreational centres).

Fourth, the development of the world economy has increased the dependency among nations, but has also supported the new economic concept that development in general (and that of the mineral industries in particular) may not depend only on local materials but also on imported resources. The local region may have little or none of the consumed materials. The new economic development pattern has proved beneficial to the arid zone when such development becomes part of a dynamic national policy. A good example of such policies is in Israel.

Fifth, the accelerated advancement of new technology has also benefited agriculture and other areas of development in the arid zone. Drip irrigation, greenhouse agriculture, and new modes of transportation are among those developments. Sixth, the improvements in political, economic, and especially health management in many of the developing arid regions have responded to national population growth. Urban growth has been accelerated by the nonorderly urbanization movement which is creating pressing needs for planning and for improvement. Developing countries which have luckily had natural resources, such as oil or copper, have channelled some of their resources to ease this pressure.

All in all, the recent three decades of development have seen an evolution of new trends in arid-zone construction throughout the world. In general, we can point to both a population and a construction growth. It is also evident that this growth pattern will continue in the future. Our study of these trends reveals that we can outline them as follows:

Heterogeneous occupancy

Contrary to tradition, it has become clearer and clearer in many countries in the arid zone that the zone's potential is not primarily to be found in agriculture because the land and the scarce water pose problems. Rather, the potential is for heterogeneous land uses with agriculture being only one possibility. In addition, it has become clear that agriculture in the arid zone necessitates highly sophisticated technology and, above all else, comprehensive planning of the entire region so that other economic activities within the region can support agriculture's revival. Among those activities can be industries related to agriculture such as services and amenities required by agriculture (including the construction of infrastructure and its maintenance and a transportation system guaranteeing supply and delivery through a marketing network).

Energy resources

Since the 1973 world fuel crisis, the world's attention has focused on the importance of world oil reserves found mostly in the arid zones of the Middle East, Iran and North Africa. Also, increasing research has been focused on the possible uses of solar energy on both household and industrial levels. The large number of cloudless days throughout the year in the arid zones make them the primary regions for solar energy use. Solar energy is the freest and most readily available energy in the arid zone. (To a lesser extent, attention has been drawn to the potential of wind energy, which is also available in many arid regions.) Although more research and technology are needed to develop sophisticated and economical systems permitting the large-scale use of solar energy in industries and services, it is already evident that we are able technologically to use solar energy on a household scale. Also, a dehydrated food industry using solar energy now exists in the United States; its existence points to the future possible uses of solar energy systems beyond the level of individual dwellings.

Recreation and tourism

This new trend has accelerated immensely in both developed countries (such as the Southwestern United States) and developing countries (such as Egypt, Israel and Iran). Although these trends are global, the arid zone's warmth throughout most of the year has provided good seasonal vacations for residents of cold climates. Moreover, the ancient archeological sites in many of these arid-zone countries have become new tourist attractions. Recreation and tourism are usually associated with high standards of services (such as hotels and restaurants) and transportation. New job opportunities and economic growth result from tourism. This important ingredient in the arid economy is expected to grow further with the increasing prosperity of the nonarid developed countries and their residents' search for moderate and warm climates for their vacations. The new decreased global air fares will also support this new trend.

New settlements

The growth of settlements in arid zones, both of established and new ones, is noticeable. Phoenix and Tucson, Arizona, for example, are the most rapidly growing cities in the United States in the past decade. Many new cities have been built in the arid zones of Iran, Pakistan, Israel, Egypt, and Australia. However, the lack of sufficient knowledge in planning and constructing cities in the stressful climate of the arid zone is extremely noticeable. Considerable research and experimentation is needed if we are to attain sufficient expertise and establish acceptable norms for arid-zone planning and building. In doing this, we have one major asset: the lessons to be learned from the cities built by the ancient civilizations of the arid zones. This trend toward increased urbanization also entails the development of a wide variety of settlement types in the arid zone: urban centres, agrarian-industrial villages, recreational villages, mining (but not isolated) towns, and, finally, settlements planned on a regional basis featuring mutually interrelated economic and social services.

National new policy formulation

To conclude this discussion of policy we should note that more than that in any other region in the world, arid-zone development or nondevelopment decisions in most countries are made on the national and regional levels. Such policies are made without a profit-maximizing intention. Rather, social goals tend to dictate the policy-making decisions on these levels. It is on the individual or family level where cost-benefit considerations become a major factor in such decisions. However, individual goals may not necessarily coincide with national or regional goals. The planner's role is to reconcile any differences with a view toward the ultimate implementation of the national goals.

Also, policy formulation for the development of arid zones differs from one nation to another since the motivations differ. The following is a listing of some of these motivations, few of which would stand alone. They could, however, command action in various combinations, depending on the nation. Some of these policies may apply only to developed countries, while others are pertinent to developing and developed countries.

1 Exploitation: exploitation of natural resources such as oil, gas, copper, and ore.
2 Growth: population growth pressure and the need for new areas for expansion, including housing needs created by urban expansion.
3 Food: food production and livestock raising.
4 Special population problems: absorption of new migrants and refugees, population dispersement in order to reorder social and economic balance, nomadic societies' revival to insure their survival.
5 Security: response to security and defence needs, including the development of camps for military training and weapons or materials storage.
6 Transportation: support systems for major transportation on water or land.

Despite the differences in national motivation for arid land development, we can generally suggest the common necessary elements to be considered when a basic policy is to be formulated for any arid-zone country, if and when development should occur: (a) public involvement in such development, (b) long-range goals, (c) resource management, and (d) land use control. If the national policy for arid-zone development is generally known to be an outgrowth of and coincidental with the defined goals and objectives, public awareness can support and unify those policies.

Direct public involvement is essential in effective, efficient arid-zone development. Development requires, from the earliest stages, regional planning and large-scale construction in which private enterprise may be incompatible with the national goals and the public good. Furthermore, the development of arid zones requires large investments of money, materials, and labour, especially when no infrastructure or transportation network is available. Private developers may be unwilling to make such investments with little, if any, guarantee of a return. Public contribution to the development can be focused on each of several areas: site selection, land assemblage and purchase, construction of infrastructure and transportation networks, regional and local plan preparation, construction of basic services, and, of course, supportive legislation. In developing more than in developed countries, public involvement is essential for the development initiation and the continuity of construction.

Social and economic considerations require a long-range comprehensive plan when an arid zone is to be developed. When the development becomes part of a national objective, long-range measures become necessary to secure the development's continuity.

Long-range goals and comprehensive planning are essential elements in policy formulation. These are not only unifying elements among the parties involved, but they also permit an overview of the contributing forces, the resources to be consumed, the impact of the development on the ecology, the joint role of the public and private entrepreneurs, and finally the total social and economic benefit of the development. Moreover, goal formulation and comprehensive plan preparation for the region will necessitate the integration of these goals and plans with those of the nation. This integration will lead to the region's designation as a priority one when a country is made up of arid and nonarid zones such as Turkey, Australia, Israel, or the United States. In others where the entire country falls within the arid zone, such as Iran and Egypt and other North African countries, the goal and the planning become an automatically essential core of the national priorities and are tied to every facet of national activity.

Resource management is an essential issue in policy formation for the arid zone, because of the limitation of these resources and, in many cases, because of the strong dependency of the national economy on a single product. Therefore, the limited basic resources, both physical and social, and the delicate environmental balance call for an effective national policy for resource management. Land, water, and financial resources are most important in arid-zone development and, thus, require national management.

Land is one of the most important resources in the arid zones and is primarily owned by public or semipublic concerns. Most of such land is virgin, and its development necessitates a high degree of coordination because of its delicate ecosystem. As in the case of Israel, public land is given to developers conditionally with a prescribed treatment in conjunction with national developmental goals. The developer must introduce comprehensive treatment of the development area throughout the construction process and after the full plan is implemented. When land is given to individuals, the public still retains some control to guarantee its proper use in order to meet the comprehensive plan's ultimate objectives, prevent land misuse by individuals and protect the interest of the general public.

Water scarcity also presents the need for control and regulation, even when water is imported. In the case of Egypt, for example, water limitation dictated centralized power and social order throughout its ancient existence.[6]

Financial resources, an effective tool which requires careful management, affect the implementation of all policies. Financial incentives can be introduced in a variety of forms to promote the ultimate goal of the development—for example, loans, grants, subsidies, abated income and property taxes, and fringe benefits. This has been the case in both Australia and Israel where private industries have been given such financial incentives to attract them to newly developing arid zones.

Resource management requires comprehensive planning and coordinated implementation. Any change in one variable will cause sequential changes in other variables. Under no circumstances can a planner afford to ignore those changes. It is, therefore, the initial responsibility of the planner to predict the changes that may occur.

Land use policy formulation is a good opportunity to ensure successful arid-zone development because most arid zone land is public, abundant, and available. The prices are relatively low. Therefore, purchase, planning and construction are relatively free from the usual financial and legal problems, and from the long fatiguing, complicated negotiations with large numbers of landowners. In addition, arid zones provide wide choices for settlement sites, for an infrastructure network, or for any other type of land use common to all developments.

Although it is in the public interest to attract private developers to ease the burden on the taxpayer, the need for each party to pursue a profit maximizing course will introduce a variety of economic forces and, therefore, a potential conflict of interest among these parties. Despite the existence of a comprehensive plan, it is necessary also to have a restricted land use control system monitored by a public agency.

PROPOSED STRATEGIES

Strategy, as I see it, is the art of marshalling resources toward a definite goal, a system of management, or the courses of action to be taken to enact policy. Thus, strategy is the employment of means, subject to fluctuations as circumstances change, to implement the principles of the stated policy.

Since strategy is sequential to policy, it deals primarily with planning and implementation. Thus, it is concerned with courses of action and the decision-making process. Naturally, strategy deals with future development over a lengthy period. Nevertheless, it also involves the study of past and present experiences in order to derive basic planning and implementation principles to be used as guidelines in line with national goals.

Strategies for arid land development vary from one society to another and differ among geographical regions. We may divide those societies, first, into those technologically advanced, but not necessarily wealthy in natural resources, as in the case of Israel; such societies constitute the minority. In the other division are those societies, not necessarily poor in resources but technologically less advanced, such as Saudi Arabia. These embrace the majority of the population areas in arid zones today.

Despite the national or geographical uniqueness of arid zones, some strategies can, with different degrees of emphasis, be common to all; they are specifically concerned with: 1 water, 2 agriculture, 3 economic development, 4 social and health issues, 5 settlement pattern, 6 site selection, 7 urban forms, and 8 research strategy.

Water

Although by definition, arid zones have limited water supplies, any one arid area may have several sources of water. In figure **1.1** the control systems, problems, and solutions for various water sources are presented. For most categories, effective control, as well as resolution of potential problems, require public regulation; despite the numerous sources, water is still a limited resource. This scarcity has a direct impact on the land use pattern and therefore necessitates a comprehensive treatment of the water supply for permanent settlement as **1.1** shows. Comprehensive treatment can ease the risk posed by the unreliability of any one water source. Settlement development, especially urban development, cannot take the risk of total water stoppage because of direct consumption by the population. Reliance on a single water source can bring any development to the edge of disaster in case of extreme fluctuations. Comprehensiveness can diminish such risk. Such a disaster happened in California where the population was dependent primarily on the winter snow. The winter drought in the mountains caused extreme water deficits in the summer of 1977. Unlike in humid regions, it is necessary to have in the arid region a national control agency with local and regional branches empowered by legislation to regulate all water resources, including ground and surface water. This one agency would oversee all collection, treatment, allocation, distribution, consumption, and pricing.

Since water is scarce and because it is primary in overall development, it is necessary that the economy of the settlements be developed to minimize dependency on water. For example, when selecting industries for development, choose those requiring less water. Water should be allocated primarily for activities which are necessary for daily human survival. Water policy will dictate, to a great extent, the land use policy, including the type, distribution and size of settlements. The intensity of the mutual impact of policies is less in the development of nonarid zones than in the arid zone. Humid areas possess greater variety and, hence, greater flexibility of resources.

Agriculture

In contrast to humid areas, arid zones can only support controlled agriculture. Such agriculture allows cultivation throughout the year to provide out-of-season crops to the market when nonarid zone agriculture cannot provide them. The need for controlled agriculture has spurred development of new production methods requiring a minimal amount of water and innovations in irrigation to improve both water conservation and crop yields.

Extreme lack of precipitation and its low retainability automatically eliminate planning for dry farming. If some water becomes available, irrigated agriculture may become economically feasible through use of drip irrigation and/or the use of greenhouses. A one-half acre greenhouse will have high initial cost but will only consume 1500 m^3 of water annually, as compared to the 12 000–15 000 consumed by an open-space traditional farm of the same area.[7]

Greenhouses, if designed properly, can certainly be integrated within the urban settlement itself. From the settlement planning point of view, greenhouse agriculture has more advantages than open farming, which requires ploughing thus creating dust storms and air pollution in an adjacent settlement on a dry day. The occasional torrential rains also leave open fields vulnerable to intensive erosion.

Other techniques for food production in arid zones are crop specialization and use of farming methods which require a minimum or no water. Such specialization may involve plants which are fast growing to minimize irrigation, self-fertilizing, water conserving, and salt tolerant (such as sugar beets, which can use brackish water) and plants that can genetically adapt and use solar energy efficiently.[8] High value and high yielding crops, those without, or with a minimum of, leaves and those requiring minimal space and water, are adaptable for economical food production in arid zones. However, it might make sense to promote agriculture in a region with more available water, on land freed for this purpose by encouraging urbanization in arid zones.

Economic development

Even if the population density of an arid-zone city is relatively high, the density of settlement distribution in a given arid zone will typically be low. This unique form of regional distribution affects all networks, infrastructure, and transportation systems on the intra-settlement level by increasing construction and maintenance costs. Under such isolation, transportation of necessary goods and agricultural products to the urban centres would be costly, infrequent and time-consuming. To ease these burdens, the arid-zone urban community needs to increase its self-containment and seek maximum possible independence in goods and services. In modern times, of course, complete self-containment is impossible for any city or country. Self-containment, however, should be possible with most, if not all, daily life needs such as food, goods, social and entertainment services, educational facilities, shopping amenities, and diversified job opportunities.

There could be three parallel phases which could lead toward the self-containment of an arid-zone city. First is diversification in the production of goods and services, including agriculture, to fulfill local needs. Second is clustering of settlements in a given arid region planned as a single economic unit. This configuration will foster mutual reliance among the individual settlements within the cluster. Third is development of sufficient storage and refrigerating systems to reduce the frequency of outside deliveries, thus reducing dependence on long-distance imports.

In any given planned settlement, the final degree of self-containment should be achieved as quickly as possible. The planning of such development requires speedy construction and may impose financial constraints on the public developer in the short run since much of his investment will take place then. The process pays for itself in the long run because the more rapid initial development encourages stability and, therefore, increases the likelihood of the settlement's success. Newly settled, newly developed regions are characterized by instability in the composition of the population, for they tend to attract a variety of groups who view their presence in a new region as temporary. Also, the harsh environment may encourage others who came to settle to change their minds after a brief residence. Both may constitute a high proportion of the population, creating an atmosphere of instability in the community and among the labour force. The achievement of the planned level of self-sufficiency for the settlement as a whole has the potential to ease such instability.

The plan of an arid-zone city should be implemented to have a diversified economic base to attract various demographic groups to the community. The interaction and socialization of a variety of social groups, moreover, contribute to rooting the new settlers and minimizing population changes.

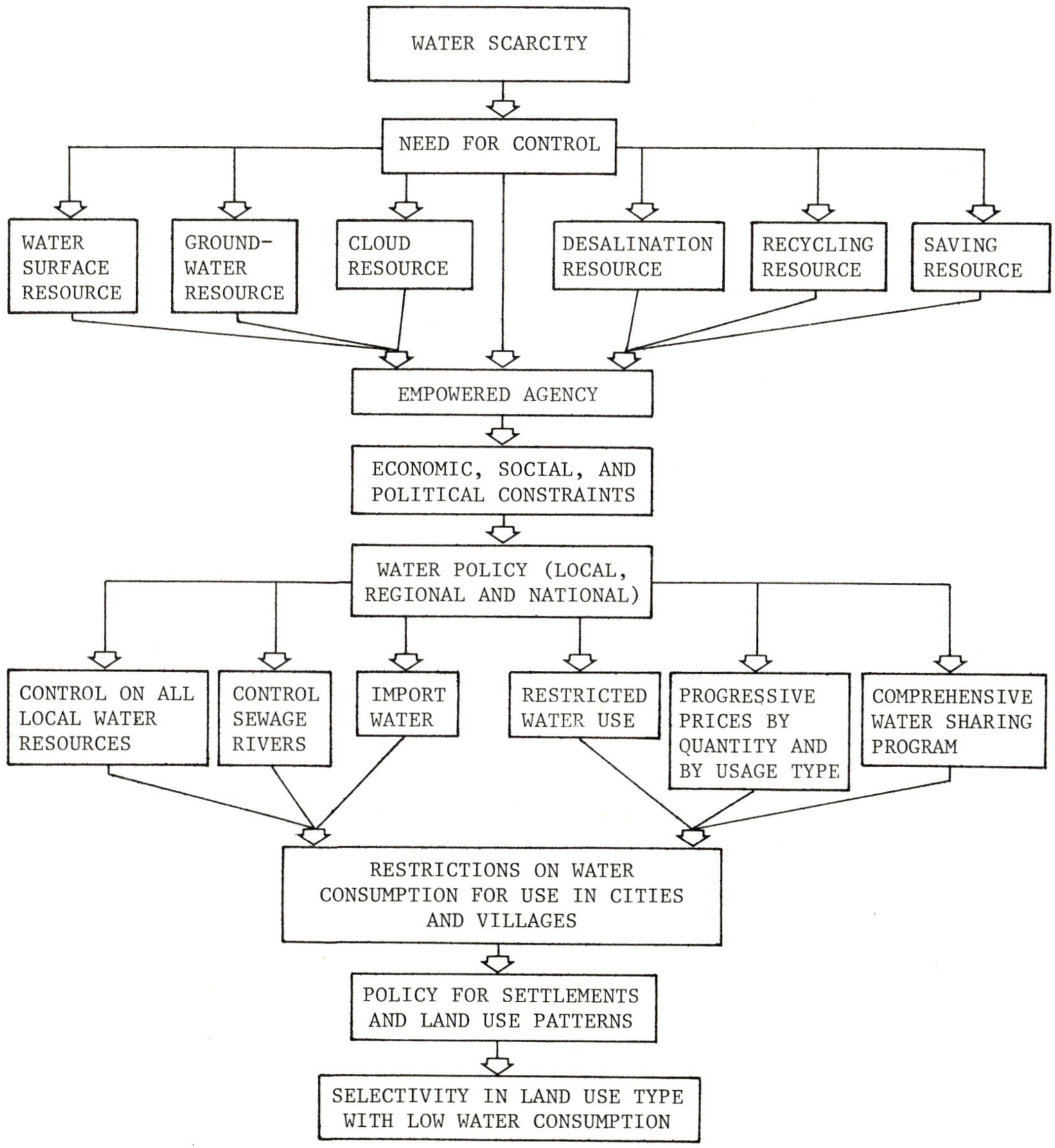

1.1 *The impact process of water policy on the strategy for settlement and land use patterns*

In short, an arid zone should be developed on a regional base, rather than by single settlements at sporadic intervals, in order to mutually support the economic and social success of the entire region, to ensure self-containment of each settlement, and to justify the economic development of a regional infrastructure and transportation network. Finally, a subregional scale of development will also strengthen a sense of unity and ease the psychological feeling of isolation.

If these self-contained arid urban centres maintain their unique dispersion apart from each other within the regional model of settlements, the high costs both of the construction of the regional infrastructure network and of the delivery of supplies throughout the zone will also be perpetuated. This likelihood only increases the need for planning for self-containment in each arid-zone city.

Several other features in different phases of the development of arid zones can increase the initial investment. Consumption of imported building materials will add to the cost and, therefore, will increase the price. The same can be said for food supplies and other daily needs until such time as those items will be produced within the region. Imported labour, both unskilled and expert, who do not plan to stay in the area will demand higher salaries. Until water networks are perfected on a regional scale, most water for daily needs may have to be imported or desalinated, two very expensive alternatives.

Finally, economically, most of the populated regions in the arid zones of the world are characterized by low *per capita* personal income and low gross value in national products. The southwestern United States and some parts of Australia are notable exceptions. The *per capita* income and gross national product of arid zones must not rely solely, or even predominantly, on agriculture which is dependent upon a largely nonexistent local resource, water. Instead, the economy should rely on industry, manufacturing, and tourism and recreation which use local talents and materials. The acceptance of these concepts leads to the acceptance of the strategy that arid-zone development must be in new urban growth centres on a regional scale rather than in predominantly rural villages.

Improvement of systems needed by both manufacturing and service industries in arid zones is important in securing a nonagricultural economic base for newly planned, newly developing arid regions. The transportation network must be made efficient and relatively inexpensive for both people and cargo. The distance between nonarid supplying regions and the newly developing consuming arid-zone cities is usually so great that the development of a land transportation system is costly. Moreover, the immediate need for an efficient transportation system cannot be met by a land system because of the time constraints on road construction. An air transportation system responds to these needs, practically and economically, and will hasten the development of the arid-zone city while the land transportation network is still under construction. A land network will lead to further development of settlements along its routes in the arid zone.

The infrastructure system, such as water supply, sewage, communication, and energy, are all necessary for manufacturing and can be planned and implemented on a regional level in the early stages of development. Furthermore, improvement of accommodations, including hotels, motels, and restaurants, will support recreation and tourism as well as attract the skilled labour base needed for the manufacturing industries and for services.

As already indicated, arid zones strongly attract tourists, especially during their own winter season. This effectively supports the national economies of many arid-zone countries such as Israel, Egypt, and Jordan and becomes an impetus for improving the transportation and telecommunication systems as well as the infrastructure of these arid regions. In addition, tourism has created new job opportunities for the indigenous populations. Tourism can have yet another positive impact on the economy, bringing foreign currency which may be needed for food and will be needed for imported goods. However, tourism must not be the dominant variable in an arid region economy because it is a new form of dependency and is therefore not secure.

Social and health services

The conditions of the harsh, stressful climate of the arid zone demand young, dynamic settlers. Accordingly, the majority of the population will be comprised of young bachelors and young married couples, with or without very young children. Consequently, the strategy of the arid-city planner must be to prepare the first phase of the development to meet the demands of this demographic group.

Some of the settlers will be in periods of transition and, as such, will be prone to move if a suitable rented apartment becomes available. Thus, housing strategy must be geared toward bachelors' and young married couples' needs in terms of size and cost. Also, educational facilities should meet the needs of a large number of nursery, kindergarten, and early elementary school students.

Strategy for environmental quality must be strict throughout the early stages of development. The initial lack of comprehensive medical facilities makes good environmental quality important. Primary health services, however, need to be of good quality at the early stages because of the community's geographical isolation. Also, the demographic composition will require specialized medical facilities to be developed in addition to the formation of comprehensive health service plans to ensure proper medical treatment and prevention of health deterioration. Initially, some specialists might be required to be in residence because of the settlement's isolation. The high solar radiation, the dehydration, and the diurnal fluctuation in temperature require special adaptations for outdoor labour and for group and individual daily activities. Comprehensive health services will encourage and attract new settlers and will persuade those already settled to stay.

Perhaps most important of all in the implementation of a development plan are the characteristics of the people who come to live in the region, for they are a strong influence on the future of the development. Indeed, the likelihood of success can be measured by the extent of the settlers' pioneering attitude and motivation (especially financial), their age group and social class, their skill and management levels, their ability to efficiently absorb advanced technology, and their expectations of a reasonably high standard of living. The quality of social and health services and the availability of labour for construction can also determine the pace of development. Lack of skilled labour will make the project more costly and will hinder the community's social and economic progress.

Settlement pattern

Agriculture has traditionally been viewed by both developed and developing countries as the main economic justification for arid-zone development. Experience in most arid countries, especially in the developed Negev of Israel, reveals that such strong emphasis needs to be changed and a new and additional focus introduced. The new strategy should rely not on scarce resources, such as water, but rather on those resources available in a particular area. Agriculture should not dominate or be the only economic activity of a settlement. Rather, several different types of business, less dependent on water resources, should form the economic core of these new settlements. Manufacturers and services such as tourism and recreation, as well as services basic to life in urban centres, are examples of possible other dominant economic activities which depend less on water consumption. Such a strategy can lead to an increase of urban centres in contrast to the traditional and primarily rural, agricultural centres in the arid zones. Israel has recently seen a greater population living in urban centres in the Negev than in rural, agricultural villages.

Two types of social organization in the settlement of the arid zone have been delineated by Erik Cohen.[9] The first type is spontaneous and occurs during adaptive penetration typical of pioneering settlements. Cohen's second group, with which we are more concerned, is sponsored. That is, it is effected through planned development and involves visionary transformation. Sponsored development is related to social and national goals, rather than only to individual interests.

Strategy for a settlement pattern is related to the infrastructure's pattern which is basic to the settlement's development. Unlike in other regions, the infrastructure's network, in turn, has an impact on the settlement's economic structure, too. Unlike most nonarid zones, the arid zones usually lack any infrastructures, as well as water networks, transportation systems, power networks, or telecommunications systems. Without such infrastructure any development on regional bases becomes questionable. As a basic planning principle, an infrastructure on a regional level should precede any development of settlements in order to secure the continuity of the development. Even in comparatively free market countries where private enterprise has been the major catalyst in the national economy, infrastructure has been the sole responsibility, at least financially through subsidies, of the public. In addition, even when a given arid region is populated, the settlement pattern tends to be sparse and, hence, increases the *per capita* cost of the infrastructure above that of an infrastructure in an equivalent humid region. This means an increase of costs—in investment, operation and maintenance—for the infrastructure, and, thus, the costs borne by the public.

Site selection

From the urban planning point of view, site selection is one of the most vital processes in the success or failure of an arid city. Strategy considerations for site selection must incorporate several needs:

1 Land sufficient to accommodate the planned population.
2 Water resources for daily consumption by homes, industry and services (cities featuring agricultural development will require water for irrigation).
3 Accessibility via standard transportation systems.
4 Local resources to employ at least part of the city's population (in some cases, the climate may be one of those resources).
5 Comfortable climatic conditions for healthy living and working.[10]

The environmental setting of the arid zone encourages the development of a scattered pattern of settlements which, we have seen, increases investment and maintenance costs. Such a pattern of scattered settlements necessitates acceleration of each settlement's movement toward maximum self-containment. Unless national, regional, security, or political reasons preclude such settlements in proximity, clustered settlements can be economically, socially, politically, and defensively a very efficient regional setting. In any case, if such a concept is accepted by the planner-developer, the site selection must, accordingly, meet the required norms for a clustered community.

Since the process of site selection is a complex one, it must involve interdisciplinary experts. Despite the generally harsh climatic stress of the arid zones, sites properly selected and planned can introduce more moderate climatic conditions, aided by physical conditions such as elevation, site position as related to topography and geomorphology, wind direction and velocity, exposure to solar radiation, degree of relative humidity, eolian transformation particles, soil characteristics, and the nature of the precipitation.

Urban forms

The lack of research and, consequently, the absence of knowledge on the unique urban forms required in a stressful hot, dry climate lead many planners to transplant planning methods of urban forms from nonarid regions to arid zones without much of an attempt to make the proper adjustments. Moreover, many of the planners, architects, developers, and decision makers have not learned or used the lessons of the impressive ancient achievements in designing urban forms in arid zones.

It is my strong belief that the new urban form of the arid-zone city should be vertical, rather than horizontal as contemporary nonarid cities are. Verticality, however, can mean both above and below ground. In an arid climate, compactness of above ground structures should be governed by characteristics gleaned from the ancient experience still displayed by many Middle Eastern cities.

1 Houses and other structures should be grouped into clusters to achieve maximum shadow throughout the day. The city should be a nest of interconnected clusters.
2 Cities should be positioned to capture cool breezes and to minimize glare and solar heat reflection.
3 Water, when available, should be used to stimulate cooling and ease the harshness of the landscape.
4 Houses and public units should include an open central court to maximize shadows and ventilation.
5 A pedestrian network should be confined by buildings or trees so as to form a protective environment.
6 Large and unshaded open spaces should be avoided by locating urban land uses near each other.

Research

Indeed, settlement planning and construction is the area which has been most ignored by researchers; yet for practical planning, it is the most urgently needed body of knowledge. Research on special settlement planning and development of special urban forms for an arid zone has been minimal and disproportionately small compared to the research which has been done in nonarid areas. In 1975, a report prepared by UNESCO on arid areas concluded as follows:

> Innovations in design and construction of habitations in urban environments [of arid zones] are still limited. On the whole, the transfer of new technology and scientific findings into action in land management and settlement design was extremely slow.[11]

Land use planning on city and regional levels is necessary and should focus on the proximity of land uses and on special and unique patterns required by an arid zone. The planning and construction processes of a new arid-zone city are also important as are cityscaping and landscaping with native plants requiring little or no water. Another unexplored area of research is subterranean settlements, compact and vertical.

If the greatest potential of arid-zone development is to be found not with the old dream of irrigated agriculture but rather in recreation, solar energy, and new urban growth, then a new set of priorities for research must be set.[12] There has been some research on arid zones, but there is still much more research to be accomplished. Such research must now focus on new disciplines, not on the traditionally accepted ones of agriculture, soil, climate or indigenous population only. Research must now probe the impact of the development of settlements and the arid environment on each other, and in study of flood behaviour, the eolian dynamics of the physical environment, and special settlement forms.

Also important is the research need for an optimal house design to meet the particular climatic constraints of the hot, dry regions. A house, as such, should provide the following:

1 Daily and seasonally controlled microclimatic conditions which will be comfortable and ease the outside harsh climate.
2 Minimum or no energy use in achieving its eased microclimate.
3 Investments equal to an equivalent house in nonarid regions in order to maintain accessibility to low- and middle-income consumers.[13]

Another area for research is the educational environment at all levels—that is, the conditions for outdoor and indoor activities of the different age groups (nursery, kindergarten, elementary, and high school (if immediately required)). Spatial requirements should be determined so as to create a school design which would ease physiological and psychological stress on both teachers and students, correct adverse teaching conditions, and provide good facilities for physical and social activities.

Social and individual entertainment also requires research, for recreation often calls for facilities requiring special planning. We can expect recreational activities in a hot, arid-zone city to be oriented toward water because of the heat. Cities located close to a seashore or perennial river can meet those recreational demands easily. Others can use brackish or recycled water for such purposes. The water can, in turn, be reused in agriculture. Research should also consider the impact of the stressful climate on socialization, settlement patterns, health services, nutrition, and the clothing necessary for community well-being in the stressful climate—aridity, extremes of temperature and dry winds.

The hot, dry climate necessitates a different labour structure than the humid environment. Outdoor and indoor labour conditions, daily working hours, seasonal occupations, wages and benefits, retirement age, technological improvement and labour adjustment are a few of the areas needing research if a citizen is to easily adapt to arid-zone living.

Research on the economics of the arid-zone city should cover the economic bases, incentives for development, priorities in investment, the role of public and private enterprise, and employment diversification.

Agricultural research should focus on helping plants adapt to the stressful climate. In addition to those areas previously mentioned, there is need to discover plants that have higher salt tolerance levels, shorter rates of maturation, drought resistance, higher heat radiation tolerance and improved grain-to-straw ratio. Improvement of soil, increased fertilization, multiple cropping, and improved agricultural management are other areas of agriculture not yet fully researched.[14]

There is much more to be learned through further research into greenhouse agriculture: a closed system scaled to avoid loss through evaporation. We know greenhouses are effective because their use of trapped solar energy which penetrates through the glass and their reuse of evaporated water in the microclimate of the sealed environment to accelerate maturation time together enable several additional crops to be grown in one year. Greenhouses can be used to control scheduled times for planting and harvesting and to meet market demands for different regions. Finally, we know greenhouses use minimum resources for maximum output. Yet, this area requires more research to ensure optimal efficiency and to devise architectural designs so that greenhouses could be integrated with the environment, especially with the urban environment.

Research on water resources is of primary importance. It should cover the sources, problems, and practical solutions outlined in table **1.2**. If solar radiation is to be made a viable universal source of energy on both large and small scales, massive research must perfect the technology we already have so that housing units are energy-self-sufficient. Lighting, heating, cooling, and housekeeping needs could be met by solar energy. Researchers should try to create new, efficient, inexpensive means to harness solar power for use in factories, hospitals, high-rise apartment buildings, and the like. Wind energy is still another resource to be investigated.

Because arid zones are unique in many ways, it is time that an international centre providing a specialized educational programme at the postgraduate level be developed. It should have as its primary goal the preparation of qualified planners for arid-zone development. Such a postgraduate programme can draw students with diversified backgrounds and experiences, including trained architects, engineers, urban and regional planners, economists, sociologists, geographers, environmentalists, agronomists, water and soil specialists, decision makers, public officials, and others. In addition, an interdisciplinary, international and specialized body of experts in diversified fields related to arid zones should form the teaching and research faculty. It should be the primary goal of this faculty to prepare the qualified expert for immediate and future planning and for implementation of arid-zone development plans. An international educational and research centre for the arid zone would be a good investment and a visionary venture. It is urgently needed.

CONCLUSION

This study of arid-zone planning and development needs leads us to the following recommendations as a basis for policy and strategy:

1 The economy of a developed arid zone should be diversified and not solely dependent on agriculture or the dominant natural resources. Diversification in employment leads to social heterogeneity and stabilizes the regional economy.
2 Resources (such as the climate), which have been rediscovered in the arid zone and are beginning to be developed, should be further exploited. Also, some economic activities, the resources for which are not available locally, should be brought into the region to support its economy.
3 Comprehensive planning is most desirable for the proper development of the region. It is necessary to introduce comprehensive planning rather than build spontaneously or plan one aspect at a time. A long-range comprehensive plan which forms a cohesive part of national goals and policy is essential on a regional and local basis. Such plans guarantee continuity and persistence in the development. The mutual ties between the national and the regional goals will not only support the development of the arid zone financially but morally and otherwise. A centralized, fully-empowered agency is necessary for efficient coordination and speedy development of the arid zone.
4 Planning and implementation strategies must be designed so as to make feasible a short initial construction time. The necessary systems for early functioning, such as an infrastructure, a telecommunication system, a transportation system, etc, must be provided. The policy for structuring the economic base of the community in an arid zone should

be based on maximizing the self-containment of the urban centre. Under some conditions, such self-containment can be planned and achieved on the local level alone or on the regionally integrated level. On a regional level, the local settlements will all be viewed as one whole—as one cohesive, self-contained unit. Likewise, settlement clustering in subregions can introduce solutions to the problems posed by isolation, the need for infrastructure network complexity and length, and economic inefficiency. Clustering, at the same time, can provide the basis for self-containment and can also introduce choices among job opportunities and social structures.

5 Neither public investment nor private investment should dominate in the development of the arid zone. Rather, both must jointly be involved in any new venture.

6 Developers must realize that the use or overuse of natural resources (oil, gas, etc) of the region is not only a depletion of these resources but, when improperly treated during processing, can be a serious threat to the environment. It is thus imperative to formulate and implement strict environmental protection codes. There must be more recognition that the arid zone is balanced at the threshold of survival as far as its flora, fauna, physical forms, and land quality are concerned.

7 Last, but not least, research on water desalinization must be intensified in order to make the production price of agriculture more economical. Also, the existing tendency to use brackish local water mixed with fresh imported water should be evaluated, as should the new method of irrigation in agriculture by dripping water and the new emphasis on greenhouse agriculture. This last innovation has brought about a revolution in the planning and implementation, in the marketing, and in cost-benefit analyses of arid-zone agriculture. This revolution may significantly raise the potential population of the arid zone. All in all, water resources still determine the bulk of planning and development decisions. It is this author's belief that the importance of the water issue makes more emphatic the need for comprehensiveness in the planning process and for more control in the implementation of any plans, even in the private-enterprise economy such as that of the United States. This control can be attained because the public sector is still a primary investor at the pioneering stage of arid-zone development.

Intensive and comprehensive research is vital, especially in the areas of site selection, planning processes, urban forms, housing designs, social and health services, and economic structure. Furthermore, the development of a unique type of subterranean settlement which can adapt itself to the harsh environment and stressful climate is a revolutionary concept which deserves special experimentation and research. Establishment of an advanced educational programme which can prepare experts to deal with the uniqueness of urban and regional planning in arid zones is urgent for most nations and should be an international enterprise.

NOTES

[1] Thomas Maddock, 'Drought is the Dilemma: The Philosophy of Water Management', in *Arid Zone Development*, ed. Yair Mundlak and S. Fred Singer (Cambridge, Mass: Ballinger Publishing Company, 1977), p. 21.

[2] Gilbert F. White, 'Unresolved Issues', in *Arid Lands in Transition*, ed. Harold E. Dregne (Washington, DC: American Association for the Advancement of Science, 1970), p. 484.

[3] Gideon Golany, *New Town Planning: Principles and Practice* (New York: John Wiley and Sons, 1976), p. 56, n. 71.

[4] Hoyt C. Hottel, 'Can Solar Energy Contribute to Arid Land Development?' in *Arid Zone Development*, p. 201.

[5] Golany, *New Town Planning*, p. 1.

[6] Ibid, p. 56.

[7] David H. K. Amiran, 'Arid Zone Development: A Case of Limited Choices', in *Arid Zone Development*, pp. 14–15.

[8] S. Fred Singer, 'A Systems Approach to Food Production', in *Arid Zone Development*, p. 178.

[9] Erik Cohen, 'Man as the Initiator of Arid Zone Development: Social and Cultural Aspects', in *Arid Zone Development*, p. 233.

[10] Gideon Golany, 'Planning Urban Sites in Arid Zones: The Basic Considerations', in *Urban Planning for Arid Zones: American Experiences and Directions*, ed. Gideon Golany (New York: John Wiley and Sons, 1978) p. 7.

[11] 'Obstacles to the Development of Arid and Semi-Arid Zones', *Nature and Resources*, 11, no. 4 (October–December 1975), 11.

[12] White, p. 489.

[13] Gideon Golany, 'Free-Energy Cooling Systems for Houses in the Desert', in *Innovations for Future Cities*, ed. Gideon Golany (New York: Praeger Publishers, 1976), pp. 246–262.

[14] Harold E. Dregne, 'Development Strategies for Arid Zone Land Use', in *Arid Zone Development*, p. 258.

SUGGESTED READING

Golany, Gideon. *Arid Zone Settlement Planning: The Israeli Experience*. Elmsford, New York: Pergamon Press, 1979.

Housing in Arid Lands: Design and Planning. London: The Architectural Press, 1980.

Urban Planning for Arid Zones: American Experiences and Directions. New York: John Wiley and Sons, 1978.

2 Town design in the arid Pilbara of Western Australia

HERBERT CALDERBANK GREEN

The Pilbara area has always been a difficult place in which to make a living, and people are not attracted to it. The main settled area, shown in **2.1**, is some 100 000 square miles; with the recent development of the iron ore resources, the population is now 33 000 scattered among 15 towns of 5500–300 people.[1] The 1961 census showed a population of only 3000 in coastal towns, isolated sheep stations, mining camps, and inland towns—some of which now are virtually unrecognizable as towns.[2]

In recent times, the economic base of the state has broadened through industrialization and the development of mineral resources. Deposits of high grade iron ore were discovered in the Pilbara in 1888, but not until 1960 did the federal government allow ore to be exported from Australia, largely in response to the expansion of the Japanese steel industry. As in many other countries, the actual mining has occurred in areas characterized by difficulties of environment. In attracting people to work in the Pilbara, the basic problem has therefore been to provide a satisfactory environment for people accustomed to suburbia with its services, choice of jobs, and range of leisure activities, and an agreeable climate. Considering the geography, developers may have a difficult time attracting and keeping residents.

INTRODUCTION

The Pilbara region consists of rugged rocky hills, outcrops, and flood plains. Vegetation is sparse, consisting mainly of spinifex and grasses on the flood plains, eucalyptus along the water courses, and acacias behind coastal dunes. This vegetation is fragile and not easily reestablished. Major river systems are the Harding, Nickol, and Maitland rivers. The headwaters of the Harding River are located in the Chichester Range. The other two rivers drain the northwest coastal plain and do not extend their water catchments into this range. Each of these rivers is dry for most of the year except for a number of permanent pools. Although artesian water is not readily available in the area, aquifers and borefields on the rivers provide for current needs on a limited basis.

Climatic realities

Even more important than geography *per se* in the development of the arid region, once an economic future has been predicted, is the climate and its effect on people. The consideration of it is thus mandatory whether designing for work or for play. Some differences of opinion still remain about how harsh the Pilbara environment is. However, a consensus seems to be forming: a survey conducted in February 1972 in the town of Newman showed a generally unfavourable reaction, as table **2.1** shows.

The Pilbara lies in the tropical monsoon area and thus has persistently high summer temperatures. The wet season features tropical cyclones, although there are often long, rainless periods even at this time. The spring and winter months generally are dry with cloudless skies and with pleasantly warm days and cool nights. Maximum humidity along the coast averages about 55 per cent throughout the year and decreases as one moves inland. The coastal rainfall in summer averages approximately 2in per month and in the spring (September to November) less than 0.12in per month. Rainfall decreases slightly as one moves away from the coast. Although there are occasional heavy rains, the region is also subject to infrequent and light southerly winter rains. Since the bulk of the rainfall is due to tropical systems, it is sporadic and unreliable. The heavy falls often result in widespread flooding with dry river beds being transformed into torrents. Plant growth then flourishes, and small fish even appear in temporary creeks and ponds. The evaporation near the coast of the Pilbara is high—90in per year, 12in in a summer month and 4in in the winter month of July. Inland evaporation is higher, approximately 120in per year (ten times the precipita-

Table 2.1 Response of Newman (Australia) residents to natural environment[a]

Subcategories	Unfavourable %	Favourable %
Climate	0.8	0.8
Temperature	6.4	0.0
Topography	0.8	0.1
Flora and fauna	1.0	0.0
Dust	1.1	0.0
	10.1	0.9

[a] All survey responses were listed under nine categories of which Natural Environment was one, and were shown as percentages of total responses. There were 63 per cent unfavourable responses; 37 per cent were favourable (see below).

	Unfavourable %	Favourable %
1 Built environment	3.0	11.0
2 Natural environment	10.1	0.9
3 Isolation and access	12.0	0.7
4 Administrative policy	7.2	1.3
5 Community facilities	18.0	1.9
6 Community dynamics	5.4	8.5
7 Economic considerations	5.7	8.5
8 Miscellaneous	0.3	2.8
9 Nonidentifiable	1.3	1.4
	63.0	37.0

Source: Brealey, T. *Living in Remote Communities in Tropical Australia I: Exploratory Study*. Commonwealth Scientific and Industrial Research Organization Report TB 27.1, 1972. P. 141.

tion). Table **2.2** presents typical climatic data for two Pilbara towns.

There are many calm periods in the region resulting in inversions. Dispersion of dust and emissions of various kinds from any future industry will therefore require careful consideration. Cyclones, however, bring winds up to 100 and 120 mph; but once they pass over land, their wind energy is rapidly dissipated. They rain themselves out or move out to sea to regenerate. Coastal towns, thus, are those mainly affected by wind damage. The general area is also subject to occasional small dust storms; much less frequently, large-scale dust storms occur.

The Pilbara, therefore, approximately six months of the year offers pleasant, warm but dry weather (June through November). For the remainder of the year, the weather is generally hot and uncomfortable, with outdoor activity sometimes difficult to endure.

Planning experiences

Growth of the new, single resource towns has occurred without relation to the existing situation. Recently, the state town planning department has been actively concerned with the Pilbara's integrated development. A government report on 'the feasibility of establishing a large industrial complex with associated infrastructure in the Pilbara' had something to say on several topics pertinent to the coordinated development of this area and is the basis for the following comments.[3]

Concerning natural resources, the report argued that the scattered development of the past decade should be replaced by a more intensive type of development where possible. More coordination, it noted, is especially required in the development of the iron ore industry from marketing, transportation, phasing, and fuel usage viewpoints. Coordination

Table 2.2 Climatic data for the towns of Marble Bar and Port Hedland

	Marble Bar	Port Hedland
Rainfall (in inches):		
average	13.00	12.70
highest	28.00	39.50
lowest	2.75	1.25
Wet days:		
average per annum	32	27
Temperature (°F):		
average maximum	96	89
average hottest month	107 (December)	95 (March)
average minimum	67.5	68.7
highest	120	116
lowest	34	38.7
Humidity:		
daily average at 3 pm	26	53
Days over 90°	248	201
Days over 100°	156	62

Source: Bureau of Meteorology, Department of the Interior. *Climatic Survey [of the] Northwest, Region 6—Western Australia.* Canberra: Australian Government Publishing Service, February 1972.

2.1 *Settlements in the Pilbara, Australia*

SOURCE: *The Pilbara Study* (CANBERRA: AUSTRALIAN GOVERNMENT PUBLISHING SERVICE, 1974), sect, 1, p. 3

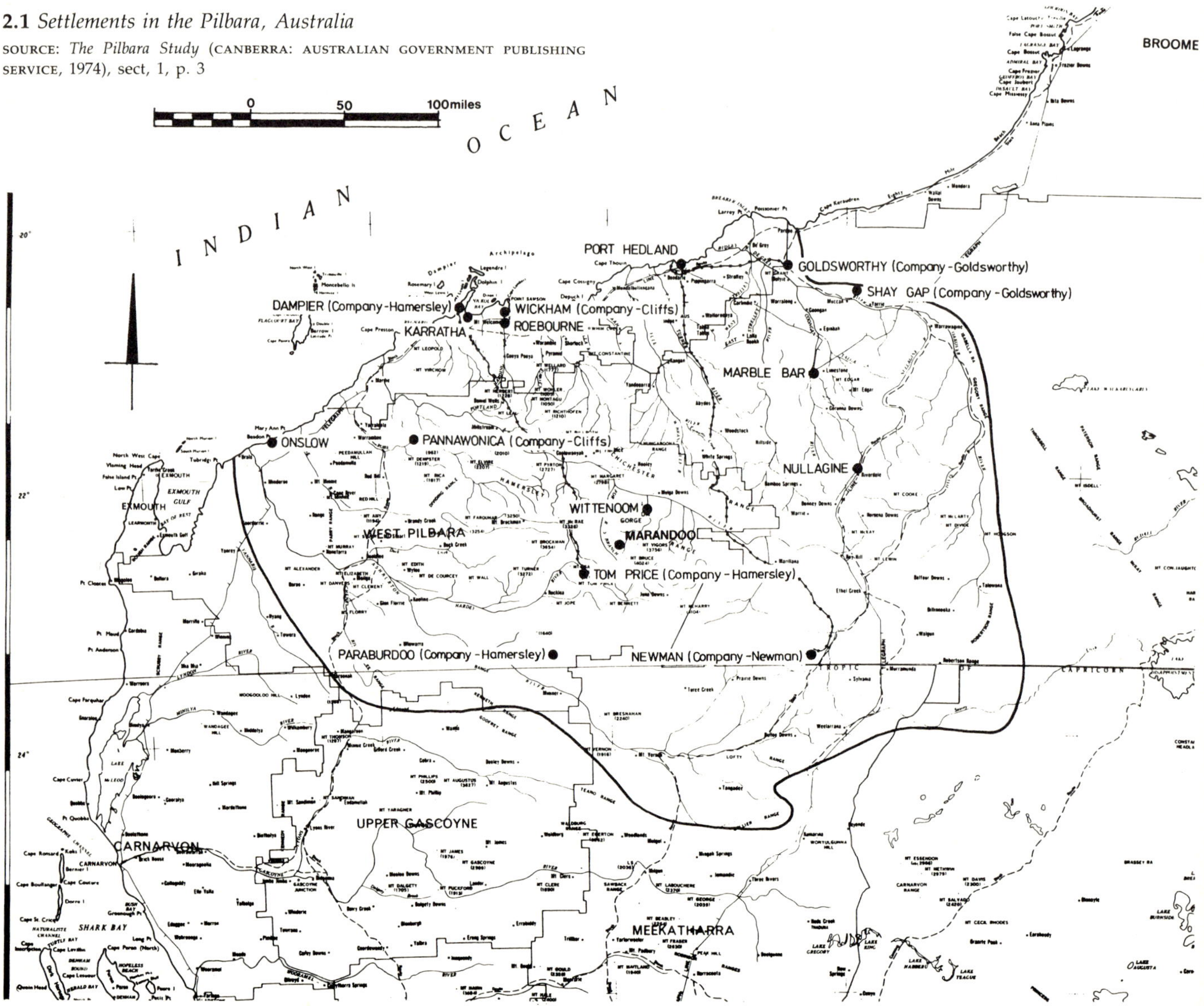

will also be required in the development of natural gas resources.

Concerning the environment, the report noted that the increased production and processing of minerals and the possible concentration of activities in one or two growth areas will place the environment at greater risk, even though the problem of control may be simplified. As for services, the report stated that there will be a growing need for the coordination of infrastructure for both community and industrial needs. However, with the large-scale industrial development expected, coordination would be infeasible without government control and financing. According to the report, scheduling of development will also be necessary to provide for the continuous availability of labour and materials from Australia and overseas.

Vis-à-vis social infrastructure, the report called attention to the wide disparity in the standard and cost of housing in the Pilbara that has not engendered social harmony. Achievement of the high level of social amenities required to obtain and keep personnel in the Pilbara at a reasonable cost will require an approach coordinated by government since it would be beyond the company resources if operating alone.

Other suggestions have been made, eg for the development of regional centres. (Karratha represents a current attempt.) Since the inland mining towns are so scattered, it is impracticable, however, to have a regional centre serve them. A suggestion was also made that lessons can be learned from Israeli arid-zone regional planning.[4] However, no analogy can be drawn directly between the Australian situation and the Israeli. In Israel, small villages are serviced by a regional centre supplying a high standard of cultural, educational, medical, commercial and administrative services and naturally aiding the social and cultural integration of the area. There are particular factors that make this approach possible in Israel. First, the distances are not comparable to those in Australia, and second, there is an ideological obsession among the Zionists to return to their native soil, despite the toil and hardships of life in such circumstances.

Although not large in terms of population, the Pilbara mining towns play an important part in the economy of the state. The mining activities in this area in the early-to-mid-1960s spawned a series of company owned and controlled townships: Dampier (3000 population), Tom Price (3500), Paraburdoo (2000), Goldworthy (1800), Shay Gap (800), Newman (5300), and Pannawonica (750). Port Hedland (5500), already in existence, was further developed as a port and did not become a company town. Other smaller satellite settlements were also established.

The state could not finance the construction of these towns and, not having any experience with such developments, did not then realize what alternatives were available. However, in 1972, these closed company towns were declared 'townsites', enabling local governmental bodies to exercise authority previously denied. Company land within the townsite limits was thereafter leased to the company by the Crown.

Table 2.3 Reason for moving to the Pilbara correlated with intention of staying

	Likely to stay	No plans to leave	Will stay up to 3 yrs	Definite plans to leave	Total	Percentage (n = 209)
Money	19.5	39.1	20.7	20.7	87	41.6
Job	25.5	33.7	19.4	21.4	98	46.9
Environment	32.7	28.8	15.4	23.1	52	24.9
Family	36.5	28.8	19.2	15.4	52	24.9

Source: Stockbridge, Margaret E., et al., *Dominance of Giants: A Shire of Roebourne Study*. Perth, Australia: Department of Social Work, University of Western Australia, 1976. P. 65.

In 1969 when Hammersley Iron decided to limit the growth of Dampier (partly because of the extremely high cost of further developing the rocky site), nearby Karratha was developed as an open town by the state government in conjunction with the company. It now has a population of 5000, and the state government is making it the regional administrative centre. The government has always been responsible for educational, medical, and police services in these towns and requires company contributions to pay for such services and for the cost of housing the necessary staff. Other new open towns are Wickham (2300) and South Hedland (5000), where the inevitable contrast between noncompany and company housing has made integration difficult.

TOWN PLANNING

The development of the remote company town emphasizes that today's social and technological changes can occur too quickly for the inhabitants to adapt, a phenomenon sometimes observed in small but older communities in other parts of the world. When the natural environment is unsympathetic to civilized man's accustomed existence, the task of the town planner is compounded. To bring people to such a barren land as the Pilbara to work requires incentives large enough to overcome the average person's need to be able to choose when, where, and how he will spend his leisure time.

Money is one of the main reasons for establishing a new mining town—basically the only one for the company. People, however, want to live in an environment that is not oriented toward work alone. Still, the two main reasons people go to the Pilbara are money and job satisfaction, as indicated by a survey (table **2.3**). Respondents who listed 'environment' and 'family' among their reasons for moving to the Pilbara tended to be readier to commit themselves to the settlement, although the 'environment' also caused most of the definite plans to leave. Some of the respondents were disappointed with the suburban appearance and life-style in the Pilbara, having hoped for a more tropical, romantic environment.

Early problems

At first, companies did not consider town planners necessary since they felt the engineer and surveyor could lay out a conventional subdivision with appropriate roads and services. However, there has been a considerable increase in the appreciation of the problems in planning a town for the Pilbara since then. Mistakes were made, and many of the early towns suffered as the need to start production overrode other considerations. The companies were at fault as well as the mining professionals employed, who, in many cases, were ill-trained to deal with the new problem. The extraction part of the industry usually performed competently since many of the mining personnel were experienced in setting up mining processes both in other parts of Australia and overseas. Unfortunately, this experience gave them authority, and those responsible for the planning accepted criteria and priorities that were at odds with principles appropriate to the situation.

Sometimes, the company contracted with a construction firm to design and build the housing and associated facilities. Housing then was provided quickly, and the time before getting into production was reduced. The initial high turnover rate of mining labour (up to 200 per cent), however, was most unsatisfactory. In other cases, the state government housing commission provided housing plans and other town planning services for the mining company. Later, companies had to pay closer attention to the problem of town design and the needs of the mining town employees in order to keep a stable, satisfied work force and to maintain efficient operations. The companies believed that houses—the standard suburban bungalows—arranged in the usual suburban patterns and air conditioned would make people psychologically comfortable and thereby minimize any adverse reactions to both company control and the environment. It was not to be quite as simple as that, although these houses were usually commended by tenants. The Goldsworthy Mining Company town at Shay Gap (to be discussed later) reflects a later attitude of wanting to do something better. Their first town at Mt Goldsworthy featured parallel streets along the contours of a hill with shops and amenities at one end.

Although it is clear there are more important factors than the built environment in influencing attitudes toward living in these towns, it has proven highly important in attracting and keeping people for as long as possible, particularly *vis-à-vis* families and women. There have been over 20 new mining towns developed in various parts of Australia within the last decade; and with this experience, planners and architects are locating, planning, and designing housing for such towns more successfully. They better understand the interdisciplinary nature of the problem of providing the best possible habitat.

Site selection

In the rush to get into production, the mining town's location was often based on inadequate data. This forced the planner to solve imagined rather than real problems, for in the Pilbara there is often little available information. Aerial photography, however, has proven most useful in selecting sites when used in conjunction with ground surveys. The priorities among a list of criteria can never be fixed absolutely, for each possible site is unique. It thus becomes a matter of weighing the advantages and disadvantages of each prospective site. There are a number of quantitative methods which can assist in this process, but for the mining towns, sophisticated methodologies were—and are—not used. A list of criteria for use in selecting a mining town site is presented in table **2.4**.

Table 2.4 Criteria for site selection

Topic	
Topography	As attractive and interesting as possible, viewed from and within the town
Slope	Not excessive, but providing view of general activities, aircraft arrivals, etc, and good drainage without erosion
Soil	Adequate for plant growth
Geology	Suitable for supporting building and establishing foundations
Climate	Not emphasizing the disadvantages of an arid climate
Water	Available for mining operations
Landscape	Lending itself to the creation of an oasis
Pollution	Removed from possible contamination, such as dust and noise from mining operations
Historic sites	Guaranteeing the preservation of any aboriginal sacred areas and historic landmarks
Transportation	Suitable for access by new roads and an adjacent air strip
Utilities	Suitable for locating effluent lagoons nearby
Land area	Allowing for conceivable expansion

One often overlooked criterion would be land necessary for growth. While many of the mining towns have no such prospects, others will experience a limited expansion. Service towns, such as ports with associated industrial operations, will be subject to some growth as known resources in the Pilbara are further developed under government stimulus. Port Hedland, for example, has expanded and become the administrative centre of the East Pilbara; in 1974 the shire had a population of about 10 500. Another town, Exmouth, just south of the Pilbara region, has been widely criticized because insufficient attention had been given to land for possible future expansion.[5]

The area required can only be determined after the density of the proposed housing and the amount of conceivable expansion are determined. Although usually only a small area is needed, it often proves difficult to find a site that is both close to the mining operation without being affected by it and meets the other established criteria. The towns of Goldsworthy and Mt Newman, for example, both suffer from fallout of red ore dust because they were located on the leeward side of an open-cut mining operation and not far enough away. Many similar mistakes duplicate those made in other parts of the world when selecting a single resource town site. Kambalda, a town further south near Kalgoorlie, was perched on top of another ore body as had happened in Canada.[6] Properly considered action, no doubt, could have prevented these repeated errors; but when time is of the essence, it has priority over thorough investigation. Sometimes it costs to gamble.

The tangible and intangible have to be synthesized in the town design. Thus, if the town planner has an ideal concept in his mind, it would be extremely fortuitous to locate a site onto which the concept would easily fit. The site, however, has to be chosen with every advantage possible. Hopefully, the many elements of a town may be organized into a plan that to some extent fulfills the designer's dreams. The plan must, however, first and foremost provide a situation into which the inhabitant can fit and live with the minimum of adaptation and stress from the natural and the built environment.

Land use planning

Housing

Site selection and the proposed type of housing are interwoven as are housing and its spatial organization on the site. The traditional subdivision concept dictates how the housing must be laid out. In the arid northwest, some latitude is allowed in meeting regulations, allowing planners to arrange the housing not necessarily according to a particular lot size, but as the planner and architect see the need for exterior space between houses and relative to the house itself. Subdivision delineation, where legally necessary, may be a subsequent operation or, alternatively, may be evaded.

Two alternatives to the conventional subdivision that have been considered are Radburn style separation, pedestrian ways with underpasses to link areas and the car parked at the house (similar in principle to the subdivision scheme used for Radburn, New Jersey, USA, by Clarence Stein); and pedestrian precinct development, a peripheral road system and the motor vehicle kept completely out of the housing area except for emergency purposes. The peripheral road system has been applied at a small settlement of 800 at Shay Gap where the size made such a solution feasible. It is inconvenient to walk from the car, often in the heat of the day; but savings in road cost resulted, however. There have been other attempts with varying degrees of success to segregate the pedestrian from the car, but none have achieved such complete separation as that at Shay Gap. Some planners doubt separation is necessary where traffic is slow, clear views and bicycle paths are provided, and car parking is not permitted on roadways.

Roadways
Most planners are aware of the standard technical and physical requirements of a town plan, such as the location of road junctions, drainage and service easements, etc.; but standard requirements are not always applicable in the case of a mining town. Road width requirements must be questioned, however. In a new subdivision where adequate area for parking is provided off roadways, roads can be minimally adequate in width and so designed that through-traffic is discouraged. Excessive road widths are a source of sand and dust problems, and roads 36ft wide are unnecessary. There is in fact no reason for roads to be wider than 18ft. In some existing cases, the roadway proper has been reduced, but the road base has remained at 36ft to permit future expansion. In these cases, selected roadways in the system should be allotted for this expansion, and these should be on the periphery of residential areas. Traffic movement has generally received precedence over other factors in most towns; occasionally however, drainage factors have, as at Karratha, dictated the street pattern.

Landscaping
Numerous examples exist of European town planning principles being applied directly to new towns in another climate and culture with disastrous results. A classic case is the new town development in the northern sector of Beer Sheva. The planners did not see the consequences of following principles established for places with abundant rainfall. The 'single storey houses with extensive plots of land, winding streets without any shade but with a generous provision of open spaces all modelled on an English garden city were bound to be doomed in the very heat of the desert . . .'.[7] The majority of Australians and most Western peoples do prefer the detached dwelling unit on its own lawned grounds planted with trees and shrubs; these areas, with adequate water and attention, are pleasant and inviting, providing space for the family to entertain and play with some measure of privacy. In the absence of available vegetation, however, the spaces become deserts; the walls of the house become the only sources of shadow.[8]

The large open spaces in arid town plans often look good on paper with plenty of trees, shrubs, and lawns; but most become dusty, hot areas which no one wants to use or look at. This is because the water and/or technology necessary is not available to maintain such greenery, or horticultural knowledge may be faulty. As a result, landscaping schemes cannot be executed as visualized. Significant efforts have been made by landscape architects to solve these problems, and suitable techniques are being developed which appear to have a high potential for success. Large green spaces will, however, always be an extravagance in an arid zone.

House design
In determining the type of housing appropriate in the Pilbara today, designers have not arrived at a particular form. The earlier efforts to imitate suburbia, as previously mentioned, were not successful. Yet, some architects are still designing modified versions of such housing in which the surrounding land area is reduced considerably. Another response has been to reduce the exposed wall area by using single-storey row housing with small front courts and rear yards. The resulting higher density creates a compact pedestrian town.[9] While a two-storey design would have the further advantages of being more compact and having less roof exposure and heat gain, the Australian preference is predominantly for single-storey housing.[10]

Some planners and architects still hold on to romantic visions based on the closely packed two- and three-storey Moroccan towns—'forms in harmony with the sun'.[11] The high population densities of these desert cities have been credited to possible population pressures and also to the continued pressure from the desert itself which has resulted in cultural adaptation to high density.[12] These people are thus able to better adapt to crowded, noisy spaces and naturally choose a close, intimate physical proximity with other people in public areas. To place such a person in the Australian suburban context would mean virtually isolating him from the physical and social interaction he enjoys. Similarly, placing an Australian in such a high density situation would not be an attractive prospect for him.

The old inhabitants of Australian arid areas have learned to live with the environment by using whatever natural methods they could find to assist in keeping the homestead as cool as possible—wide shaded verandas, thick walls, and orientation to control the sun and to catch breezes. At one blow with today's technology, the major problem of climate control in housing design for the arid zone is solved—at some expense. For example, the older residents at Roebourne, the 'old capital' of the region, indicate that they manage the heat problem fairly well without air conditioning. Most newcomers, however, cope with the climate with the aid of an air conditioned house. However, in Roebourne, very little business is done between 1 pm and 4.30 pm when people take the siesta traditional in many hot countries. This appears to be a sensible idea, but it is not amenable to the mining town's routine.

With the application of virtually unlimited technology (limited only by cost), the designer of a mining town has a fairly free hand. All sorts of subtle influences affect the designing of a house for the employees of a large corporation in a remote arid area, but the designer still has a wide choice of alternative directions. Success or failure becomes a matter of his talent, experience and insight. Design of the house then is a problem of form, modified by the social, political, economic, moral and aesthetic aspects of the culture. Consideration, nevertheless, has to be given to materials, shading and orientation in order to reduce the cooling load and provide an attractive view. Planning thus becomes a matter of integrating building science requirements with design ideas. There are always, of course, lessons to be learned from history; such lessons, if they are relevant, can be applied in original ways by the gifted designer. Thus, given sufficient money, nearly ideal physical solutions are possible—once the problem is properly defined. Advantages offered by a design, however, have to be carefully balanced against the cost incurred; therefore, practically all final housing designs represent in some way a compromise with the ideal.

Social planning
Without substantiation, a variety of assertions have been made regarding the problems of living in a company town in the Pilbara. Although this does not make the assertions necessarily incorrect, there is room for doubt. Recently the results of research work done in the Pilbara have become available.

In a survey made during February 1974, a majority of informants thought the major social problem was marital breakdown.[13] About 75 per cent of respondents, male and female, felt the Pilbara was hard on marriage, although only 40 per cent reported having experienced difficulty themselves. People come to the Pilbara with their hopes and problems: many couples are already in conflict. Establishing some financial security can solve problems, but this does not always happen. Furthermore, the shortcomings of life in an isolated arid-area mining town can intrude more and more into the continuing hopes of financial reward. This intrusion of economic aspirations is doubtless why living costs also are high on the list of disadvantages of living in the area, even though housing rentals are subsidized by the company and usually represent less than 10 per cent of a worker's gross income. A major cause of marital problems cited was the presence of different behavioural norms such as heavier drinking, freer interaction between married and single persons, and greater promiscuity. Other contributing factors were the limited kinds of entertainment, the limited shopping areas, the absence of older people, the different adjustment speeds of spouses, the demands of shift work, the long hours, and the involvement of the husband in all-male groups to the neglect of the wife and family activities. The climate and physical environment were lesser considerations, but they were felt to restrict family activity, causing short tempers and some health problems.

Results of another part of the same survey in the Pilbara are shown in table **2.5**.[14] The overall indication is that quite a few people felt the 'country life' of the small town was an advantage. Disadvantages listed were three times the number of stated advantages, and some respondents obviously found it difficult to list any advantages at all. Companies had hoped that by mixing the housing for staff workers and labouring personnel, one particular social problem, stratification, would be avoided; but they are finding that their plan did not work since the staff personnel living among the workers seem to maintain a social distance to reinforce their authority. Attitudes are always of more consequence than appearances. Although the physical environment may be well-planned with the very best of intentions, it cannot ensure that people will be happy and productive even if it does offer them a better chance to be so. Further stratification was due to perceived differences between company and noncompany homes in open towns, between migrants' and English-speaking persons, and between single men and married persons.

Table 2.5 Advantages and disadvantages of life in the Pilbara

Reasons	Percentage (n = 53)
Advantages	
Country life	26.4
Good amenities (including housing)	22.6
High living standards	15.1
Climate	7.5
Beautiful country	5.6
Free way of life	3.7
Good promotional opportunities	1.8
Challenge	1.8
Disadvantages (excluding marital problems)	
Social stratification	45.2
Living costs	22.6
Long work hours	18.8
Climate	16.9
Isolation	16.9
Intertown jealousy	15.1
Impermanence	13.2
Sexual imbalance	13.2
Age imbalance	11.3
Paternalism	11.3
Lack of choice	11.3
Schooling	11.3
Gossip	7.5
Deviant behavior	7.5
Lack of outside interests	5.6
Venereal disease	5.6
Medical services	1.8
Aborigines	1.8
Bad mail services	1.8
Instability	1.8
Gambling	1.8

Source: Stockbridge, Margaret E,. *et al. Dominance of Giants: A Shire of Roebourne Study*. P. 19–20.

The long work hours were reported as a problem largely by the women. In time, the towns with a foreseeable future will probably assume a more normal work pattern; but where the mining activity is comparatively short-term, overtime will remain the order of the day with the worker caught between the opportunity for extra money and the loss of time and energy for the family. Women also appear to suffer most by moving to the Pilbara; it is, therefore, for women that adequate compensating amenities should be provided. The normal social environment for families also no longer exists. In the unique circumstances of the Pilbara, adaption can be slow and stressful, depending on the personality and quality of home life the person brings with him or her. The first 12 months are the critical period for adjustment. It is much easier for the man since he has his work. Loneliness, boredom, and ill-temper, are all symptoms of the woman's experience alone.[15]

There have been several reports that in the larger towns which have a hotel, many wives spend much of their day drinking. This may be due to the above factors in addition to the lack of opportunity for work with a company which typically can only employ about 10 per cent of the wives. The more enterprising women start small businesses such as hairdressing and selling jewellery, cosmetics, etc. It is difficult to say how many are in the habit of spending most days at the hotel, if one exists; but where the habit is pronounced, it is symptomatic of the other problems in their lives.[16]

The single man is another special problem. For him, life in the Pilbara is generally a matter of working, eating, and sleeping, with the remaining time spent playing sports or pool, drinking, or reading. This constant pattern of life understandably becomes monotonous over a long period.[17] The imbalance of the sexes is another difficulty and will remain one unless the Australian culture accepts more women as manual workers. No conclusions were reached in Stockbridge's survey concerning the single man and his view of living in the Pilbara because of the statistically poor sample. The intention to stay in the Pilbara, however, was clearly shown to be associated with marital status.[18] A number of possibilities can explain why the single man does not stay. One is that many are unskilled, and to them the job is an opportunity to work intensely for a two or three year period and accumulate some capital.

The single men's quarters can also significantly affect their attitude. Usually, they are housed in dormitory-like facilities. If this is necessary, a group of eight is probably the optimum number.[19] If there is a larger number, the group tends to fragment. If it is very large, the men do not register ownership; the common lounge becomes 'the' restroom not 'our' lounge. Some tentative attempts have been made to integrate the single men's quarters with the main housing areas, but they are unlikely to be successful considering the different life-style they have compared to married couples with families. The inclination is likely to continue to be to isolate the single men's accommodations from other housing. Some other new approaches are dividing the single men's accommodations into single person bachelor flats and single men's quarters.[20] One or two companies have provided their single staff, as opposed to the workers, with motel-type accommodations, but this has only been in one or two towns for which a substantial length of life is forecasted. Where these rooms exist, they are much preferred, despite a considerable increase in the weekly rate.[21]

Another survey of mining towns was made in February 1972.[22] The aim was to chart favourable and unfavourable responses to living in such towns. The results for the Pilbara town of Newman (population 5300) are shown in table **2.6**. Among the favourable aspects of living at Newman, it is interesting that no 'country life' category emerged. This has always been a quality of life consciously appreciated as such, given its limitations (as the first survey showed). Where there is the possibility of a town's future, it is likely that the major core remaining to form a small stable community will consist of those who happily accept this way of life. As with all country towns, however, staying also means that many will also have to accept their children's leaving, either to go to a tertiary educational institution or to make their own life elsewhere. Faced later by these circumstances, some families are finally prompted to leave the region. Perhaps as a result, the mining towns have very few old people in the community; those over 55 years of age represent not much more than 1 per cent of the population, while they form over 26 per cent of the state's population. The demography also shows that approximately 85 per cent of the population ranges in age from 21–40 years.

One recent study argues that a new town has little chance of survival without the expectation of future growth and the future-oriented attitude of residents, in spite of all the attractions and incentives offered.[23] This view, however, does not account for the small town where people are able to establish a sound financial future and can appreciate what the particular life-style offers. As cities in Australia become larger and more inconvenient and uncomfortable places in which to live, it is possible that more small stable communities of this type will form in the Pilbara. Periodic vacation visits to the metropolis must be subsidized, however; for country folk have always made their visits to the big city to make special purchases and to supplement to some extent the limited social activities of their towns.

Not only in their analysis of such trips but in their design of the small towns, planners and companies must realize the importance of social amenities. It is the architect's responsibility then to design interesting and attractive commercial and recreational buildings in a suitable setting. Today, such community facilities are provided by companies and operated under contract (subsidized as necessary). They invariably include a shopping centre (with the number of small shops dependent on the size of the town), a service station, a community centre, a social club, possibly an outdoor cinema, and—for larger towns—a hotel/motel. Television is also provided by those mining companies who have private transmitters. Church sites are provided, and often the company will contribute towards building costs. Recreational amenities generally include most of the following: swimming pools, playing fields, tennis and basketball courts, a bowling green, and (where possible) a small golf course. Night lighting is provided so that the courts, etc, can be used in the cool of the evening. The forming of social and recreational groups is encouraged by companies; they assist in a number of ways without assuming too large a role in the operation, usually under the aegis of a recreational association of elected members and company nominees.

Generally, the amenities are minimal, but they are adequate and compare well with those in the usual small country town. The main difference is the isolation of the mining town. Since in many cases people are not able to get to another town easily, much more emphasis should be placed on the quality and extent of the amenities and services provided. What seems especially important is the need for more entertainment facilities and a restaurant or (at least) a coffee shop. Better community amenities would assist in giving the town a sense of identity, even though an increase in shops and services could not be justified economically.

While some of the indicated problems are intractable, others are subject to modification. Cost-benefit analysis techniques typically used have never satisfactorily dealt with the intangible areas of life, and decisions remain with the mining company. Compensatory benefits are practically always limited to the amount of money that can be allotted so that a stable work force may be maintained. The planner, if involved, can never make the isolated small mining town a utopia, of course. What he can do is push the companies to allow for as much choice as possible in areas of social activity and for the development of a social structure in the community. It is extremely difficult to change a person's behaviour or attitudes, but the context can be organized physically and socially so that he or she can adapt more easily. In doing this, the planner must have the advice of the experienced social scientist.

Table 2.6 Responses to living in Newman (ranked)

Reasons	Percentages
Favourable	
Buildings	9.5
Wages, savings	6.7
Work opportunities and relationships	3.3
Social relationships	3.2
Unfavourable	
Entertainment and recreation	10.1
Temperature	6.4
Undefined feeling of isolation	5.4
Cost of living	5.4
Shopping and commercial services	5.4
Social relationships	4.7
Transportation	4.5

Source: Brealy, T. *Living in Remote Communities in Tropical Australia. I: Exploratory Study*. P. 133.

Environmental considerations

In arid areas, water and vegetation have always been elements prized in counteracting the oppressiveness of the climate. Practically every Pilbara new town plan has demanded successful landscape development for its fulfilment. Previously, all the planner could do was to indicate that planting should proceed, without any confidence as to the results. However, with a better understanding of the cohesive, harmonious order of the natural landscape, he can do more.[24] When man enters the natural scene, equilibrium is upset. A successful landscape plan, therefore, creates a new order.

The landscape architect should initially make an ecological and taxonomic analysis of the indigenous flora of the area. Secondly, he should review other native species and finally, exotic species in order to isolate those most suitable for cultivation under existing edaphic and climatic conditions influencing plant growth. The exotic species which one may find desirable to introduce into the landscape design will require water, fertilizer, and maintenance throughout their lives, whereas adapted native species can be self-maintaining within three years. Shrubs and trees in some cases may require onsite trials before a major expenditure is undertaken.

Plants are acquired from commercial nurseries when possible and, if necessary for acclimatization, grown on or close to the site. Often the majority of the plants required are not commercially available. Adapted propagative material from the locality is then, if possible, supplied to the nursery in order to produce the chosen plants. Most trees, shrubs, and ground covers are most sucessfully propagated by seeding. The availability of seed, however, sometimes restricts the selection of species for initial planting. It is therefore essential that steps are taken at an early stage to ensure the collection of seeds of desirable species at the times when they are mature. Arid species are renowned for being desultory in their dissemination habits and thus need constant surveillance. The seed of a few of the adapted native species are sometimes available from commercial sources; but some seeds, if available at all, may only be obtained from noncommercial seed collectors such as the forestry department, botanical gardens, and wild flower nurseries. Where cuttings are used for propagation, the knowledgeable selection from the best available plants of a species ensures the reproduction of uniform, true-to-type progeny of high quality. The collector must understand the type of material most suitable for propagative purposes for each species and the optimal time for the gathering and striking of these cuttings.

Arid areas of the Pilbara have little available water; where an irrigation system is necessary, very special consideration is required if maximum water economy is to be attained. By keeping surface evaporation to a minimum (mostly by below-ground methods and by controlling the time of watering) and regulating the supply of water to suit the particular requirements of individual species, it is possible to strike a balance between optimal growth and water conservation. To attain this balance, the developer must have an understanding of the physiology of the plants involved and the effects that particular microclimates and soils have upon their water requirements. We should note that by using adapted species to reinforce the existing landscape, we need to water the plants only during the first three seasons. After that they are self-supporting in the climate of the area. Maintenance is also particularly important in establishing these plants' lives, and staff has to be available and trained by the time planting starts. During this 36 month period, there must be sufficient water, protection from damage or destruction, control of weeds and pests, and any necessary remedial measures.

The new landscape should complement the existing topography, accentuating the site's positive qualities and creating the impression of a man-made oasis. Where housing meets large-scale open spaces, the transition should be made more natural by the use of the same species of tree as exist naturally in the surrounding area. The landscape of focal points should include a careful selection of suitable exotic plants which lend vitality and life to these areas by diversity and contrast. They are also useful in creating a desirable atmosphere at points of interest.

The landscaping of housing areas should normally echo the general character of the area. It is considered unwise, however, to have a detailed scheme for each house, for people need the opportunity to express a sense of personal identity through landscaping. Tree planting should follow the contouring of the residential block. Athletic fields produce the need for the largest open grass areas; with special soil preparation and the proper selection of grasses and drainage controls, they can be practical with a mobile irrigation system providing the minimal required water.

Utilities and resources

Water

The basic distinguishing factors of an arid area are scarcity of water and high evaporation rates. Although water systems are normally installed by the mining company, extraction from natural sources is regulated by the state government. At present in the Pilbara, total water usage is fairly well divided between industrial and town use. However, by 1992 it is expected that industry will need nearly 75 per cent of the total required. This water demand is expected to be met by the full use of existing natural resources, future dams on a number of rivers, and the desalination of bore water.[25] Using natural gas and waste heat, good quality water from desalination could be supplied for special industrial purposes or for general water supply at prices comparable to that of treated natural water.

Domestic water consumption will have to go down. Considerable savings in the use of water for domestic purposes are possible through high penalty rates for excess water consumption; limited hours for lawn watering; restrictions on the use of water-wasting automatic washers, dishwashers, etc; ring-main hot water systems; thermostatically controlled mixing valves on showers; and the reduction of water pressure at the house connection. Recycled treated effluent is currently being used at Port Hedland for the watering of large grass areas such as playing fields. Although there is a musty odour associated with it at this time, it is not offensive; the procedure effects a considerable saving of water and is therefore recommended. Unfortunately, large losses due to evaporation decrease the amount of effluent available for reuse in irrigation.

With domestic conservation, relatively large water schemes are feasible in this arid region, but they are costly by south-western Australian standards. General water conservation should thus be encouraged by pricing policies with incentives for water economy in industrial processes and by the determent of the establishing of water intensive industries unless they offer a high 'value added' factor.

Power

Towns now receive power from their respective powerhouses connected with the town's industrial operation. Generation has been by diesel engines, but alternative fuel generation is being considered for the near future. The alternative fuels are natural gas, which is now being produced off the northwest coast, and coal from either the southwest of the state or the eastern states. The production of nuclear power normally requires a plant of 500 megawatt capacity: this would be too large, at least for the earlier stages of Pilbara development.

Many towns feature surface power lines, but these towns are mostly inland. Dampier and Karratha, coastal towns, have their power cables underground to avoid cyclonic damage. Where poles must be used to support power reticulation, the better schemes place these in an easement at the rear of allotments; in an environment where trees are not easily cultivated, it is an error to thrust wires, poles, and transformers forward unsympathetically into the landscape.

Air conditioning

The universal decision in the Pilbara has been to air condition new houses in order to give positive relief from the heat—particularly during sleep. While this decision makes the design of houses easier in some respects, they often lose a certain character from no longer needing to respond directly to the climate.[26] There is no consensus whether evaporative or refrigerated cooling should be used, although refrigerated cooling has been recently much preferred. It gives positive temperature control at all times and uses no water but is much more costly. An evaporative cooling system can be very effective in cooling in most of the inland areas of the Pilbara featuring comparatively low humidities, but it uses about 4–5 gal of water per hour in peak periods. Research engineers of the Commonwealth Scientific and Industrial Research Organization have devised an improved evaporative method of cooling that does not appreciably increase the humidity of the cooled air and is therefore more widely applicable.[27] This system, however, has proven a good deal more expensive than the existing evaporative cooling units. The technology is being refined and may yet be on all counts acceptable.

Generally a packaged refrigeration unit of approximately 5 hp is employed in cooling. An interesting exception is at Shay Gap where a central chilled-water system is used. A central system is more economical, but it involves more planning problems. The houses, for example, have fixed windows. Problems have occurred, although infrequently, when the air-conditioning service has been interrupted. Since people do not normally enjoy the feeling of being closed in, psychological problems were also feared. With views from the upper floor and large windows onto the central court, families seem to have adjusted easily to the situation.

Solar energy

The only application of solar energy to date is the use of solar heat collectors on house roofs for domestic hot water purposes. Since the Pilbara general or global radiation rate is 260 watts/m^2 in summer and 170 watts/m^2 in winter (averaged over 24 hours), it can be expected that this potential energy will be used eventually to a considerable degree. Scientists at Sydney University recently noted that a commercially feasible method of solar heating can now deliver heat at 150°C at an operating temperature of 300°C. This method would be applicable in water distillation and industrial water heating, space cooling and heating, refrigeration, and the generation of electricity.[28] Since there are an increasing number of domestic water heating installations using solar technology, the future will probably see these high temperature collector units on roofs and linked to a community system and a central energy station. Advantage may then be taken of variations in demand, and chilled and hot water may be recirculated for cooling and heating. Such a house would have to be designed with thermal considerations having a high priority, as would the town itself. As distribution and reticulation become more and more costly, the question of self-sustaining units for each house should be raised. In the arid country it is difficult to see a house ever being self-reliant for water, at least at the rates of consumption commonly accepted today. Likewise, in the smaller towns of the Pilbara, the traditional organization and distribution of utility services present very few problems. Therefore, the question is really not pertinent.

Sewerage

In the Pilbara, the treatment of sewage generally uses an Imhoff tank with secondary and tertiary ponding systems and sludge drying beds. Such a traditional sewage disposal system is dependent on a large volume of water acting as cleansing agent and transport medium and on gravity flow through the sewage system. This system has proven to be most successful. Its main disadvantages are the consumption of water, especially where this is in short supply and/or is expensive; and the necessity of minimum gradients, which in level or near-level land requires expensive excavations and/or pumping stations, especially in earth with a high rock content or a high water table.

The system used at Shay Gap is different. It uses the vacuum transport technique with consequent water savings and increased flexibility in planning (since normal drainage considerations do not apply). For other liquid wastes, a separate system was necessary. This duplication of controls, pumps, and emergency power supply meant that there was no overall cost savings, but the water economy remained. One disadvantage is that special acoustical precautions are necessary to minimize the noise associated with the vacuum operation. Another alternative would be each house's having a self-contained system in which aerobic organic breakdown of wastes occur. It does not, however, appear as suitable for townships as for very small, isolated settlements.

Drainage

During thunderstorms excessively large runoffs can occur and present special problems. In the town layout, any natural drainage system that exists is normally utilized and extended as seems prudent. Beyond this, since normal street drainage is usually not effective when deluges occur, one government proposal suggested the use of overland flow wherever possible.[29] The streets would be designed with inverted crowns to remove some surface water, but the main system would use open channels and temporary compensating basins designed to hold runoff away from surrounding housing while the area is being drained by a small pipe system. Since the soakage capacity of the Pilbara soils generally is very low, soakage wells cannot accept large quantities of water. Although water is a precious commodity in the area, the expense of trying to store flood water is prohibitive, considering the amount that could be collected and the evaporation rates experienced.

NEW DESIGNS FOR THE PILBARA TOWN

Shay Gap

The town of Shay Gap (population 800) is at the edge of the Great Sandy Desert approximately 6½ miles from the Shay Gap mine. It is enclosed to the south and west by cliffs varying from 25–120ft in height and is defined by three isolated mesas. The existing vegetation of spinifex with eucalyptus and acacias along water courses is typical of the Pilbara area.

The town plan is a departure from the traditional design of an arid-area mining settlement in Australia, and it has been successful in two ways. First, it functions better than conventionally planned towns, and second, it has demonstrated to planners and mining companies that other approaches are indeed possible. There were a number of ideas behind the specific scheme, one being the shaping of the town's housing to create the physical circumstances in which a community of possibly 1600 might form. Another goal was to use the latest technology in ways natural to a town in a remote area. A final goal was to build a virtually demountable town. The iron ore reserves are expected to last only ten years (depending on the demand), and this precludes a conventional township on economic grounds. These goals gave the architect-planner (Lawrence H. Howroyd) an opportunity to put some of his theories into practice.

The basic philosophy behind the plan was to create a particular physical environment in which community spirit could develop. The designer assumed that the principal characteristic of a community is the capacity of people in the group to both recognize strangers and become aware of the loss of a member. Thus, the successful creation of a community depends upon each member's knowledge of and identity with his fellows. However, most definitions of community include the components of area and common ties in addition to social interaction.[30] Howroyd and others expected that social interaction would take place in the town given the opportunity, but nothing definitive has as yet been published on the success of the Shay Gap design. Some general studies certainly have shown, however, that social relationships have been influenced and indeed at times determined by a particular site plan.[31] The built environment alone cannot be depended upon to influence people's behaviour or social relationships as much as many designers may like to think. This does not mean, however, that the planner should not design for opportunities for social interaction.

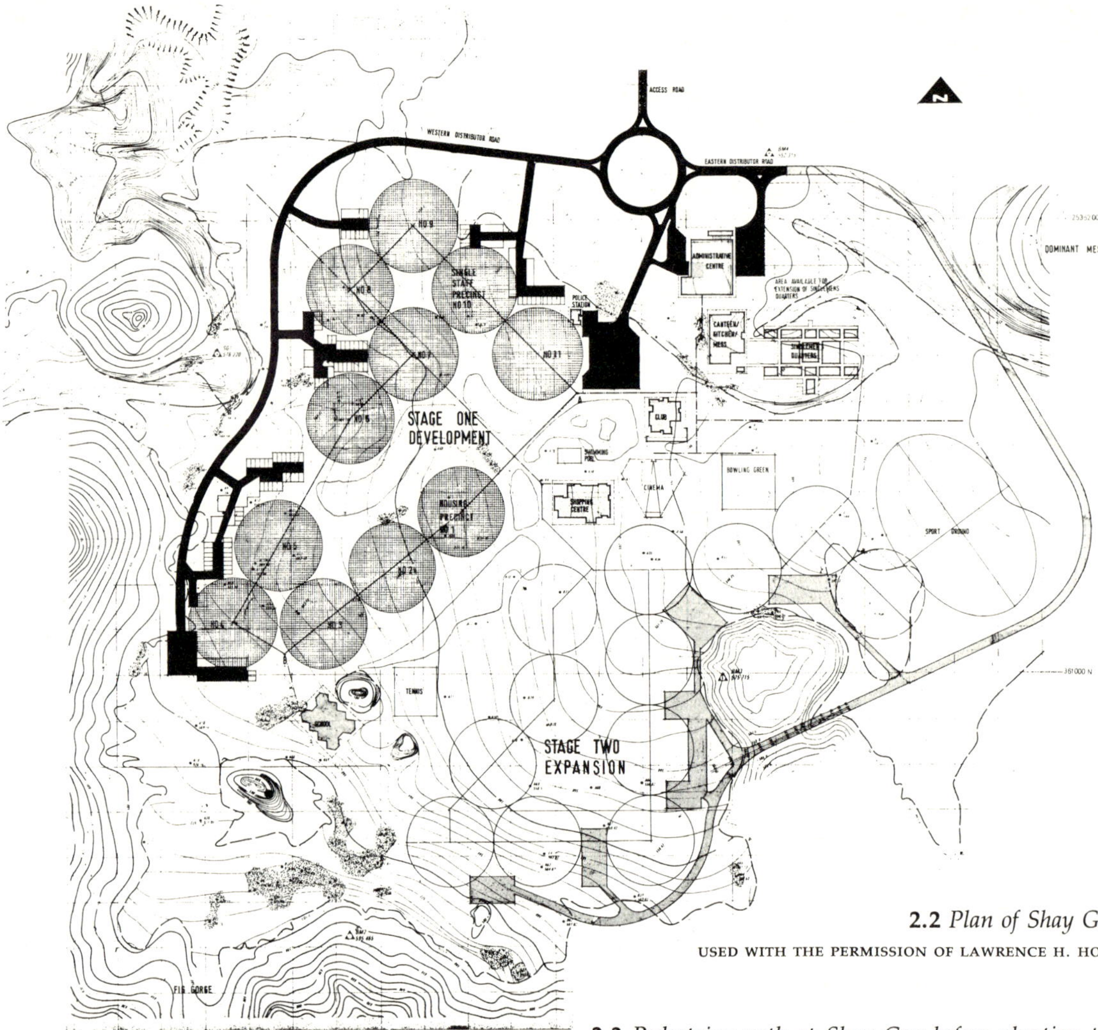

2.2 *Plan of Shay Gap, March 1977*
USED WITH THE PERMISSION OF LAWRENCE H. HOWROYD, ARCHITECT

2.3 *Pedestrian path at Shay Gap before planting took hold. March 1977* USED WITH THE PERMISSION OF LAWRENCE H. HOWROYD, ARCHITECT

The objective of creating a community resulted in a town plan for Shay Gap with the following characteristics:

1 The town features a number of pedestrian precincts with cars confined to a peripheral road system as **2.2** shows. The car was seen to be a cause of isolation in such a community. With the automobile gone, acquaintances are made while walking along shaded walks and in the common activity areas, and the town is also free from the stress of traffic dominated streets. A view along one of these walkways is shown in **2.3** before planting began.
2 Clusters of 10 or 11 houses form small precincts, and these combine to form the larger town precinct with the amenities, schools, etc. No house is no more than 330yd away from any public facility.
3 Pedestrian links within the town are directed to community foci, with the school in particular serving as a social centre.
4 Housing allotments are as small as practicable. Small courts (about 600ft^2 each) link the house to the public area.
5 The small precinct is identified with areas of public space designed for common social activities, for pedestrian use, and for children's play areas.

The car was thus not ignored but given a more relevant place in life at Shay Gap, as the designer saw that life. Because of the reduced road system, it was built in greater harmony with both shade trees and service areas. The peripheral road features communal garaging and parking bays between the road and the residential areas. About 40 per cent of the houses which are close to the roadway have direct automobile access to it, but others are up to 100yd away. In an informal discussion with a number of residents, the consensus was that an approximately 60ft separation from house to car was the maximum they would like to see; at the same time, the traffic-free dirt and cement pathways were highly commended.

The small residential precincts are 200ft in diameter with a density of 50 persons per acre. The two-storey houses have been arranged to prevent loss of privacy and to minimize the viewing of neighbouring courtyard areas. The two-storey design also creates shaded outdoor activity areas. The housing is in general an integral part of the town's character, and the design stems largely from the designer's use of technology.

The housing area has been landscaped to create discontinuous ground levels throughout, as is shown to some extent in **2.3**. Numerous insect pests and native fauna have been attracted to the town's greenery, and the insects are a particular nuisance at night, the time when many recreational activities occur. The fauna does not present a major problem since a perimeter fence was erected before construction began to keep out animals and thereby protect young plants. Plants in the shared open spaces have thus developed most satisfactorily, and maintenance is carried out by a corporate body of which each resident is a member.

The adopted planning principle results in the land being used predominantly for public open space (compared with the predominance of private open space in the conventional subdivision). Some tenants have not surprisingly expressed a desire for more outdoor privacy since they feel the private and public areas merge together too closely. No demarcated plot of land is now assigned to each residence except that defined by each residence's courtyard fencing which is not all that enclosing. Some additional fencing or screening thus might be considered, although this would not promote the designer's goal of encouraging social interaction.

Regardless of its social success or failure, Shay Gap's technical achievements, some of which were noted earlier, are significant. The houses were fabricated on an assembly line at the site from factory-prepared components. Walls were formed of prefinished asbestos-faced, polyurethane foam sandwich panels attached to a light steel frame, and all windows were factory fixed and glazed. Ceilings were similarly constructed with a steel skin replacing the asbestos for the roofs. Floors were composed in panels of vinyl-surfaced plywood and aluminium sections with pressed-steel floor joists. All plane external surfaces were furthermore post-stressed to ensure rigidity and airtightness. At 1974 prices the houses were said to cost approximately \$24ft^2, the lowest of all the new Pilbara houses; comparable cost for a similar size conventional house made of concrete block walls and a tile roof was \$32/ft^2.[32] Furthermore, whereas on-site labour costs account for 60 per cent of all costs for a conventional house in the Pilbara, this figure was 13 per cent for Shay Gap. The use of lightweight insulated panels, however, unfortunately resulted in a loss in privacy by allowing too much transmission of sound.

Utility 'lines' are contained in a common trench which carries chilled water (for air conditioning), electrical power, television, potable water, vacuum sewage lines, irrigation water, and telephone lines. A circular module serves each housing precinct where the services branch off and are controlled, as is seen on the side of the walkway in **2.3**. Maintenance and emergency service within the town is provided by all-electric service vehicles. Gardens are irrigated according to a timed programme during periods of least evaporation and not during peak demand hours. Water is supplied from borefields which tap underground water in the alluvium of the Turner and Yule Rivers.

Despite these technical innovations, Shay Gap does suffer as all small towns do, due to limited shopping, educational, and health facilities. Although foodstuffs are subsidized, many people maintain that direct buying from Port Hedland is nonetheless cheaper and more satisfying; a free, air conditioned bus operates three days a week to the town (66 miles away). A boarding high school has also been established at Port Hedland to serve the entire Pilbara. There is only one police officer stationed in Shay Gap. The only problems are associated with the single males, who break cues and pool tables, etc., no doubt with the expectation that the company will continue to freely repair or replace such damaged items.

Shay Gap is now within a council area, and the town has a right to be represented on the governing board of the council. Council policy is to encourage local sporting groups to improve their facilities by assisting them in the raising of loans. The company provides a licensed social club, community hall, tennis courts, swimming pool, bowling greens, a free outdoor motion picture theatre, and a sporting oval. More than 80 per cent of the families have a motor vehicle, and the ocean beach is about 40 miles away. It was estimated that 75 per cent of all townspeople make regular visits to the coast in their leisure hours.[33] Thus, the town is not deprived of social services.

Many of the sociological problems already outlined for remote mining towns do apply generally to Shay Gap, although its built environment has made living in arid-area conditions better. The single men's quarters, for example, are isolated, and the town's design accepts that little can be done about this problem. These men are housed in single rooms in one large complex. Due to the temporary nature of the town, the company could not even consider providing motel-type accommodations since the cost of building and operation would have been prohibitive.

The general impression of Shay Gap is that, while the housing precincts are integrated, they are not in turn integrated as part of the town—with its school, community buildings, and single men's housing, etc. The better qualities of design and planning obtained within the small housing development are thus lost when we turn to the larger scale buildings and their exterior spaces. Therefore, although the town falls short of being a classic model of a contemporary working community in the semidesert, the town's housing is a positive expression of man's attempt to provide a dramatic, interesting, and appropriate environment for living.

Marandoo

The design for Marandoo, another town in the Pilbara, is still on the drawing boards. It represents a major change in that row housing is used in an imaginative way. The houses are single-storey and formed in groups of an organic character. Each house has its own private courtyard with an unimpeded view, thus giving it a quality associated with the detached residence; yet the overall design makes a much more compact town possible. The pedestrian is separated from the road system, but not as decisively as in the smaller Shay Gap scheme. The car is brought much closer to the house. A schematic layout is shown in **2.4**.

The site is located about 25 miles northwest of Tom Price and is 2600ft above sea level. The two planners who made initial site investigations had differing recommendations for the town's location. The first proposal was to build on a flat area flanking a large creek system, but this would have meant the distinct possibility of flooding and offered no striking vistas. Construction, however, would have been simplified. The planner of the selected site, Julius W. Elischer, chose elevated ground out of range of flooding. This site had the additional benefits of a downhill gradient for an on-site sewage system; thickness of alluvium-colluvium cover on the spurs; spurs open to the cooling effect of breezes; a pleasant view, avoiding a flat, monotonous town layout; and no salting problem due to good drainage.

The planning study for the town was based on a population ratio of 50:50 between married and single people. The eventual ratio is expected to be about 65:35, and the initial population estimate is 1650, including 640 children. The main planning principles were as follows:

1 The size of the town was to be minimized in order to decrease all distances and to accentuate social life and activity.
2 The town was to be basically for pedestrians, and walking in the town was to be both an aesthetic and a social experience along narrow, shaded walks.
3 Priority was to be given to the safety of children.
4 Walking distances to school, town square, etc, were to be no greater than 500yd.
5 Areas needing to be landscaped were to be reduced to a minimum.[34]
6 Allowance was to be made for some expansion.

The broad objective was that the visual character of the town should be similar to the small Mediterranean hill towns, attracting people to work and stay there, living in a unified town with narrow shaded walks, greenery, and atmosphere.

2.4 *Schematic town layout for Marandoo, February 1977*

SOURCE: THE OFFICE OF J. ELISCHER, ARCHITECT, PERTH, WESTERN AUSTRALIA. USED WITH HIS PERMISSION

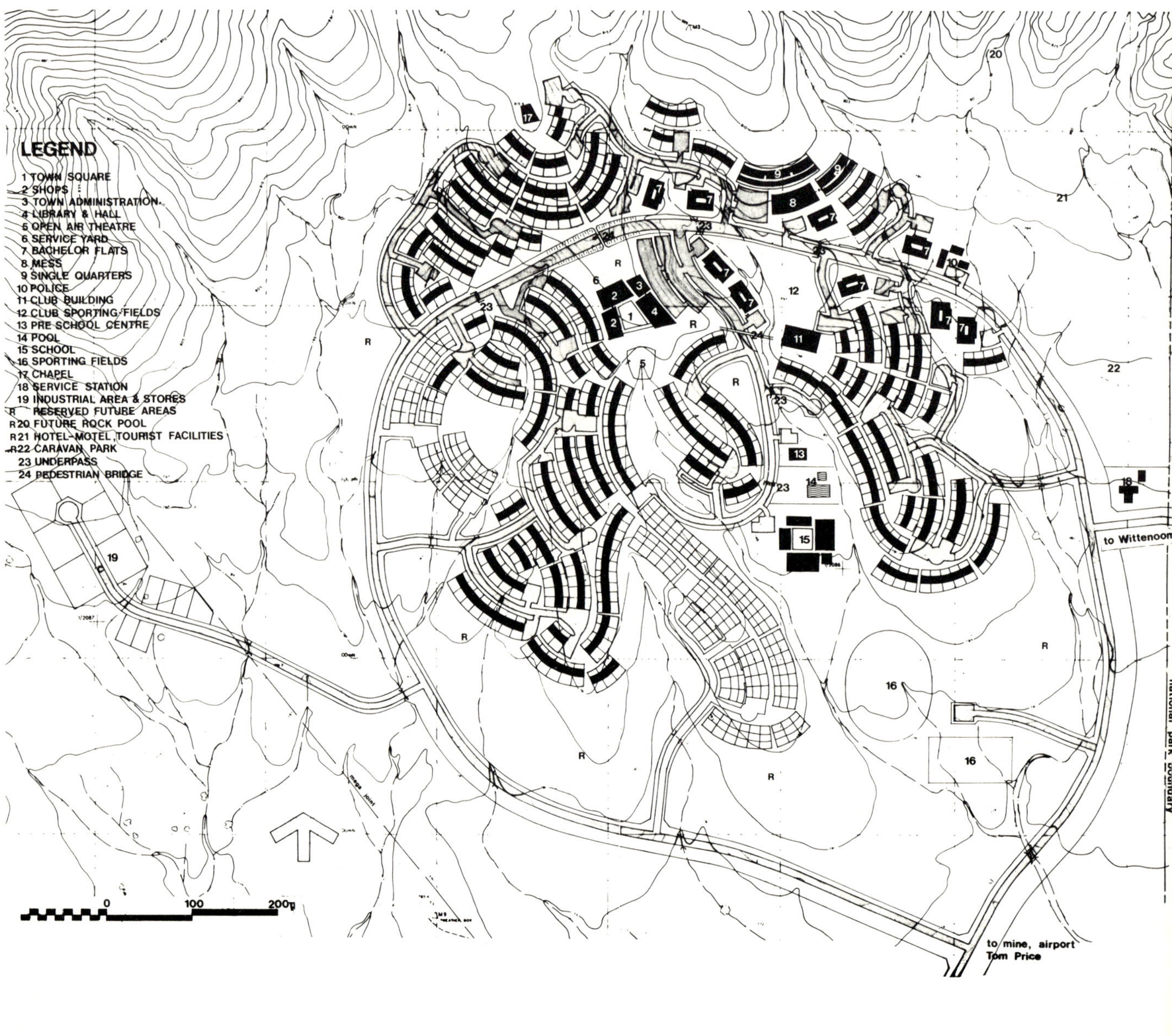

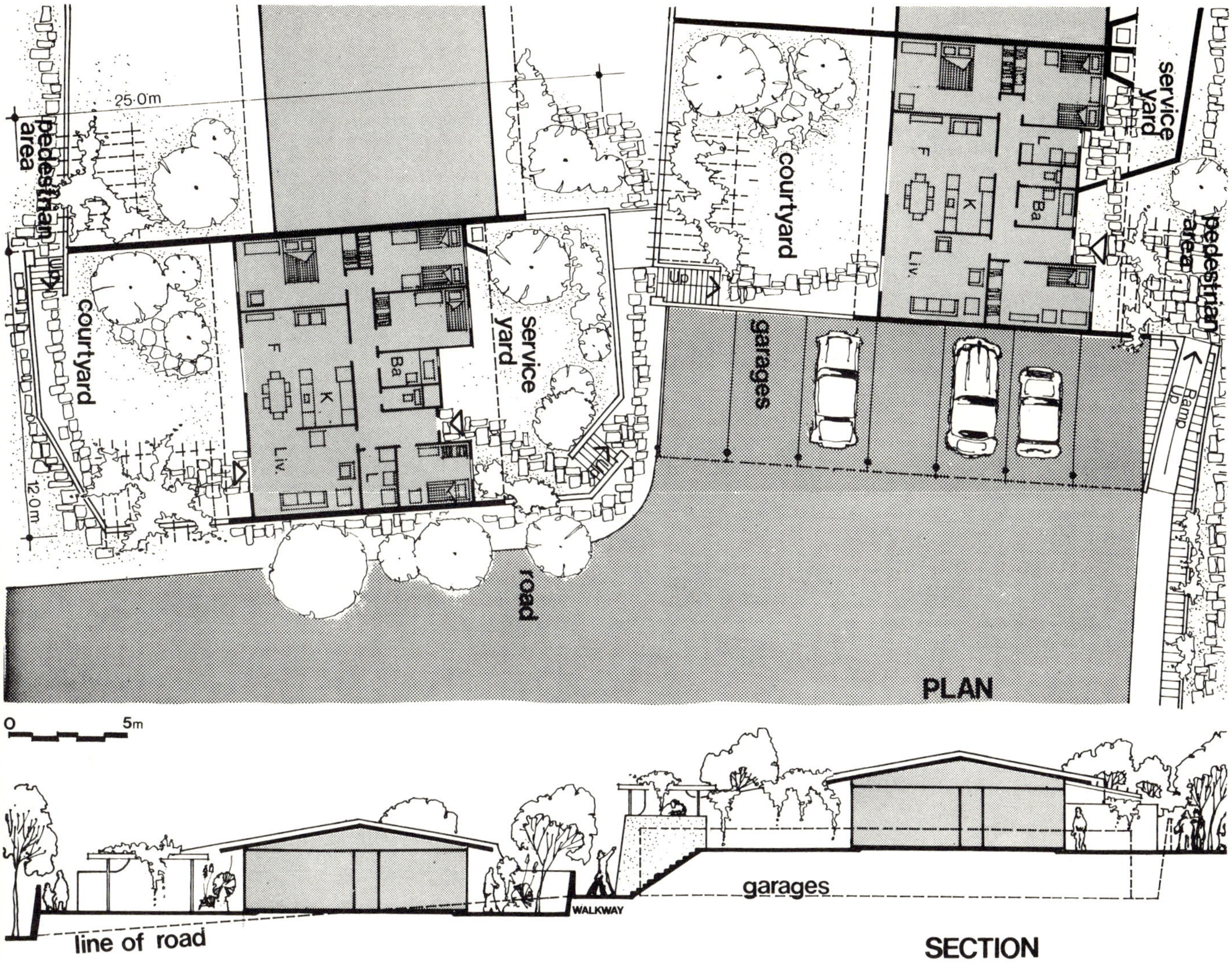

2.5 *Typical section of housing development proposed for Marandoo, February 1977*

USED WITH THE PERMISSION OF J. ELISCHER, ARCHITECT

Tourists, artists, and nature lovers are expected also to be attracted to Marandoo in due course, providing some relief from the monotony of the social structure that inevitably goes with a small, remote, single-resource town.

A peripheral main road, 24ft wide for the most part, serves and encloses the major portion of the town. From this road cul-de-sac roads penetrate into the housing areas. The remainder of the town climbs the steeper slopes to the north of the road, and access to the town centre and amenities there is maintained by a pedestrian bridge and by underpasses, making a separation between the car and pedestrian possible. The planning in its present state, thus, very nearly achieves its object of a natural and complete separation of cars and pedestrians, and further detailed study should make the solution complete.

Walking is encouraged by the closeness of most housing to amenities, the maximum walking distance being less than 500yd. If the walkways can be properly shaded for most times of the day and are as pleasant as envisaged, then a majority of people will probably prefer to walk, or at least ride bikes. However, with cars parked under cover near the residence and not superheated from standing in the burning sun, this preference might be doubted. Yet, since there is to be no cover provided for the cars at the shopping centre parking lot, the scheme might just work. Approximately 80 per cent of the families will have cars in the town. Half of the houses will have an attached garage or carport, and of the other houses, the majority will have their garage or carport directly joining the allotment. The rest will have parking facilities 60 or 100ft from the house.

A small but compact town centre with square, shops, and civic buildings is the heart of the town. However the town square does not assume the traditional place that it occupies in the long-established Mediterranean town, and the residential areas do not radiate from it as might be expected. The parking lot serving the centre is hidden from view (it is cut into the slope of the hill), but it still has the effect of isolating the town square activities from the housing, physically if not visually.

The housing design had to meet the following requirements:

1 Medium density to give a smaller town area.
2 Provision of small, private (both visually and acoustically) garden courtyards.
3 Houses and courtyards with unobstructed views and providing privacy, shelter, shade, and a sense of protection.
4 Houses with a low cooling load for economical air conditioning.
5 Residential block sizes conforming to the minimum dimensions prescribed for row housing by state regulations (a frontage of 40ft [12 m] and depth of 82ft 6in [25 m]).

About 320 houses for families and 364 single units are to be provided, and room for a further 70 houses has been set aside. In addition, since it is expected there will be a gradual increase in the married work force while the number of single men will decline, initially, one-third of the single men will be housed in family residences, one person per bedroom. These houses will revert to family accommodations as the work force changes.

The character of the town's design cannot easily be appreciated by looking at drawings alone, and a model has not yet been prepared showing how the housing is situated in relation to the land contours. The design does, however, capitalize on the site's spur ridges, enabling housing to have the required vistas. The housing is, to be more specific, built on steps going up the hill site, and thus, the courtyards are private and have a view, as **2.5** indicates. To achieve the desired housing density, the use of row housing on the rather small allotments was necessary.[35]

An important element in the design is the walling material to be formed from no-fines concrete. It is cheap and readily available on site and presents a natural, interesting texture that is most appropriate for the locale. The walls of the courtyards and the walkways made of this material will dominate the appearance of the housing and will give the town a distinctive appearance while visually anchoring the housing to the site's slopes. It will be used sculpturally, with curves, various thicknesses, holes, and recesses that introduce variety while maintaining harmony of appearance. It was established by the architect-planner that the cost will be about half that for normal dense concrete.

Housing has been designed to only withstand a wind velocity of 90 miles/hr since the area is not subject to full-scale cyclonic winds. While inter-unit, courtyard, and retaining walls will be of no-fines concrete, external walls will be made of load-bearing, insulated, concrete panels, faced with asbestos cement. Internal walls will be of plaster panel construction. All roofs will be insulated, and all framing will be of steel, clad with low-pitch sheet steel roofing with a baked enamel finish. This row housing, properly designed for visual and acoustical privacy, appears to be a good solution to the problems of maximizing housing density and limiting heat gain. There is a minimum exposure of walls to the environment, and these will be shaded where exposed. With the insulated roofs, heat gain is reduced to a practicable minimum.

The decision was made to use evaporative air-conditioning systems for housing since these systems have worked well in similar climatic conditions where the summer relative humidity at 3 pm averages only approximately 30 per cent. It does not provide the same degree of temperature control as does refrigerated air conditioning, but this loss of precise control is not critical for housing. Further, the basic evaporative cooling unit will cost less than one-third as much, with running and maintenance costs less than one-quarter of those for refrigerated air conditioning. Costs will be much lower, but the system's water consumption could represent at least 10 per cent of the total annual water consumption per family.

Water is supplied from the Marandoo Mine site, which in turn is supplied from an underground source at Millstream approximately 100 miles away. Unfortunately, this source is now approaching its long-term safe yield limit. Base water cost is 6¢/gal; but since it contains a number of salts, treatment is necessary before general use. It is proposed to provide industrial-type units for each house, and this would be the most economical method since only the water required within the household would then need to be treated. The proposed sewerage treatment will use Imhoff tanks and lagoons. Studies showed that this method would be cheapest and that the recycling of effluent would not prove economical.

Electrical power is generated near the mining operation, as is usual, and from there will be supplied to the town site by an overhead line. At the perimeter of the town site, it will be transferred to a 22kv underground ring main. Unit substations will be padmounted supplying a 440v underground service with distribution pillars in suitable positions.

The amenities and facilities proposed are similar to those found in country towns of a comparable size. There is no hotel since a town of this size would not be able to support one without a surrounding rural population, but a motel with a liquor licence is proposed for the future. Marandoo will operate as a company town with the government responsible only for the police, educational facilities, and mine, health, and building inspection. These government facilities will include a nursing station, medical and dental clinics, and preschool and primary schools with extensive athletic facilities for community use.

Shopping facilities will be provided by the private sector, but it is expected that the more specialized shops at Tom Price (25 miles distant) will also be patronized by the Marandoo population. There will be daily bus service taking children to the high school at Tom Price, and this will also provide transportation for shoppers and visitors and outpatients going to the hospital. This bus link is considered a most important factor in the town's social interaction, particularly for the women of Marandoo.

A tavern is proposed, and a social club is to complement this restaurant. These will be associated with the nearby sportsfield, basketball and tennis courts, bowling greens, and a swimming pool. A gas service station will be located at the town's entrance from Wittenoom. An open-air amphitheatre is to be provided with a small stage and dressing rooms to permit live performances as well as films. A cultural centre for the arts and crafts is also to be built, along with specially equipped hobby workshops for activities such as boat building or motor maintenance requiring heavier duty equipment.[36] An interdenominational chapel has been located at a high point in the town, rather than at the town square.

In addition to the motel, a caravan park has been planned for the future since Marandoo is on the tourist route to the Wittenoom Gorge, a well-known scenic attraction, and the planner believes tourism will be an important factor in the town's social activities. The town is also at the very edge of the Hamersley National Park. Since there are no first-class accommodations near Wittenoom at present, Marandoo could well serve it, especially with the national park airfield only 6 miles away. It cannot be denied that outside stimulus such as this would be beneficial.

The important factor in the planning of Marandoo has been the topography of the site. One of the main purposes in the site selection was to be able to give all housing a view and avoid the monotonous appearance of a town on the flat plains, and this has been achieved by the careful design of the housing. The site plan features a geometric pattern of arcs superimposed on the basic contour plan, and these curves will soften the usual angularity of housing: the geometric form will make a strong but harmonious statement on the natural site. However, in addition to modifying the usual form of residential layout, the housing arcs should also have been a unifying factor in the total design, but this has not occurred and probably was not possible considering topographical factors. The design thus offers the great advantages of vistas and privacy while having the general character of a Mediterranean town. It seems, however, that the town could have been made even more compact with advantage.

The plan appears to meet the objectives set; but as with all plans, the acid test will come after this town is built. Despite the quality of the proposed built environment, it would be most unusual if Marandoo does not suffer from the strikes and turnover of labour that occur in other mining towns. The plan nonetheless indicates an excellent solution based on research, experience, and insight—all resulting in an original town form using the selected site in an ingenious way.

CONCLUSION

Many of the issues in town planning for the arid zones of Australia are now better understood, but intractable problems remain inherent in the development of these areas. Water dictates the number of people who can be supported in any arid region, and the number will remain comparatively small in the Pilbara. There seems now to be no conceivable and economically viable way to produce or bring sufficient quantities of water to the region to raise the population limit appreciably. While future town designs will doubtless be more concerned with the conservation of energy and other resources, present indications are that such conservation will not be of real concern for possibly a decade except in the use of water.

Although technology makes life reasonably comfortable in arid areas, it cannot assist with the human need for richness in the range and variety of social activity and intercourse, something which only numbers of other people can provide. A town of some 30 000–50 000 is required for people with a variety of interests to meet in a satisfactory pattern of activity and to support a centre with an adequate range of amenities, but it is unlikely that the Pilbara will have more than one such town in the forseeable future. It can be expected that most towns will have populations less than 10 000, often much less.

When creating a township, planners must exploit any circumstance that offers favourable possibilities to its full advantage. For example, site selection is extremely important, for the setting gives interest and stimulation to the townscape. The uniqueness of the site contributes immeasurably in forming the visual character of the town, and people will be more apt to identify themselves with a distinctive habitat. Location gives a sense of place which, combined with the form and layout of the housing, may establish a town that fits naturally into its environment, respecting the topography when a grid pattern would be of little or no use.

Medium density housing appears to offer many advantages for the small arid-zone town. A density generally of about 40–50 people per acre in the residential area is an appropriate figure since it helps emphasize the place of activity and promote interaction without crowding or threatening privacy. Reducing the town to this size from what has been customary in the Pilbara makes a pedestrian town logical since distances to shopping, schools and amenities are reduced. The town design can be geared to the pedestrian and bicyclist with the car relegated to a servant's status. Under such circumstances, a green town can be more easily established and maintained. The town also holds together better perceptually. The street can 'belong' to the residents; shaded by trees and housing, the street becomes a positive functional design element. Landscaping can now be designed with more confidence. Although much can be done for the early development of some plant growth, only time brings the maturity and character looked for in landscaping just as in the town itself.

Western culture does not easily accept the street as an extension of the house. Nonetheless, the spaces between buildings must be carefully considered and consciously designed, not just left to be filled in some way. A variety of spaces should be linked in different ways to provide space for incidental activities and places for pausing, whether for children or adults. These spaces may be dynamic lineal areas serving as walkways of varying widths and levels similar to those in Shay Gap but taken further, or areas made restful by the planner's balancing of their size and treatment. Walkways should not be too wide, and home privacy should be protected from the walkways visually and acoustically.

Despite the use of air conditioning, housing should be oriented or designed to reduce heat gain. The east and west walls should have little or no openings, and the roof area also could be reduced with some advantage. A two-storey house is advantageous, although a housing mix is desirable.

To date residential planning has received most attention in the design of Pilbara towns. Their centres have not been so thoroughly considered. There, vertical forms could be used for emphasis and articulation, incorporating residential accommodations for single persons and young couples. Possible expansion could then occur vertically without having to leave large open spaces for growth, and adjacent housing could then be brought closer to the centre. Shopping facilities should be in this centre and not scattered, with meeting places provided for teenagers as well as places for general social activities including at least a restaurant and a coffee bar. Where isolation and smallness of scale are parts of the design, the scope of amenities must be enlarged, even though enlarging them will require subsidizing. Housing is subsidized, and it is equally important to subsidize amenities.

An isolated town should present as much variety and choice as is feasible while maintaining the essential cohesiveness of the town, both visually and socially. The static situation of the small arid-zone town is in the end more discouraging to many people living there than the environment. Pioneers in these areas had a long-term purpose which helped them to overcome the social disadvantages. Those persons who by nature appreciate the quiet life of the country and accept its particular limitations are those who can be encouraged to stay in a well-planned and designed town. They can form the nucleus of the Pilbara towns being developed and consolidated as stable communities. People with an urban life-style who go to an isolated arid area to make a living rather than a life are inclined to dissatisfaction no matter how well the town is planned and designed, and they eventually move on.

NOTES

1. The statutory area of the Pilbara is 190 000 square miles.
2. Marble Bar, population 300, was, for example, a gold mining town of 5000 in 1895.
3. *The Pilbara Study* (Canberra: Australian Government Publishing Service, 1974), sect. 10, pp. 4–5.
4. Balwant Singh Saini, *Building Environment: An Illustrated Analysis of Problems in Hot Lands* (Sydney: Angus and Robertson, 1973), p. 20.
5. Russell Taylor, 'New Towns in Western Australia', *Australian Planning Institute Journal*, 7, no. 4 (1969), p. 117.
6. Ira M. Robinson, *New Industrial Towns on Canada's Resource Frontier* (Chicago: University of Chicago Press, 1962), p. 38.
7. Erika Spiegel, *New Towns in Israel* (New York: Frederick A. Praeger, 1967), p. 140.
8. It has been shown that bare ground temperatures cool by as much as 40°F in 5 min when shaded. See C. F. Kelly and N. R. Ittner, 'Artificial Shades for Livestock in Hot Climates', *Agricultural Engineering*, 29, no. 6 (1948), cited in Saini, p. 78, and not available for my inspection.
9. Such a design has been proposed for Marandoo township, which we will discuss later.
10. H. Calderbank Green, 'State Housing Commission Tenants in Two-Storey Flats', unpublished study, School of Architecture, University of Western Australia, 1971.
11. J. Paynter, 'Town Form in Hot Arid Areas', paper presented to the Australian New Zealand Association for the Advancement of Science, 45th Congress, Perth, Australia, August 1973.
12. Edward T. Hall, *The Hidden Dimension* (London: The Bodley Head, 1969), p. 147.

[13] Margaret E. Stockbridge, et al., *Dominance of Giants: A Shire of Roebourne Study* (Perth: Department of Social Work, University of Western Australia, 1976), p. 21.

[14] For this part of the survey, a smaller sample was used—53 residents of Dampier, Karratha, Wickham, and Roebourne, all coastal towns and not isolated.

[15] Not being able to own one's own house in a company town might force the woman to make other attitudinal adjustments.

[16] Heavy drinking occurs throughout the Pilbara. It does not appear to be particularly associated with new towns, although it is certainly more visible in a small town. Hot weather particularly encourages drinking among workers everywhere, and it is hot in the Pilbara. There is furthermore no firm evidence that alcoholism is more prevalent in the Pilbara than in the rest of the country for men or women. At the same time, Hamersley Iron Properties Ltd, has asked the Alcohol and Drug Authority to investigate alcoholism among its workers following a $500 000 train crash.

It was reported by Kim Murray in the *Daily News* (Perth) of 1 October 1976 that Mr J. Williams of the Alcohol and Drug Authority reported to a Senate Committee on drug abuse that a rehabilitation programme will soon be established to help workers with alcohol problems in the Pilbara and that only the lack of funds prevented such a programme being established for all towns of Western Australia. He believed that a lack of community spirit in many mining towns resulted in heavy drinking. He suggested that deficient community facilities were at the root of the problem.

[17] The mining industry has suffered its share of work stoppages, many, no doubt, due to the environment of the Pilbara and the high demands of the companies. Union officials point out, however, that the situation is no worse than in other parts of the country. One interesting strike occurred because the company delayed fixing the washing machines in the single men's quarters.

[18] Only 25 per cent of the single men indicated they either had no definite plans to leave or would like to stay.

[19] The Western Mining Company of Kambalda showed after some experimentation that two units, each housing four men, with a common lounge worked best.

[20] The flats are necessarily smaller in scale and are usually located on the periphery of the main housing areas.

[21] A typical weekly rate in 1976, however, was only $26.

[22] T. B. Brealey, 'Living in Remote Communities in Tropical Australia', paper presented to the Australian New Zealand Association for the Advancement of Science, 45th Congress, Perth, Australia, August 1973.

[23] Ferdinand Zweig, *The Cumbernauld Study* (London: Urban Research Bureau, 1970), p. 19.

[24] The discussion here is indebted to the work in the Pilbara of Marion Blackwell of Blackwell and Cala, Landscape Architects, of Perth.

[25] The cost of water reticulation dictates that the source of supply shall not be more than 150 miles from the town.

[26] The housing designs adopted at Shay Gap did give character to air-conditioned dwellings.

[27] R. N. Morse, 'A Rock Bed Regenerative Building Cooling System', *Transactions of the Institute of Engineers Australia*, 4, no. 1 (1968), p. 55.

[28] The necessary development and marketing will take four years. See Adam Tarock, 'Australian Advance in Solar Energy Technology', *Journal of the Institution of Engineers Australia*, 49, no. 8 (1977), 8–9.

[29] *The Pilbara Study*, appendix 8.2, p. 176.

[30] See C. Bell and H. Newby, *Community Studies* (London: George, Allen and Unwin, 1971).

[31] See Herbert J. Gans, *The Levittowners* (New York: Vintage Books, 1967), p. 156.

[32] *The Pilbara Study*, appendix 8.2, p. 139.

[33] I. G. Ellison, 'Shay Gap', in *Man and the Environment: New Towns in Isolated Settings*, proceedings of the Australian UNESCO seminar, Kambalda, Western Australia, 18–23 August 1973 (Canberra: Australian Government Publishing Service, 1976), p. 272.

[34] Landscaping proposals have not been detailed as yet. The small valleys which are to be used as natural drainage channels will undoubtedly be maintained as a natural landscape feature, and current house plans do indicate extensive but easily maintained planting.

[35] The number of persons per acre will probably be approximately 40 in the residential area.

[36] These latter activities in other towns often cause untidy yards and unwelcome noise.

SUGGESTED READING

Brealey, T. *Living in Remote Communities in Tropical Australia. I: Exploratory Study*. Commonwealth Scientific and Industrial Research Organization report TB 27.1, 1972.

Man and the Environment: New Towns in Isolated Settings. UNESCO Seminar. Canberra, Australia: Australian Government Publishing Service, 1976.

McHarg, Ian. *Design and Nature*. Garden City, New York: Doubleday, 1971.

Negev, Avraham. *Cities in the Desert*. Tel Aviv: E. Lewin-Epstein, 1966.

Rudofsky, Bernard. *Streets for People*. Garden City, New York: Doubleday, 1969.

Rogers, Tyler S. *Thermal Design of Buildings*. New York: John Wiley and Sons, 1963.

Stockbridge, Margaret E. et al., *Dominance of Giants: A Shire of Roebourne Study*. Perth, Western Australia: Department of Social Work, University of Western Australia, 1976.

Way, Douglas S. *Terrain Analysis: A Guide to Site Selection Using Aerial Photographic Interpretation*. Stroudsburg, Pennsylvania: Dowden, Hutchinson and Ross, 1973.

3 Urban and architectural planning in warm, arid zones

LÜTFI ZEREN

Since man began to build his shelter, he has considered climate or the effect of the components of climate. Throughout history, his aim has been to maintain his comfort. In the transition from isolated to cooperative living, neighbourhood units have been built with a similar approach and in accordance with climatic effect as well as some other factors. Observation of the settlements forming the nucleus of many historical cities shows this very well. Generally, settlements have been fairly well organized and from the point of view of site selections, orientations, building shapes and building shells, all constructed in relation to climatic conditions. Man's approach to problem solving, by making use of the technological level of his time, has been intuitive and based on his past experiences and customs.

These appear to be the facts on the organization of the nucleus of a city, but one may ask about the effectiveness of the climate control in the later organization that surrounds the nucleus. The initial settlement may be in balance with the climatic character of the region; but in the course of time in some sections of the city, it is possible to observe a deficiency of performance. This may be caused by a change of environmental effects, where a change in topography is the main reason. Therefore, the expansion of an urban organization is of primary importance when we consider climatic and microclimatic conditions.

BIOMETEOROLOGICAL APPROACH TO URBAN AND ARCHITECTURAL DESIGN

It has become clear that evaluation of the climatic character of a region becomes necessary for design purposes. For a long time, climatology was studied mainly from the agricultural point of view. Thus many of the classification systems for climatic character were related to the vegetation of a region. But, as man is the basic measure in urban and architectural design, the true way to evaluate the climatic character of a region should always be in accordance with human requirements. A relevant approach was used in 1953 research sponsored by the Housing and Home Finance Agency, Washington DC; it was carried out by A. Olgyay, V. Olgyay and Associates and published under the title *Application of Climatic Data to House Design*.[1] Their concept of treating problems and the applied methodology are specific and very feasible for urban and architectural uses.

In organizing a built environment with the aid of artificial climatization, it is possible to obtain a climate balance within that environment by using today's technology. But the adaptation would be on buildings to a great degree and would require excessive energy consumption. Such a solution could not be assumed as optimal. A building, a neighbourhood or a city can never be conceived without surroundings. Therefore, the design elements *and* their surroundings must be considered as design units in climate control. The size of the built environment depends on the size of the design.

To reach a peak of performance in climate control of a built environment, natural climatization should be integrated with artificial climatization. Natural climatization is essential to climate control, and the resulting design is called a climate-balanced environment. This balance must be achieved at different design levels as urban, architectural, and building elements design; it should be organized according to design parameters based on human climatic needs of a region. The aim is to define the true regional character in design.

Making decisions on the climatic character of a zone

All the relevant investigations of a zone's climatic character are based on a bioclimatic chart where the coordinates are respectively the percentages of relative humidity and the dry bulb temperatures (see **3.1**). Without going into a detailed description, we can say that the most significant element on the chart is the shading line which divides it into two main parts. The lower portion shows the underheated period where wind protection is necessary; different quantities of solar radiation are needed according to the values of the coordinates. The upper portion of the chart depicts the overheated period. Here, shade is necessary; and in some zones, wind, or wind plus moisture, or moisture may be required. Right over the shading line lies the comfort zone where no additional protections are needed. These evaluations of human climatic needs are independent of time for the maintenance of human comfort (see **3.1**). Therefore, the problem is to define these climatic needs as read from the bioclimatic chart in relation to the time of the year, the 'timetable of human climatic needs' (see **3.2**).

If these investigations are made in several locations within a region, correlation among them makes possible significant decisions on the climatic character of the zones in relation to the human climatic needs for urban and architectural uses. Such a process has enabled me to establish a method of classifying climatic zones in Turkey. First, I selected 35 locations and determined their timetables of climatic needs. By weighing the results at a few critical locations, it is possible to map the boundaries of the typical climatic zones in Turkey.

Climate control design parameters

I have already indicated that climate control combines natural and artificial climatization. The basic principle is to solve problems at different design levels with natural climatization and to integrate it with artificial climatization. Of course, the ideal solution would be a climate-balanced environment without the use of artificial climatization in order to conserve energy. The design of different phases of a built environment to an appropriate scale requires development as a design unit with consideration given to various dimensions, one of which

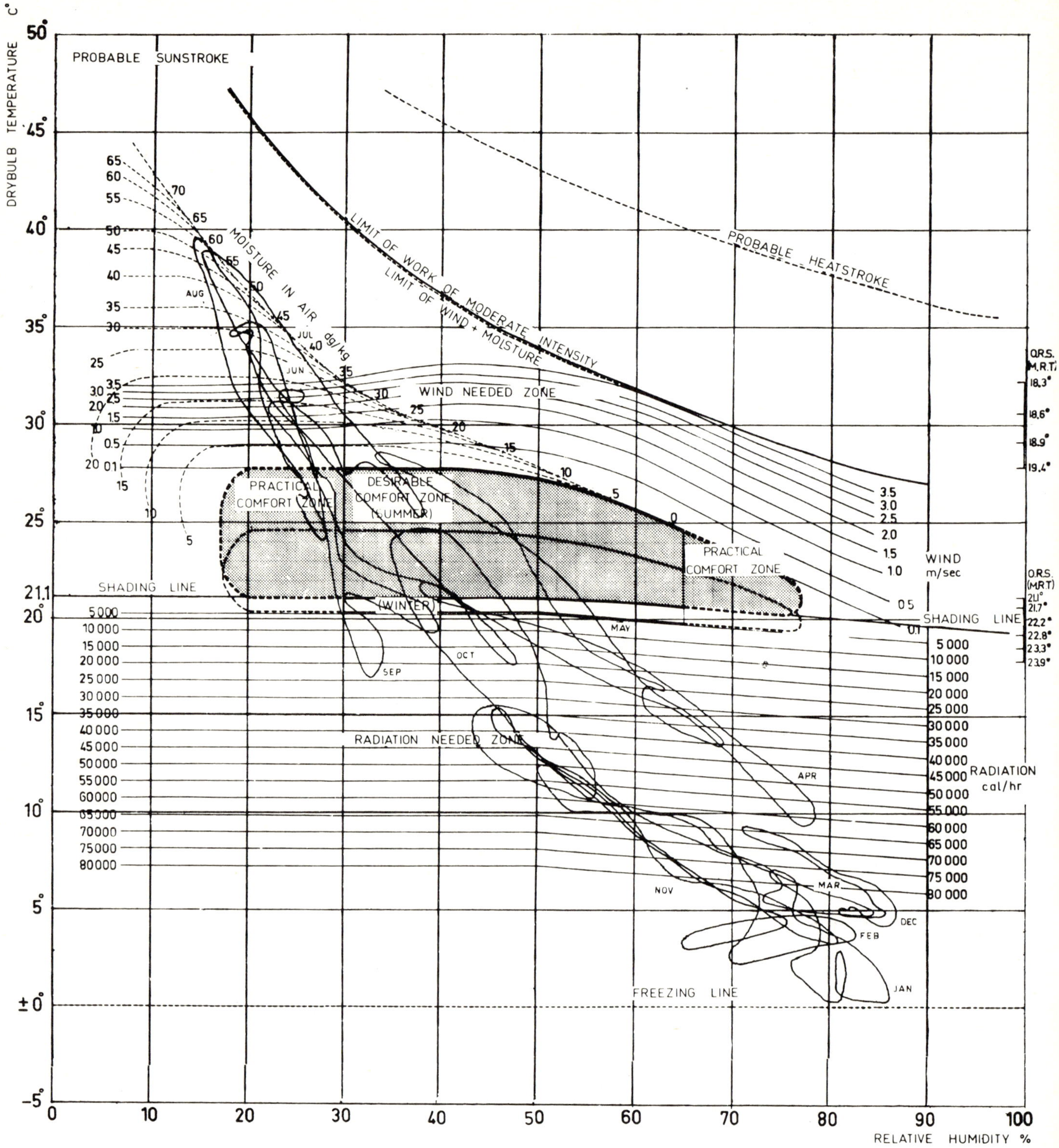

3.1 *Bioclimatic chart and registration of climatic data for Diyarbakır area, Turkey*

SOURCE: ALL CHARTS AND DRAWINGS ARE BASED ON THE AUTHOR'S OWN INVESTIGATIONS. THIS BIOCLIMATIC CHART IS ADAPTED (IN METRIC MEASURE) FROM VICTOR OLGYAY, ALADAR OLGYAY, AND ASSOCIATES, *Applications of Climatic Data to House Design* (WASHINGTON, DC: US HOUSING AND HOME FINANCE AGENCY, 1953)

is the climatic dimension whose components may determine such design parameters as the orientation, site selection, building shape and building shell. Certainly, these are the basic or imperative parameters to be used in a climate zone. Other factors (sun and wind, for example) requiring preventive measures in climate control can be mentioned.

Orientation

In orientation we must consider both the solar and wind effects. Usually, insolation is of primary importance. From that point of view, several theories have been developed. It is worth considering the 'Sol-air Orientation Theory' where human stress is the objective measure of solar impact.[2]

According to this theory, a building oriented to a particular axis receives maximum radiation during the underheated period and minimum radiation during the overheated period (see **3.3**). The designated axis gives optimum orientation; the values near to that, weighted by values for the overheated and underheated periods, would be good orientations. The main purpose of the theory is to obtain a balanced insolation for a building. For that reason, some flexibilities were proposed in relation to the use of buildings. A weighted correction should be made, if necessary, considering wind effects obtained from the wind analyses chart of the area (see **3.4**). By these analyses one could be informed about the velocity, prevalence, and direction of the wind; but the character of the wind, an important factor in orientation, can be found by studying general trends. Since buildings are the main elements of an urban unit, the determined optimum axis and the good orientation bearing angles are valid for the orientation of the city itself. Its main axis must be normal to the determined orientation axis.

Site selection

For successful settlement in a climatic zone, proper site selection is an essential consideration. The influence of the topography is of major importance because the ground is rarely even. In above-ground structures, we must consider the effects on their orientation, inclination and altitude. Solar and thermal impacts and wind cause many problems requiring study.

It is very well known that sloped surfaces receive higher solar radiation depending on their particular orientations. A study carried out for 40° N latitude, gives basic knowledge concerning areas near to that latitude.[3] However, since the problem is to select a site for settlements where buildings are to be built, the optimum and good orientations determined according to Sol-air Orientation Theory are suitable. As for inclination, the slopes considered reasonable for urban design could be used since they are within the limits delineated by the above mentioned study.

A difference in altitude creates a difference in the amount of solar radiation received in relation to the cloudiness. During daytime, the intensity of the solar radiation received on the earth's surface increases with the height above sea level. It is similar at nighttime for the outgoing long-wave radiation, but inversely proportional to the cloudiness and the relative humidity at that time. Therefore, the heat gain and the heat loss of the earth's surface, at different altitudes of different characters, initiate some periodic air currents, which are broadly explained by R. Geiger.[4]

Study and comparison of these mutual effects makes desig-

3.2 *Timetable of human climatic needs for Diyarbakır area*

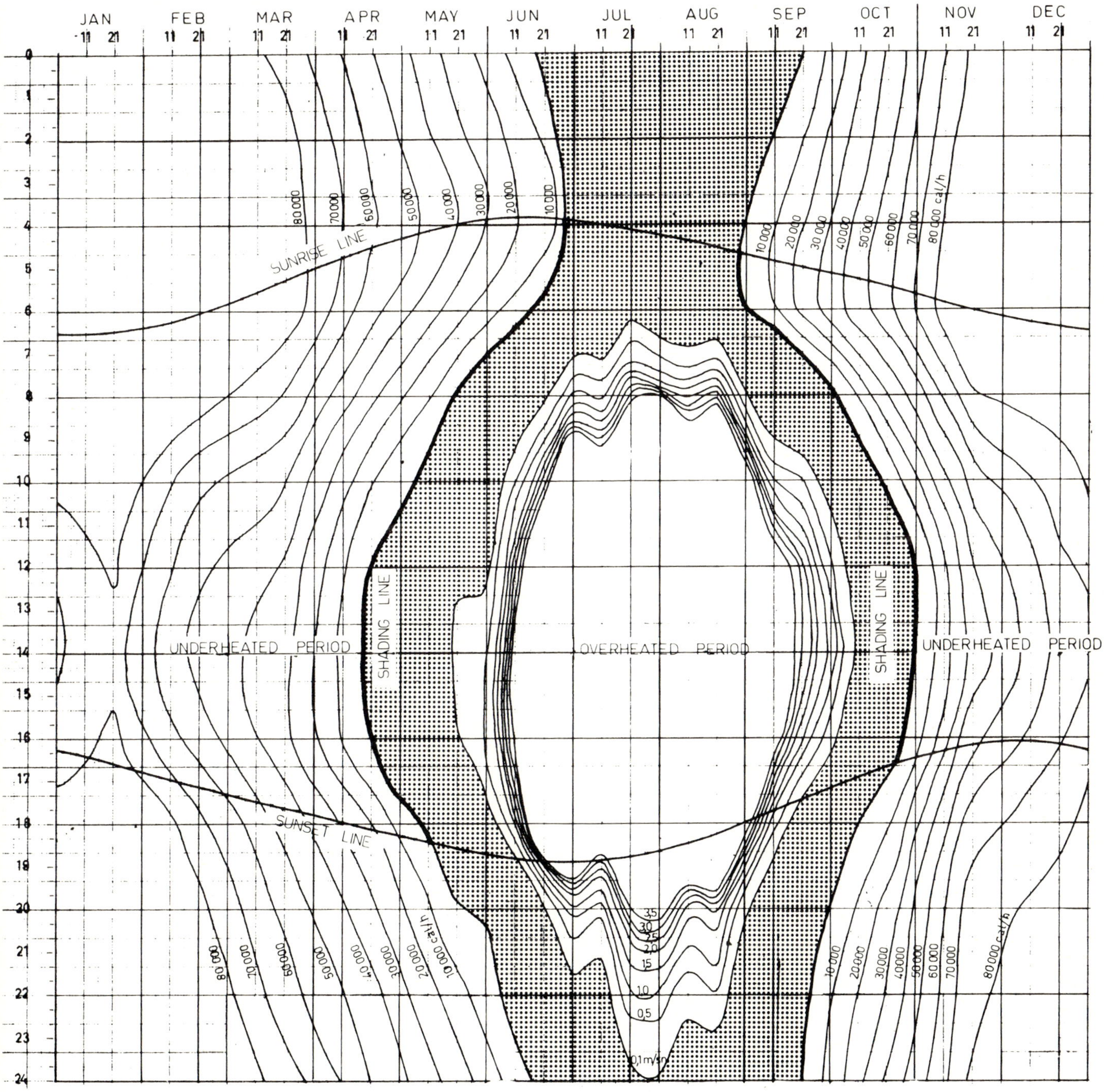

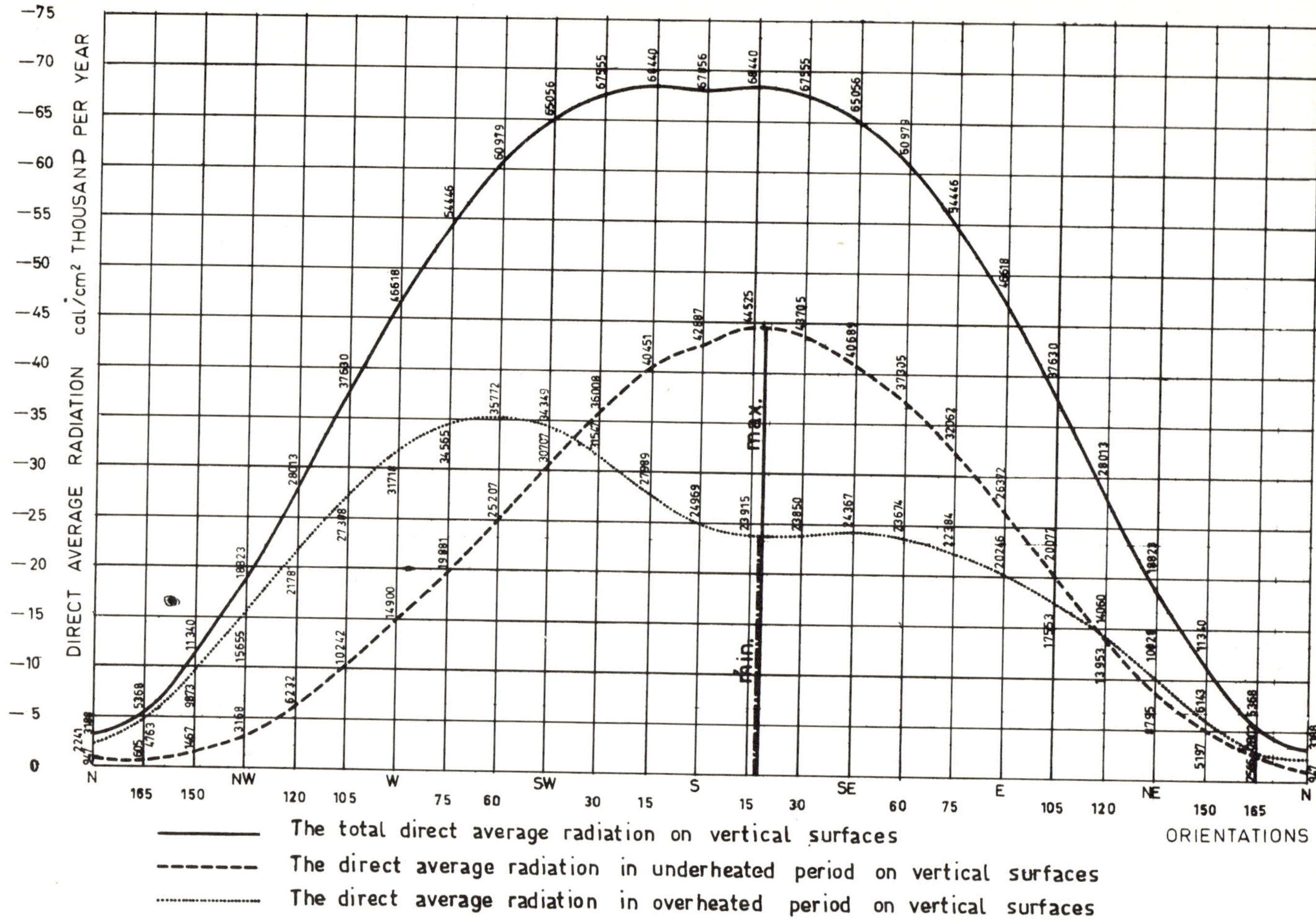

3.3 *Yearly radiation received at different orientations in Diyarbakır area*

nation of a thermal belt possible. If a theoretical cross section of land is considered—its valley, slope and plateau—the thermal belt is the intermediate zone on the slope between the plateau and the valley where there is optimum climatic performance (see **3.5**). At different locations on the cross section mentioned, there are different effects occurring from climatic elements; it would be possible to correlate them with the human climatic needs of the corresponding climatic zones. The results of many investigations have proved that for each particular climate zone, the optimum human settlement would be at a different elevation, according to the topography and in relation to the thermal belt. Therefore, the locations to be considered are: in temperate climate, at the thermal belt; in cool climate, below the thermal belt; in warm, humid climate, on the plateau; and in warm, arid climate, in the valley's vast flat lands (see **3.5**). Thus, in any climatic zone, correlation of the above-mentioned effects of topography with the determined orientation would give the properly climate-balanced site for human settlement.

Building shape

Because of their widely exposed surfaces, buildings and their masses are influenced more by thermal climatic elements (the combined effect of air temperature and radiation) than by humidity, wind, etc. As these thermal climatic elements give rise to an amount of heat (which differs according to the climate zone), the problem of heat exchange of the building mass in that climate arises. When the heat exchange of building mass (a function of its surface areas) and the human heat needs (deduced from the timetable of human climatic needs) are considered, correlation of the two would solve the problem for a climatic zone. A method has been developed by V. Olgyay for the determination of climate-balanced building shapes.[5] His basic premise is that in any climate zone, the building shape which loses a minimum amount of heat during the underheated period and gains a minimum amount of heat during the overheated period is the optimum. His method also includes a comparative application for the purpose of observing differentiations in building shapes in different climate zones.

For the purpose of investigation, the house type usually selected has a constant area but a variable ratio of sides. Calculations of the heat gains at the overheated period and the heat losses at the underheated period are made for different shapes, with the square shape values being used as reference points. The optimum basic shape is found with the mentioned method, and with flexibility the proportion of a plan can be determined which can be considered as generally good (see **3.6**). The results of such a quantitative investigation are used as a parameter for design purposes in a particular climatic zone. However, the wind effects must also be considered in order to state general information about the building shapes for each of the typical climate zones.

We have already mentioned how proper orientation of buildings leads to proper orientation of the city. Suitable climate balance in the building shapes has a similar ameliorative effect in an urban unit. Buildings are the main organizing elements; and if their shapes are climate-balanced, the unit tends to be developed in a similar shape, with a particular textural pattern of buildings in accordance with the climate of the region.

Building shell

The building shell is the building element which separates the indoor environment from the outdoor environment. Thus, the indoor climate is created by the intervention of the building

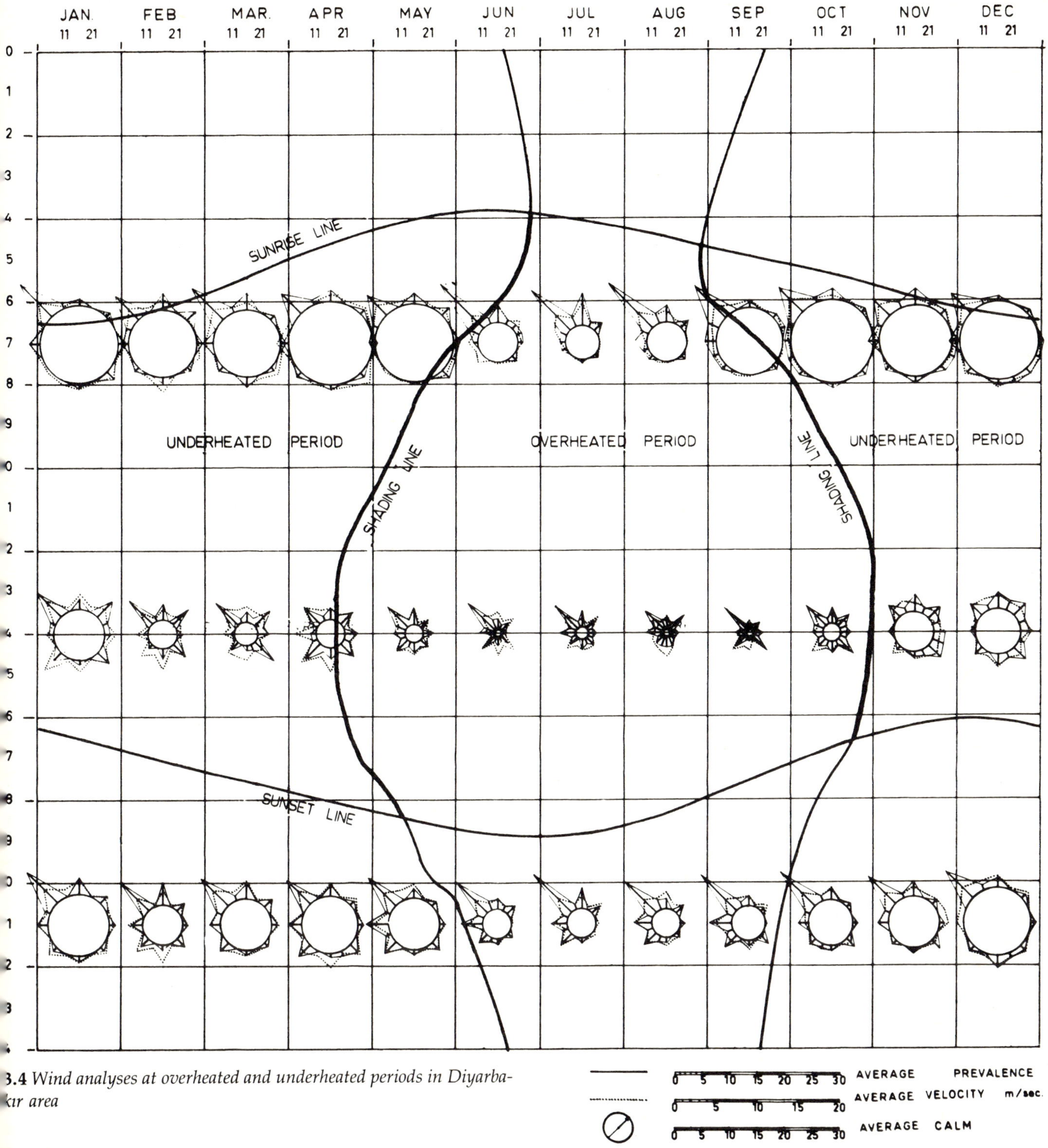

3.4 *Wind analyses at overheated and underheated periods in Diyarbakır area*

shell. Or to put it another way, the function of the building shell is to maintain the indoor bioclimatic comfort conditions despite the outdoor climate.

The building shell is always under the influence of the climatic elements, especially the thermal elements represented as sol-air temperature. This sol-air temperature impact creates a heat on the outside surface which flows into the inside surface in amounts consistent with the properties of the building shell and dependent on the thermal diffusivity resulting from both capacity and resistance insulation.

V. Olgyay has proposed a method of evaluating attempts to obtain a climate-balanced building shell.[6] The calculation of

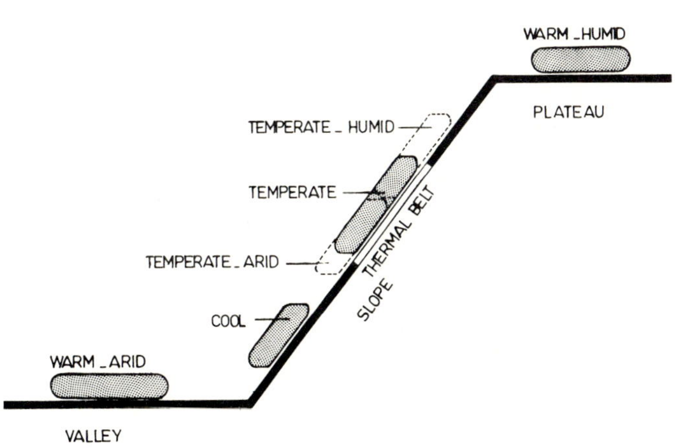

3.5 *Thermal belt and suitable settlement site elevations for different climatic zones*

the capacity insulation or time lag requirement has been determined with the aid of sol-air temperature graphs of different exposures. The translation of them onto the axis of the abscissa, by having the maximum impact coincided with the minimum, makes it possible to read the shifting time. Calculation of the resistance insulation or the overall heat transfer coefficient requirement has also been based on sol-air temperature impacts. The hypothesis is that the desired amount of insulation is in direct relationship to the difference between outside thermal conditions and inside comfort requirements. To account for these differences, two thermal conditions were considered: (a) the 'balanced insulation' which represents the average deviations of sol-air temperature impact at different exposures from the indoor comfort conditions, determined by relative insulation values; and (b) the maximum deviation from the design temperatures of different localities, represented as the 'insulation index'. This method suggests interpretation of these values according to the varieties of heat load experienced by the different exposures of the building shell. Olgyay's method does not propose a definite result for the resistance insulation of the building shell, but it is interesting from the general approach point of view.

In my investigation in Turkey, I developed the following method: (a) for capacity insulation, I used the previous method to reach some general conclusions; however, I used flexibility in regard to the time of use and the function of the concerned space in the building; and (b) for the resistance insulation, I found that the 'balanced insulation' (at relative insulation values) could be determined with the same principle. However, for the determination of the overall heat transfer coefficient, I proposed another procedure.

Since the living spaces in dwellings are basic elements and since their dimensions are limited, the radiation from inside surfaces on the users becomes of major importance. This is observed when users in a space may feel uncomfortable while the comfort conditions are maintained. However, the human body can counteract this effect to a limit, without any effort.[7] Therefore, the influencing inside surface temperatures of a building shell could be obtained by augmenting the comfort temperature for the overheated period and by diminishing it for the underheated period according to the body's limit value without any change in comfort conditions. Then the resistance insulation (the overall heat transfer coefficient, the 'U' value) could be calculated by means of the appropriate formula for different exposures and for the roof of buildings in a climatic zone.

The resulting 'U' values (combined values for opaque and transparent surfaces of a building shell) are effective, and values for the overheated and the underheated periods can be calculated. The weighted averages according to the percentages of the yearly duration of each period help us to find the yearly balanced effective 'U' values. Since the surface of a building shell is composed of certain percentages of opaque and transparent surfaces, represented by their 'U' values, the sum of the same percentages of 'U' values equals the balanced effective 'U' values.[8] With this relationship in mind, we tentatively can assume 'U' values in regard to the required capacity insulations of the opaque surfaces and can assume that the transparent surfaces are shaded. Then it is possible to determine the percentages of the opaque and transparent surfaces suitable for different exposures in a climatic zone.

By these methods we are able to find out definite information concerning our climate-balanced building shell. As a general rule, we have learned that resistance insulation is required in humid zones, while capacity insulation is best in arid zones.

APPLICATION IN A WARM, ARID ZONE

The approaches and methods I have mentioned have been applied in my investigations in Turkey. Diyarbakır was selected as a pilot area for the warm, arid zone. (The location of the meteorological station was at 37° 55′ N latitude and 40° 12′ E longitude with an elevation of 660 m (2165ft) above sea level.) My objective in this chapter is to demonstrate the values determined for this warm, arid area in order to give the reader and the investigator the opportunity to compare them with information on other areas. I shall base my suggestions on the quantitative values.

Evaluation of human climatic needs

In order to evaluate human climatic needs in the Diyarbakır area, we first established the overheated and the underheated period because this determination is essential in climatological studies for urban and architectural usage. We superimposed charts covering a ten year period of average dry bulb temperatures over those showing average relative humidity percentages. This enabled us to draw another chart showing a shading line, representing the overheated and the underheated periods, according to coordinates on the bioclimatic chart (see **3.1**). On such a chart, we placed the days of the year on the abscissa and the hours of the day in standard time for Turkey on the ordinate (see **3.2**).

When the percentages are computed of the diurnal yearly durations in the Diyarbakır area, we find: 47.64 per cent for the overheated period and 52.36 per cent for the underheated period. The duration of the overheated period is usually shorter in warm, arid zones than in warm, humid zones due to the extremes in diurnal conditions. Generalizing, we could say that during the 47.64 per cent of the diurnal yearly duration (the overheated period), shade is needed; and during the remaining 52.36 per cent (the underheated period), both radiation and protection from the wind are needed. The chart clearly shows the time when shade or radiation is needed. We need to have this information clearly in mind when we consider solar control. When the overheated and the underheated periods are plotted on the sun-path diagram of 38° N latitude, it is possible to overlay the shading mask and to read the efficiency of a shading device for that area.[9]

The second step is the registration of climatic data on the bioclimatic chart. The averages of dry bulb temperatures and relative humidity percentages for a typical day in each month (the 21st) are plotted on the bioclimatic chart. When they are connected, closed curves indicate the monthly variations (see **3.1**). These monthly variations give a clear idea of the climatic needs in this area. For warm, arid zones generally, the yearly situation of the curves is almost diagonal according to the intensity of the climate.

This evaluation of the Diyarbakır area can also be transferred to a yearly chart as a timetable so that the varying climatic needs of any days in the year can be read simultaneously (see **3.2**). In other words, the timetable of human climatic needs indicates which element, at what time, and what intensity is needed to maintain the feeling of comfort. When the timetable of human climatic needs is interpreted, the general character of the curves shows that the daily variations in intensities of all elements are great. Late in the afternoon before sunset, a decrease of values starts; this continues till sunrise, and then the abrupt increase starts. This is characteristic of the warm, arid zone.

During the underheated period, there is a rather high need for heat which cannot be satisfied by direct natural means in some parts of the period. During the overheated period, the area lying between the boundaries of the shading line and of the 0.1 m/sec (20ft/min) wind velocity curve is the comfort period; only shade is needed. The wind area, where curves show velocities from 0.1 m/sec to 3.5 m/sec (700ft/min), lasts for a very short period. Even high wind velocity cannot restore comfort for a rather long period when additional artificial climatization is needed. Note that the velocities would be annoying when they are over 1.5 m/sec (300ft/min) for indoor conditions and over 3.5 m/sec for outdoor conditions. The suggested interpretations for the Diyarbakır area also can be represented by the percentages of the diurnal yearly duration. The average yearly needs are: 52.36 per cent heat (but 43.59 per cent of this need can be met with the sun's radiation); 47.64 per cent shade; 52.36 per cent wind protection; and 10.15 per cent breeze period. During 19.22 per cent of the time, additional artificial climatization is required; 18.27 per cent of the time is a comfortable period.

Determination of the design parameters

I do not intend to give detailed solutions for the determination of design parameters; rather I shall state briefly the results obtained by the application of methods previously discussed. This will give an idea (or lessons) based on the values found in the Diyarbakır area.

Orientation

Wind effects would not be taken into consideration for orientation because, in an arid area (such as in Diyarbakır), the wind's character (or its relative thermal property) is generally warm during the overheated period and cold during the underheated period. Therefore, some protective measures are needed during both of the periods. To determine the exact protective measures required, we need a wind analysis of the area. A satisfactory solution is to plot the monthly average wind roses of different observation hours on the chart of overheated and underheated periods (see **3.4**). This enables us to make a simultaneous evaluation of the needs. As a general rule, the wind's character is not available from official meteorological reports; however, in most cases, very general information will be sufficient.[10]

For orientation in this area, the solar effect is of primary importance. Where Sol-air Orientation Theory has been applied, the radiation received by a unit area of a vertical surface exposed at different orientations was calculated for the overheated and underheated periods. We considered the insolation factor, which is the ratio of the yearly average insolation duration to the yearly possible insolation duration (IF=0.673 for this area). The values calculated are graphically represented in **3.3**. With the aid of the theory's criteria, we determined that the optimum orientation is 18° east of south; good orientations are from south to 40° east of south.

Site selection

I have previously mentioned why the effect of topography was of such major importance in site selection. The site is best located in a valley or at a lower and flat part of the land so that there is a noticeable decrease of wind effects, but an increase of helpful cold-air currents. As the characteristics of these cold-air currents are known, attention must be given to retaining their cooling effects as long as possible. With such a location, orientation of the site becomes less important because of the flatness. Practically speaking, the very low inclinations can be considered as flat because the sun's radiation on these surfaces is approximately similar to radiation on horizontal surfaces.

Therefore, I selected a site (an experimental unit of 5000 inhabitants) on the flat and very large basin of the Dicle (Tigris) River lying in an east-west direction within one of the wide bends of the river. The elevation is around 565 m (1854ft) and the geographical location, 37° 49′ N latitude and 40° 26′ E longitude. This would make it possible to change the wind's character from all directions except from the north.

Building shape

To determine the optimum building shape for this area, I applied the method already discussed. Essentially, we have a heat exchange problem; according to the method, the heat gains during the overheated period and the heat losses during the underheated period have to be calculated. The purpose of the method is comparative, and a reference house type was selected for investigation in pilot areas of all climate zones of Turkey.

For the calculation of incoming and outgoing heat, thermal impacts had to be analyzed. Therefore, we made an auxiliary investigation into the sol-air temperatures for vertical surfaces exposed in four main directions and for the horizontal surface. We selected typical days for the overheated and the underheated periods. The values of heat gains and heat losses, calculated by the periodic heat flow method, were plotted on a graph in order to make the minimums of both of the represented curves for the overheated and the underheated periods coincide in accordance with the method (see **3.6**). For the area in question the following proportions—1/1.16 as optimum and 1/1.4 as adapted flexibility—had been determined for a climate-balanced basic building shape. This led us to the general conclusion that the building should be compact. However, if the wind effects are taken into account, the patio type of building is the most suitable in this region.

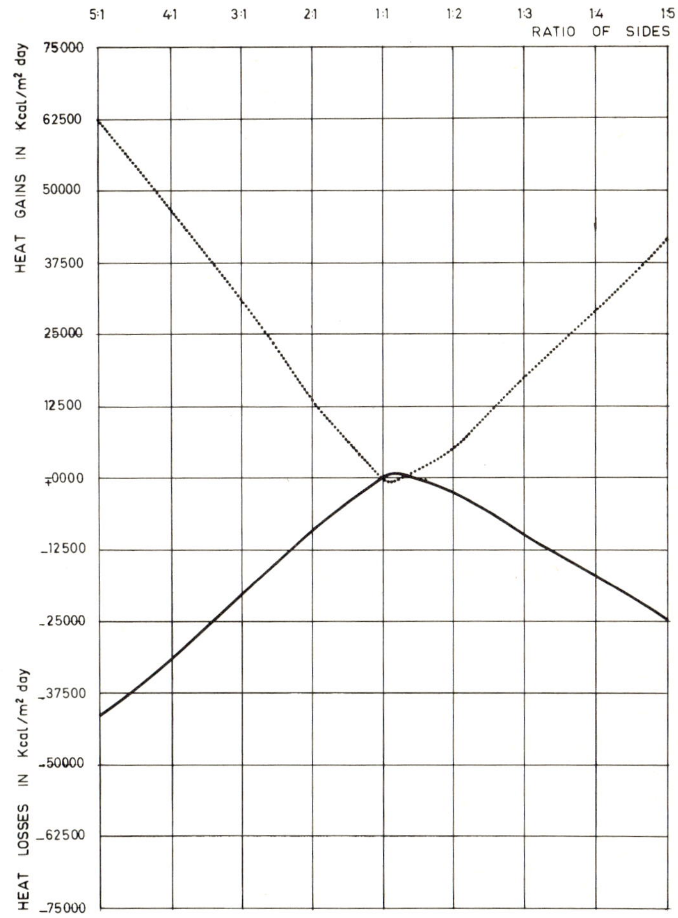

3.6 *Graphs of calculations for determining basic building shape in Diyarbakır area*

Building shell

We determined the climate-balanced building shell by application of the method mentioned earlier in the chapter. We considered the four directions relative to the optimum orientation already determined and decided the roof should be flat. To calculate the capacity insulation needed, the graphs of sol-air temperatures were used. We translated them onto the axis of the abscissa; the coincidence of the minimum values with the maximums gave the shifting time or the necessary time lag for the building shells, exposed in four directions, and for the roof. The values obtained were as follows: 11 hours for 18° west of north, 12 hours for 18° north of east, 11 hours and 45 minutes for 18° east of south (the optimum orientation), 11 hours for 18° north of west, and 12 hours and 15 minutes for the roof. These are general values; for that reason, flexibility is possible. Therefore, some change in the values above could be tolerated in regard to the appropriate time of use and the function of a space. In some cases, it might be useful to decrease or to minimize, or vice versa, the time lag of the shell of a space. Next we considered the amount of resistance insulation needed. We determined the relative insulation values, the deviations from the comfort temperatures, for the overheated and the underheated periods of each of the main orientations related to the optimum orientation and of the roof so as to give measure for 'balanced insulation'.

The inside surface temperatures of the building shell were determined as previously described. Deviation from the comfort temperatures of both periods was taken as 2.7°C (5°F) as concluded from the values proposed by many investigators. As for the averages of sol-air temperatures, the daytime (7 am to 7 pm) averages were considered for the above-mentioned orientations and for the roof.

We calculated the effective overall heat transfer coefficients for the four orientations, the roof, and the two periods. Their weighted averages in relation to the durations of the overheated and the underheated periods are as follows:

1.39 Kcal/m^2 h °C (0.29 BTU/ft^2 h °F) for 18° west of north orientation;

1.33 Kcal/m^2 h °C (0.27 BTU/ft^2 h °F) for 18° north of east orientation;

1.49 Kcal/m^2 h °C (0.31 BTU/ft^2 h °F) for 18° east of south orientation (the optimum orientation);

1.35 Kcal/m^2 h °C (0.28 BTU/ft^2 h °F) for 18° south of west orientation; and

1.23 Kcal/m^2 h °C (0.25 BTU/ft^2 h °F) for the roof.

The effective 'U' values might then be used for the determination of the recommended percentages of opaque and transparent surfaces with the aid of the mentioned relationship, according to the materials to be used for the construction of the building shell. The 'U' values obtained show that the resistance insulation of the building shell is also important because of the high thermal impacts and their variations.

PAST EXPERIENCES

In urban and architectural studies, one should never neglect the early practices. From the climatological point of view, man has always striven to build his dwellings or cities so that he has had most favourable conditions for his comfort. In all cases, he has found some fairly good and some perfect solutions. Close examination of buildings or cities gives rise to some questions about these solutions, however. The task of a designer or of an investigator is to ask himself these questions, to determine the problems, and to search for modern solutions. Indeed, this was the purpose in my investigation.

After the determination of human climatic needs in the selected locations, a survey of the climate zones in Turkey was completed in the summer of 1965*. The research included documentations, studies and measuring drawings, and it has been the basis of a continuing study. Within these regions, the southeastern area has a warm, arid climate; many cities were founded there in early historical times. There were cultural centres that influenced the art and architecture in many of the surrounding countries—for example, in the city of Diyarbakır, selected as a pilot area for this region, where the first settlement was in 3500 BC.[11] We will try to demonstrate with some samples from cities of the region the approach to and the solutions for problems by considering the design parameters in a general sense.

We found that the cities and the buildings in them were placed so as to take advantage of the optimum insolation for human needs. For instance, when the city map of early Diyarbakır is examined, two main axes in the east-west and north-south directions are visible. The city's orientation is approximately south (see **3.7**).[12] The buildings have the same orientation, which is in accordance with the scientifically determined good orientations. Some exceptions could be observed in patio-type buildings which were oriented according to customs and function. In this early city of Diyarbakır, solar control was provided by the construction of an arcade, tall garden walls, and shaded, narrow streets.

It is interesting that these ancient cities had climate-balanced sites. In most cases the early settlers built on flat land

*In 1980 Professor Zeren completed a new bioclimatic chart which was published by NBS.

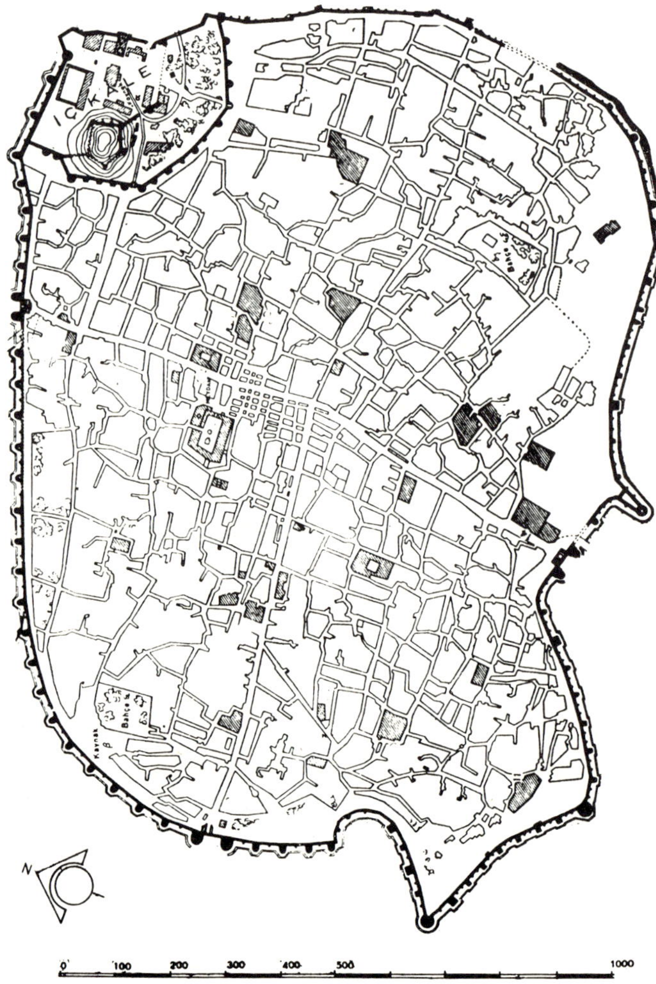

3.7 *City map of early Diyarbakır*

SOURCE: MAP ISSUED BY THE CITY OF DIYARBAKIR. ALSO SEE METIN SÖZEN, *Diyarbakır'da Türk Mimarisi* TURKISH ARCHITECTURE IN DIYARBAKIR (ISTANBUL: GUN MATBAASI, 1971)

instead of on slopes or plateaus, unless there were other influencing factors. The selection of the site was intuitive and based on experience. However, we note the sites actually are in accordance with the criteria previously mentioned. Therefore, these early settlers chose to profit from cold-air currents and to make use of protection from the wind effects which are warm in overheated periods and cold in underheated periods.

The form of the early historical cities can be learnt by tracing their peripheral walls which are generally similar to the form of the city because of the wall's function. The effects of climate might not have been considered in building them; however, since city walls were built to surround the city, at least the part forming the nucleus had to be organized. Investigations have concluded that in most of these organizations, the climatic conditions that influence the form were taken into account. Thus, the formation of city walls was at least indirectly influenced by the city form. If the city map of Diyarbakır is examined again, we can find the proportion of its form by the ratio of the two main axes, approximately 1/1.4 (see **3.7**). The result is within the adapted flexibility proportion determined to be satisfactory for this area. While this may be a fortuitous coincidence, it improves the hypothesis.

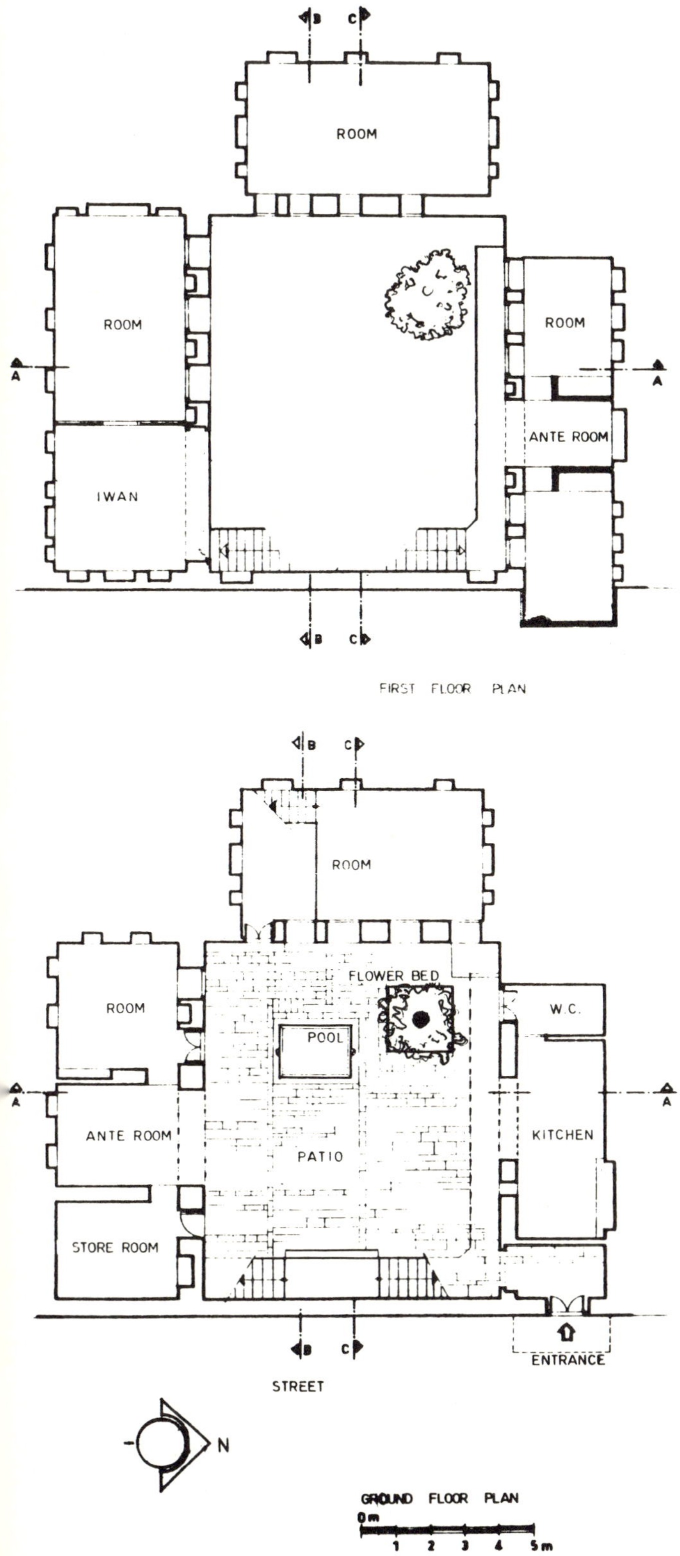

3.8 *Plan of a house in Diyarbakır*

In warm, arid zones, the expansion of cities is a more important consideration than in other zones. The expansion generally takes the form of concentric circles and results in a single massive unit. Therefore, the expanded cities lose their responsiveness to climate, and this becomes one of the weakest parts of the later organization. In such a climate, for the organization of cities and for their expansion, I suggest the form of grouped compact units because of the climatic and microclimatic effects previously mentioned.

The house shapes coincide quite closely with the determined proportion for the area. Drawings of one house give a clear idea of my hypothesis (see **3.8**). In general, the houses are compact and have a patio. The size of the patio depends on the size of the house which is proportional to the owner's social level. Usually, a large area of the patio is stone covered, and there are deciduous trees and a pool on the remaining part. The houses are placed next to each other in a row which creates the pattern of the streets. The entrance to each house is through a gate and into the patio. This eliminates direct inlet of outdoor conditions to the house itself. The above descriptions and evaluations prove that the shape of these very old houses was climate-balanced.

During my visit to that region, I noted the interesting construction concept of the building shell. The vertical building shell, in other words the exterior walls, were always thick masonry, mostly of stone because of its abundance. There are very few and very well protected window openings on the street side. Most windows open on the patio side and are at two levels. The second register of windows, called head windows, are near the ceiling and decorated, and they are smaller in size for ventilation purposes with some fixed open. Such a disposition of window openings is climatically favourable in both the overheated and the underheated periods. However, the desire for privacy is another highly influencing factor in the window placement. The timber ceilings are supported by strong joists, sometimes very nicely decorated. The roofs are flat in most cases, and are supported with similar constructional elements covered with adobe.

From the above description, it can readily be seen that in construction of the building shell, attention was given to capacity insulation (in this case a natural climatization) essential in warm, arid zones. But the applied construction system provides a rather high resistance insulation, too, Such an intuitive solution for the building shell is, of course, the result of experience. The time lag obtained allows the outside daytime thermal impacts to reach the inside at night for a daily period. The accumulations of these delays gives rise to a yearly pattern, wherein one can feel cooler in the overheated period and warmer in the underheated period in a building.

SUGGESTIONS FOR FUTURE CONSIDERATIONS

When the main purpose of an urban designer or an architectural designer is to express regional character in his design, a climate-balanced organization can be quite appropriate to his concept. Other influencing factors may change in the course of time, but the climatic character will never change. Variations can be expected in the intensity of the microclimate, however.

These can be balanced by means of preventive adaptations in the built environment. Classification of the climatic character of different regions must be made in the way cited in this chapter in order to give the designers an opportunity to find definite urban and architectural expressions for them. A climatic zone might be warm, arid, or hot, arid, or very hot, arid; the architectural and urban expression would be similar while showing the gradation of intensity with the aid of the delineated quantitative approach. The general characteristics and the differentiation in the details would present the true specific regional character. We are now ready to generalize the information based on the analyses and the evaluations cited and to make suggestions for design considerations in warm, arid zones.

Shade and water to provide moisture are the two main needs and are essential in warm, arid zones. Winds are welcome when their characters are changed. These three elements would be basic to organizing a built environment, and all the constituent elements and their correlations must be developed from this point of view because of the intensity of the climatic conditions during the overheated period. The process of shading a built environment should very deliberate. We must break the sun's rays in the overheated period and admit them in the underheated period—a principle in solar control. Once the periods are determined for an area, the efficient and definite shading devices could be designed accordingly. These shading devices could be natural or artificial. As natural devices, deciduous trees and climbing plants are preferable because they can show good performance in both periods for this area. However, when the yearly duration of the overheated period is excessive, evergreen plants become more efficient. The artificial shading devices are man-made design elements and must be built to permit air convection and to hinder the accumulation of warm air. When all the necessary investigations have been completed for an area, the manipulation of shading devices becomes easy with today's technology.

Environmental design in warm, arid zones requires that the ground surface be shaded as much as possible. Buildings, garden walls, or hedges must be organized to provide abundant shade on nearby surfaces and to prevent heat gain. Pedestrian ways—public squares, walkways, sidewalks etc—must all be shaded. Nowadays, streets and roads cannot be narrowed to get shade as they were in the early historical cities. Shading could be accomplished by means of trees with tall trunks which would allow air movement beneath them. These trees could be planted between wide lanes, with repeated planting patterns to help solve the problem. It should be noted that in warm, arid zones it is important to organize the design elements in groups instead of in large single units. Parking areas also provide a relevant application. When they are large, they are affected by direct and reflected solar radiation and by long-wave radiation. Heat islands with high emissivity develop and influence the surrounding elements. Therefore, like the building units, the parking areas should be reduced in size and be very well shaded. We can achieve the size needed by repetition of parking areas. The glazed surface of buildings should always be shaded in accordance with the principles of solar control. Materials with high reflectivity and low absorptivity are preferable in making shading devices. Even the suggested compact building shape offers shading possibilities if there are some recessed parts on the surface of the facades to provide a shading effect. Roofs of buildings could be shaded and ventilated or covered with materials having high reflectivity and low absorptivity to minimize heat gain.

In warm, arid zones, waters providing moisture (lakes, rivers, ponds, pools, etc) function as a regulator of thermal conditions. Therefore, we should very consciously use water so as to profit from this property. As early as the site selection phase, we must search for an area where water exists. Water is essential to the design of a built environment, and it must be considered a design element and must be combined as much as possible with other elements. Perhaps we could use water in all the common sections of the built environment in narrow waterways that widen at some places and sometimes form ponds or pools. Shading enhances the effects. Water pools should be requisite elements in residential sections to be used in patios and even houses. Since water influence on the microclimate of the area is considerable, the cited suggestions seem to be a very good idea and not impossible with today's technology. Evaporation of water moistens the ambient air which is necessary in such a climate according to the timetable of human climatic needs. On the other hand the temperature of water is warmer than the ambient air in underheated periods and cooler in overheated periods, and this gives rise to some local convection. In some warm, arid areas it may not be possible to provide sufficient water to carry out all the above suggestions. In that case, shading becomes more important.

Air movement is essential in all warm climates; but in arid zones, it is undesirable because of its character. The purpose of a designer, therefore, must be to change the character of winds by using the principles of wind control. The information obtained from wind analyses enables the location of preventive measures according to prevailing wind directions. This could be done on a large scale outside the built environment and on a smaller scale closer to that environment. To change the character of wind, evaporation from the ground is needed. Having determined the prevailing wind direction, the ground surface layers should have vegetation which will aid evaporation; water in some form could be provided on the windward side of the elements and units in the built environment. We should consider air movement in buildings, too, by means of cross ventilation but always in relation to outside shaded and moistured areas.

The cold-air currents, as previously mentioned, initiate from the higher elevations of the land and flow down, like a fluid, to the lower parts during the night. Therefore, the built environment must be so organized that the path of these cold-air currents is not obstructed and they enter within the elements as much as possible.

Future planning in warm, arid zones must correlate the above suggestions with the climate control design parameters, based on biometeorological analyses and quantitative evaluations. Otherwise cities in warm, humid zones would be built like those in warm, arid zones or vice versa. I have seen this happen in some areas along the Mediterranean coast, a warm, humid zone which had the character of warm, arid zones.

In my investigation in Turkey, I designed for five pilot areas experimental settlements of 5000 inhabitants so as to show the structure of different types of climate-balanced built environments. By using equal elements in each of them and by applying all designated climate-balanced design parameters and all the suggestions made here, the surface areas of these settlements turned out to be different.

I hope the conclusion above and the approaches mentioned in this chapter may cast light on future investigations concerning primarily the density of cities in different climate zones.

NOTES

[1] Victor Olgyay, Aladar Olgyay and Associates, *Applications of Climatic Data to House Design* (Washington, DC: US Housing and Home Finance Agency 1953).

[2] Victor Olgyay and Aladar Olgyay, 'The Theory of Sol-Air Orientation', *Architectural Forum*, March 1954, 133–137.
[3] Olgyay, Olgyay and Associates, *Applications of Climatic Data to House Design*, pp. 90–104.
[4] Rudolf Geiger, *The Climate Near the Ground*, revised edition (Cambridge: Harvard University Press, 1965), pp. 393–417.
[5] Victor Olgyay, *Design With Climate* (Princeton, NJ: Princeton University Press, 1963), pp. 84–93.
[6] Ibid, pp. 113–125.
[7] Olgyay, Olgyay and Associates, *Applications of Climatic Data to House Design*, p. 7.
[8] Aladar Olgyay, Wayne F. Koppes, John H. Callender and Robert W. McLaughlin, *Thermal Behavior of Curtain Walls*, Curtain Wall Research Project, School of Architecture, Study No. 6 (Princeton, NJ: Princeton University Press, 1957), p. 22.
[9] Lütfi Zeren, *Türkiyenin Tipik İklim Bölgelerinde En Sıcak Devre ve En Az Sıcak Devre Tayini* [Determination of the overheated and underheated periods in the typical climate zones of Turkey] (İstanbul: İ.T.Ü. Mimarlık Fakultesi Matbaası, 1967), p. 35.
[10] Imanuel S. Wiener, 'Solar Orientation: Application of Local Wind Factors', *Progressive Architecture*, 36 (February 1955), 112.
[11] *Meydan Larousse* (Istanbul: Meydan Yayınevi, 1970), 3: 766.
[12] Metin Sözen, Diyarbakır 'da Türk Mimarisi [Turkish architecture in Diyarbakır] (İstanbul: Gün Matbaası, 1971), p. 21.

SUGGESTED READING

Aronin, Jeffreye. *Climate and Architecture*. New York: Reinhold, 1953.
Dreyfus, Jacques. *Le Comfort Dans l'Habitat en Pays Tropical*. Paris: Editions Eyrolles, 1960.
Fanger, P.O. *Thermal Comfort*. New York: McGraw-Hill, 1972.
Geiger, Rudolf. *The Climate Near the Ground*. Cambridge: Harvard University Press, 1965.
Givoni, Baruch. *Man, Climate and Architecture*. London: Elsevier Publishing Company, 1969.
Housing and Home Finance Agency, Housing Research, *Application of Climate Data to House Design*. Washington, DC: US Government Printing Office, 1954.
Olgyay, Aladar, and Victor Olgyay. *Solar Control and Shading Devices*. Princeton: Princeton University Press, 1957.
Olgyay, Victor. *Design With Climate*. Princeton: Princeton University Press, 1963.
Zeren, Lütfi. *Mimaride Günes Kontrolu* [Sun control in architecture]. İstanbul: Fakulteler Matbaası, 1959.
Türkiye'nin Tipik İklim Bölgelerinde En Sıcak Devre ve En Az Sıcak Devre Tayini [Determination of the overheated and underheated periods in the typical climate zones of Turkey]. İstanbul: İ.T.Ü. Mimarlik Fakultesi Matbaası, 1967.

4 Feasibility of the Australian desert coast for future urban settlements

ERNEST GORDON HALLSWORTH

Although some of the great civilizations of the ancient world developed in desert areas in the irrigation communities of Egypt, Mesopotamia, and also Peru, the great population increase after the Industrial Revolution was an urban one. The cities which consequently grew have been in the better watered lands of the temperate zone.

In the last few years, the loss of productive farm land associated with the expansion of cities into great conurbations has attracted attention. In the United States, some 30 million hectares (ha) (72 million acres (a)) of land suitable for dry farming are now covered by urban settlements, roads, railways, and airports, and another million ha are being transferred from agricultural production every year.[1] The situation in Europe is much the same. In both North America and Europe, the hot desert area is relatively small (less than 10 per cent of the total area). By contrast, in Australia where European settlement began in an almost empty continent just over 200 years ago, 532 million ha out of a total of 769 million ha receive insufficient rain for dry farming or improved pastures. Consequently, two-thirds of the country must be considered desert.[2] Furthermore, in the United States and in the Union of Soviet Socialist Republics the deserts occupy inland areas, but in Australia the desert extends to the sea.

There are two large sections of the Australian coastline that are desert. One, in the south, is along the Bight coast; and the other, in the northwest of Western Australia, is on either side of the Tropic of Capricorn (**4.1**). If either region could be used for urban settlement, Australia would find an alternative to the continuous loss of productive land caused by the same population growth that bedevils the rest of the developed world. There are advantages and disadvantages.

CLIMATIC RESTRICTIONS

Lack of water

The two major restrictions on the development of human settlements in hot deserts have been the lack of water and the high daily temperatures. Because of the lack of water, all major settlements in desert areas have been on the banks of rivers rising outside the desert itself. However, for those parts of the desert which border the seacoast, modern technology can now remove this restriction by desalinating sea water. For instance, an oil-fired plant is now supplying the city of Hong Kong with water at $5.00 per 1000gal. J. Simmonds calculates that a nuclear power station generating power for a city of 500 000 people would supply water at $.80 per 1000gal.[3] Where pumping through a long pipeline is involved, the cost would be comparable.[4] Alternative procedures, such as freeze desalination—thermodynamically more attractive than distillation—or reverse osmosis, may further reduce the cost of desalinization.

Temperature stress

The second restriction cannot be considered in terms of temperature alone, for high humidity exacerbates the effect of high temperature. There are now several indices used to compare these effects. In Lee and Henschel's study, the critical figure for discomfort is when their index rises to more than 0.3 at 1500 hours.[5] In the southeastern and southwestern corners, where the bulk of the Australian population lives, the index rises above 0.3 on less than 25 days during the year.[6] On the northeastern coast (Queensland), it rises above 0.3 on less than 75 days during the year (**4.2**). Cold stress also affects southern Australia. Nix's index of cold stress shows the proportion of the year when the temperature at 1500 hours is below 15°C in Australia (**4.3**).[7] Here we see that the plateau to the east of Perth in southwestern Australia and much of the densely settled southeastern regions suffer cold stress for more than 25 per cent of the year.

By comparing **4.1**, **4.2** and **4.3**, we can see that the desert coast of the Bight, showing both cold stress for less than 25 per cent of the time and heat stress for less than 25 days a year, looks particularly favourable. The desert coast of the northwest on either side of the Tropic of Capricorn shows no cold stress; but the number of days on which the heat discomfort

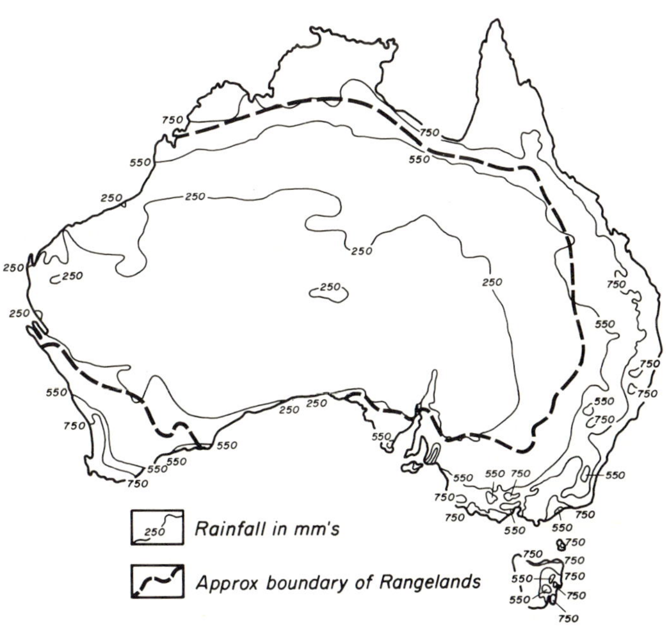

4.1 *Map of annual rainfall and area suited for dry farming, Australia*
DRAWING BY R. A. PERRY IN E. G. HALLSWORTH ET AL, *Principles of a Balanced Land-Use Policy for Australia* (MELBOURNE: CSIRO, 1976)

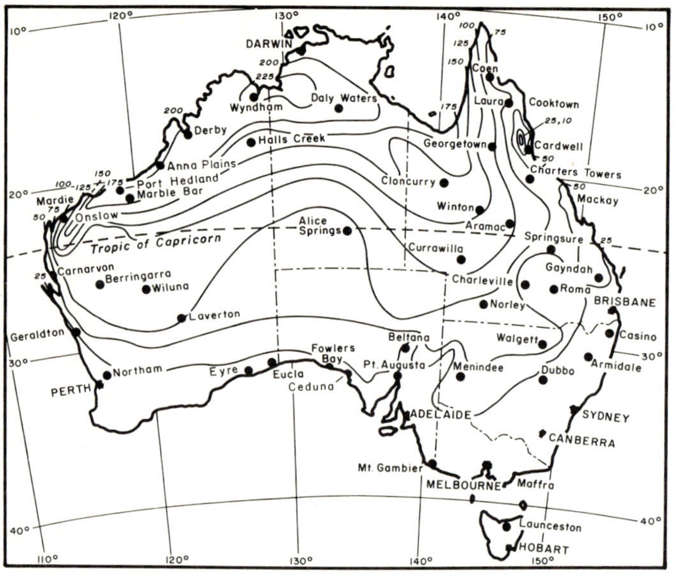

4.2 *Discomfort index for temperature and humidity as indicated by Lee and Henschel: the number of days when the index rises above 0.3 at 1500 hrs*

FROM C. E. HOUNAM, 'CLIMATE AND AIR-CONDITIONING REQUIREMENTS IN SPARSELY POPULATED AREAS OF AUSTRALIA', IN *Building Climatology* (GENEVA: SECRETARIAT OF THE WORLD METEOROLOGICAL ORGANIZATION, 1970), p. 182

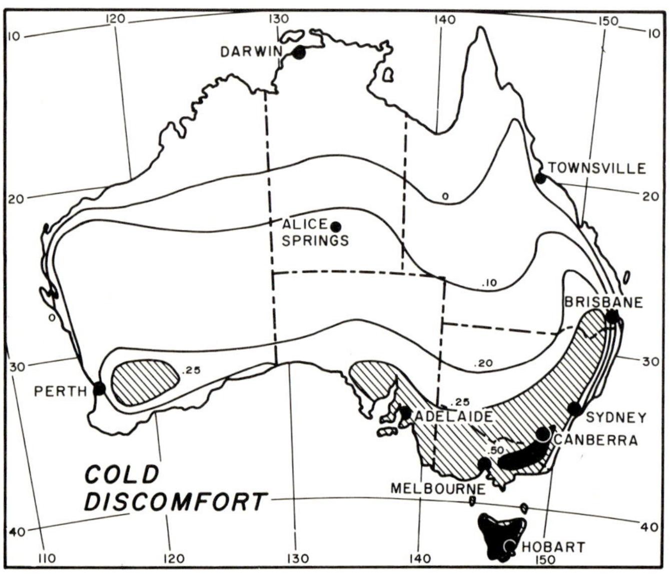

Areas experiencing cold stress 25–49 percent of the time

Areas experiencing cold stress 50+ percent of the time

4.3 *Proportion of the year when the Australian temperature at 1500 hours is below 15°C*

DRAWING BY H. A. NIX

index exceeds 0.3 increases northwards from less than 25 at Carnarvon to 75 at Onslow, and more frequently as one moves north (**4.2**).

To overcome the restrictions of high temperature, one must provide cooling mechanisms. This can be easily accomplished by electric refrigerated air conditioning. In principle, this is the same as removing cold stress in winter in cool temperate lands by heating through air conditioning (a more sophisticated approach to heating than by the familiar hot water pipes, gas fires, or open fires). However, the cost of making comfortable conditions in houses, offices, and factories in a city on a desert coast largely depends upon the extent to which solar energy can be used for heating, and rock piles for cooling. Solar powered air cooling units have now been designed but are not yet in commercial production in Australia. There are two variants of the rock pile system—one in which heat is transferred from the air to a rock pile, another in which water is sprayed on the layers of rock for evaporative cooling. Both units require more space and are heavier per unit of lost heat than a conventional air conditioner. For the rock cooling system to be more economical than a conventional air-conditioning unit, the minimum night temperature must not exceed 21.5°C (70°F), and a large diurnal variation should exist. The regenerative system also works better with lower humidity. Consequently, rock cooling systems will operate more effectively in climatic conditions typical of desert areas. The effectiveness of such a system in the Melbourne climate is illustrated by **4.4**.[8]

Since both night temperature and diurnal fluctuation are related to cloud cover, it is possible to assess the relative comfort value of various possible city sites from simple meteorological measurements. Further selection can be made by taking into account the third component of the evaporative potential, the direction and strength of the wind.

The climatic limitations of two specific sites on the Bight coast and two on the northwest coast are illustrated in **4.6**. Both Ceduna (which is not a desert area) and Eyre (which is) have mean minimum temperatures not exceeding 15°C. The difference between maximum and minimum is in excess of 10°C (11.35°C at Eyre and 13.0°C at Ceduna) with a low coefficient of variation. Since this difference is also greatest in the summer months, the Bight coast is suitable for rock cooling systems.

The sites on the northwest coast show certain notable differences. For example, the mean minimum reaches 25°C at Port Hedland and 22° at Onslow, while the difference between maximum and minimum is 10°C in December at Port Hedland but rises to 12°C or more for most of the year. At Onslow, the difference falls to 10°C only in March. It exceeds this figure on all other occasions and rises above 15°C in November and December. At no time at either centre is the rainfall adequate for the growth of commercial crops; the relative humidity of the atmosphere is always very low.

From these figures it seems unlikely that simple rock-bed cooling and breezeway-ventilation at night could keep the internal temperature of buildings on the northwest coast a suitable 25°C. Consequently, on this part of the desert coast, one would have to use a regenerative or a refrigerated cooling system. The costs of maintaining satisfactory working conditions would be higher than on the Bight. That part of the northwest desert coast south of Onslow shows similar conditions to the Bight coast, however.

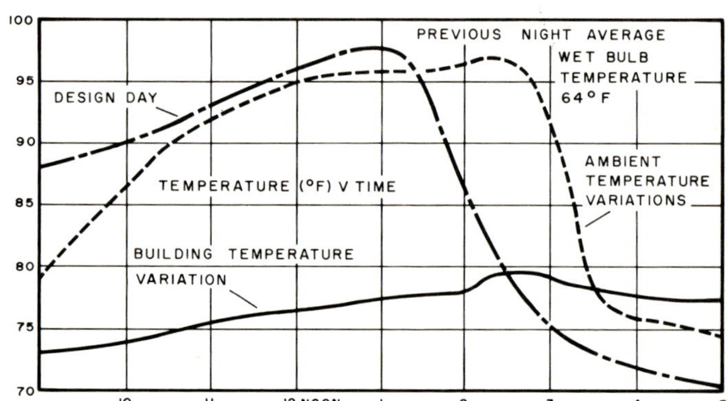

4.4 *Building and ambient temperature records for high cooling load, Melbourne*

FROM D. J. CLOSE, R. V. DUNKLE, AND K. A. ROBESON, 'DESIGN AND PERFORMANCE OF A THERMAL STORAGE AIR-CONDITIONING SYSTEM', IN *Mechanical and Chemical Engineering Transactions*, INSTITUTION OF ENGINEERS (AUSTRALIA) REPORT MC4, 1968, pp. 45–54

ADVANTAGES OF THE DESERT COAST

Optimal size of a city

With the availability of cooling systems and water, one can consider the desert coast for human settlement. In doing this, one should determine the optimal size of the settlement. As city size increases, there are better educational and medical facilities, improved facilities for theatre and concerts, and a greater variety of organized sports. In the early and middle stages of growth, there is also greater employment opportunity. In response to the pressures in developed countries throughout the world, big centres of population have grown larger, and small ones have grown smaller. Of the great metropolises of the world, only London, which began to appreciate the disadvantages of great size earliest, has made a successful attempt to limit growth and to reduce its population.[9]

The disadvantages of increasing size are:

1 Increased time and cost of moving from home to place of work.
2 Greater distance between production sites of perishable foods and their point of consumption.
3 Increased pollution arising from the concentration of industry and traffic.
4 Greater distance required to travel to enjoy the recreational resources of nonorganized sports and the countryside.
5 Pressure of the metropolitan population on recreational resources—in Australia, the waterfront.[10]
6 Social problems that arise from the loss of a sense of identity; (for in the larger and more mobile community, the sense of neighbourliness, or belonging, tends to be lost).
7 Partly as a result of all these, an increase in tension, in what is commonly called the pace of life (Bornstein and Bornstein have shown that this is literally true: the bigger the city, the faster people walk on their daily business **4.5**).[11]

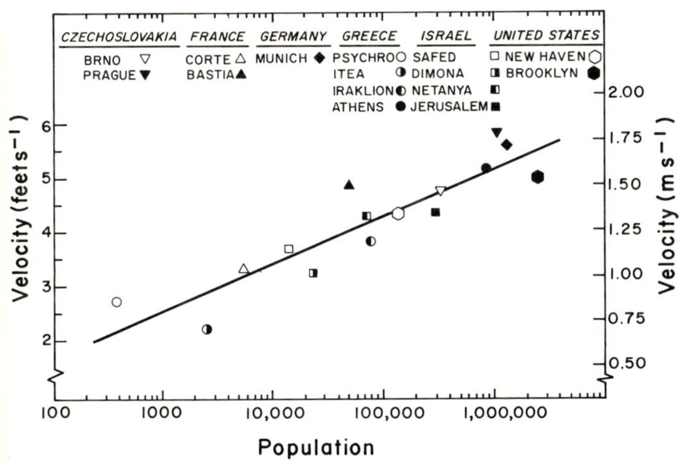

4.5 *The pace of life*

FROM L. M. BORNSTEIN AND H. G. BORNSTEIN, 'THE PACE OF LIFE', *Nature*, 299 (1976), 557–558

There is no need to comment further on the first disadvantage. It is obvious to anyone who has commuted in London or New York, Sydney or Los Angeles. The second disadvantage may seem less important to people living in the colder climates of the major conurbations of Europe and North America; they are accustomed to 'fresh' fruit and vegetables arriving at the market either jaded or preserved by snap freezing. It is more obvious in Australia, where Adelaide, Perth, and Brisbane (in spite of mushrooming metropolitan sprawl), still draw most of the fresh fruit, vegetables, and milk from within a radius of 50 miles of the city centre. To supply perishable foods to cities in the desert may seem more difficult, but this need not be so.

The third disadvantage, the increased pollution arising from the concentration of industry and traffic, is now attracting attention all over the world. There are two aspects. One is the increase in the concentration of airborne pollutants toward the centre of the city as illustrated by the data for Melbourne.[12] The other is the increased volume of urban refuse and sewage that has to be removed from the urban area; for this removal, transport costs will increase with any increase in the length of the network although economies of scale may reduce operating costs.[13]

The other disadvantages require no further comment. Whilst not begging the issue as to what is the ideal size for a city, my comments below pertain to a conurbation of 500 000 people.

Reuse of water

One of the problems of the present Australian conurbations, in common with those found elsewhere, is that they have grown up without reference to resource use. Without exception they rely on water carriage for disposal of their sewage into the sea. This practice results in the loss of increasingly expensive water and of phosphates and nitrogen in the sewage. It has been calculated, for example, that an amount equal to 5 per cent of all the phosphorus used in fertilizers in agriculture in New South Wales each year is discharged through the Sydney sewers into the sea.[14]

If a new city were to be built on the desert coastline, it would be possible for the sewage treatment plant to be sited to allow reuse of the effluent water. Assuming that the quantity produced by a city on the desert coast is similar to the dry-weather flow at Adelaide—75 US gal per person per day—the sewage outflow from a city of half a million people would supply 24in a year to 20 000 acres. The city would thus generate enough water to irrigate its own market garden area, while part of the power needed for pumping to the distribution areas could be generated by fermentation of the sewage during the treatment process.

The reuse of water from sewage for irrigation and the total abandonment of running the sewage into the sea would also minimize pollution. This process would keep the beaches clear of the pollutants from the effluent which is often washed to shore by winds and unusual tides. There would also be less pollution in the atmosphere. Increased organic matter content of the ocean water near the sewer outfall contributes to aerial pollution. As the Lamontagne study has shown, the action of sunlight on organic matter in seawater releases carbon monoxide into the atmosphere.[15]

When we consider city pollution, in general we find dispersion the most satisfactory way of disposing of biodegradable wastes. Logically, it is easier to disperse the wastes of a city of 500 000 than one of 5 million inhabitants. Consequently, in the last decade, the half-million people that have poured into each of the city conurbations of Melbourne and Sydney would have been better situated in two separate cities of 500 000 inhabitants, each away from the metropolitan centres.

Actually, the fundamental advantage of building on the desert coast would be low land cost. This would encourage people to construct buildings far apart. Several advantages stem from this ability to develop a low profile city.

Dispersal of air-borne pollutants

The materials of atmospheric pollution in city areas arise from motor car exhaust gases (concentrating toward the centre of the city owing to the concentration of motor traffic there) and from the fumes and particulate matter from industrial and domestic fuel consumption. The increase in concentration of

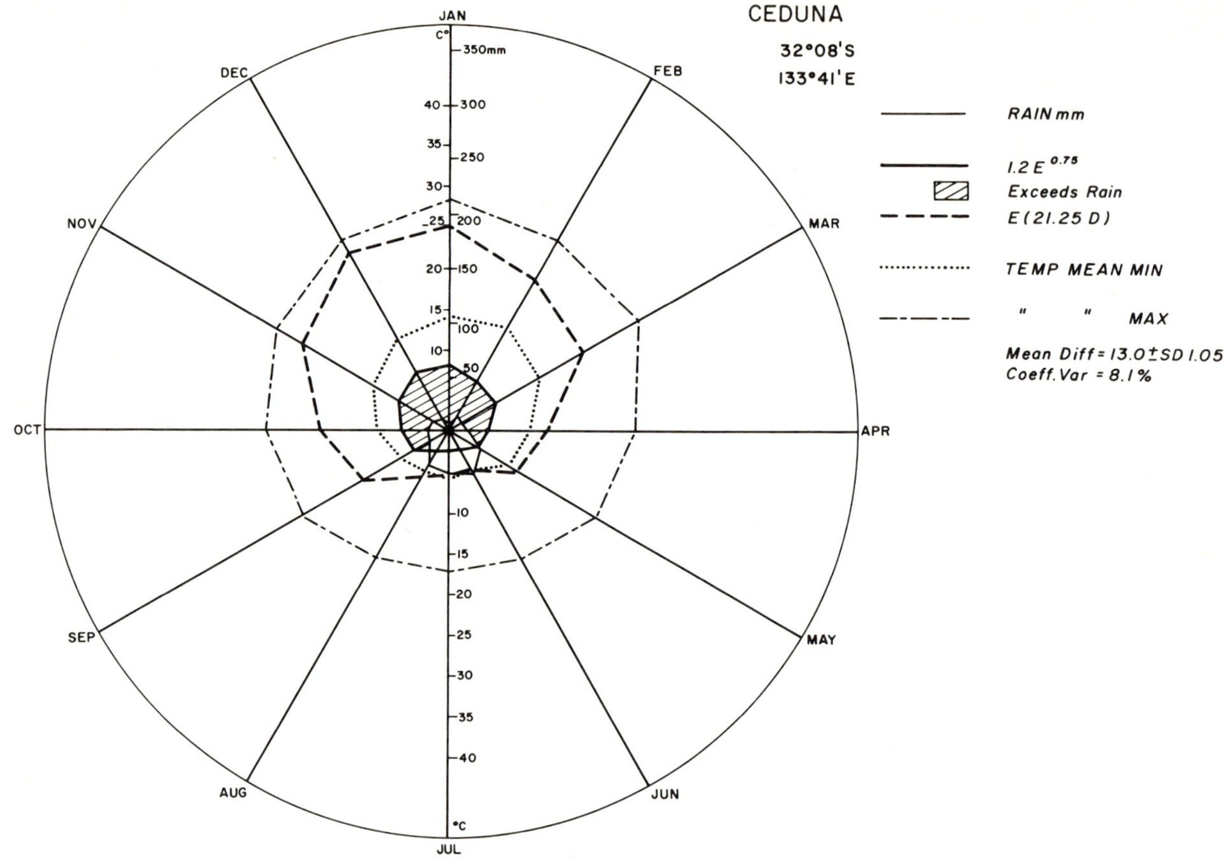

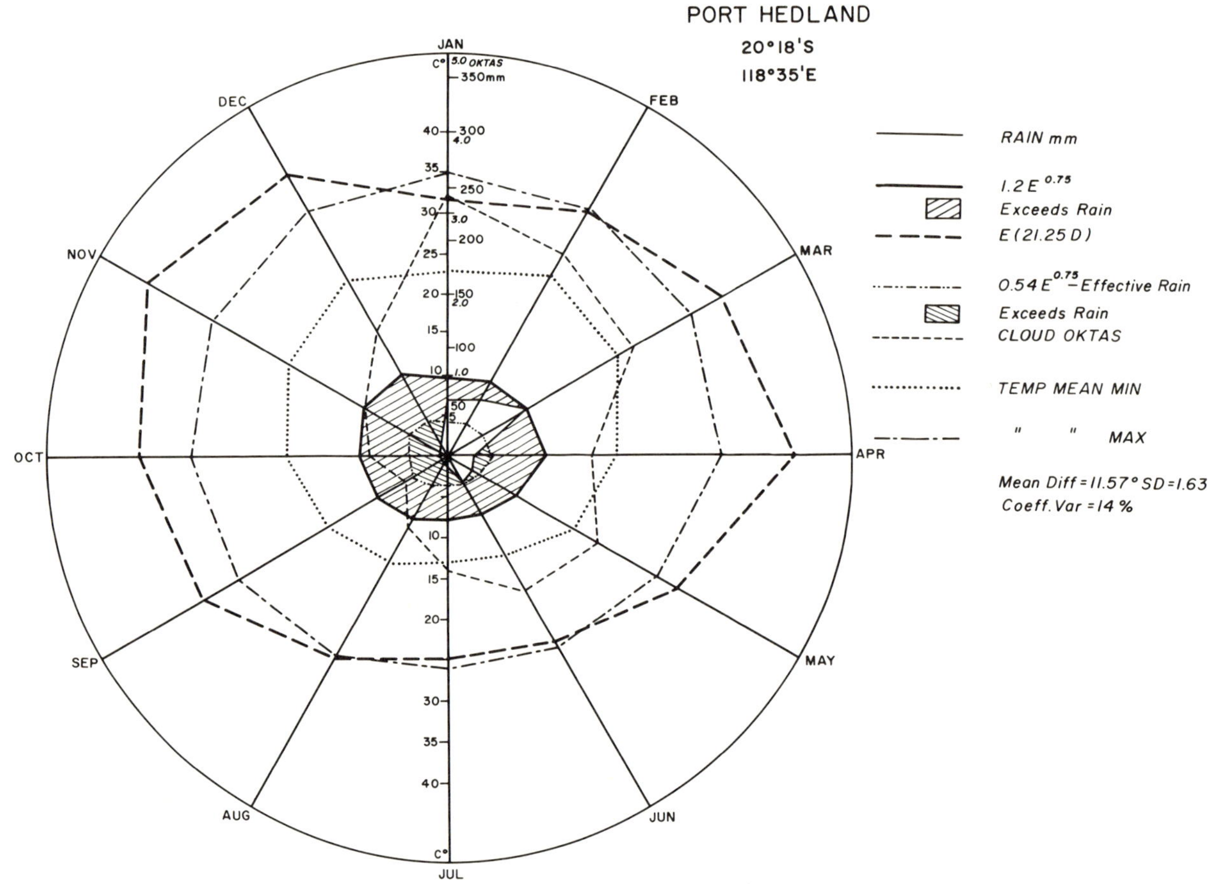

4.6 *Climatic parameters of four localities, two on the Bight and two on the northwest coast. Ceduna v. Port Hedland; Eyre v. Onslow*

DRAWINGS BY J. ARCHIBALD AND E. G. HALLSWORTH

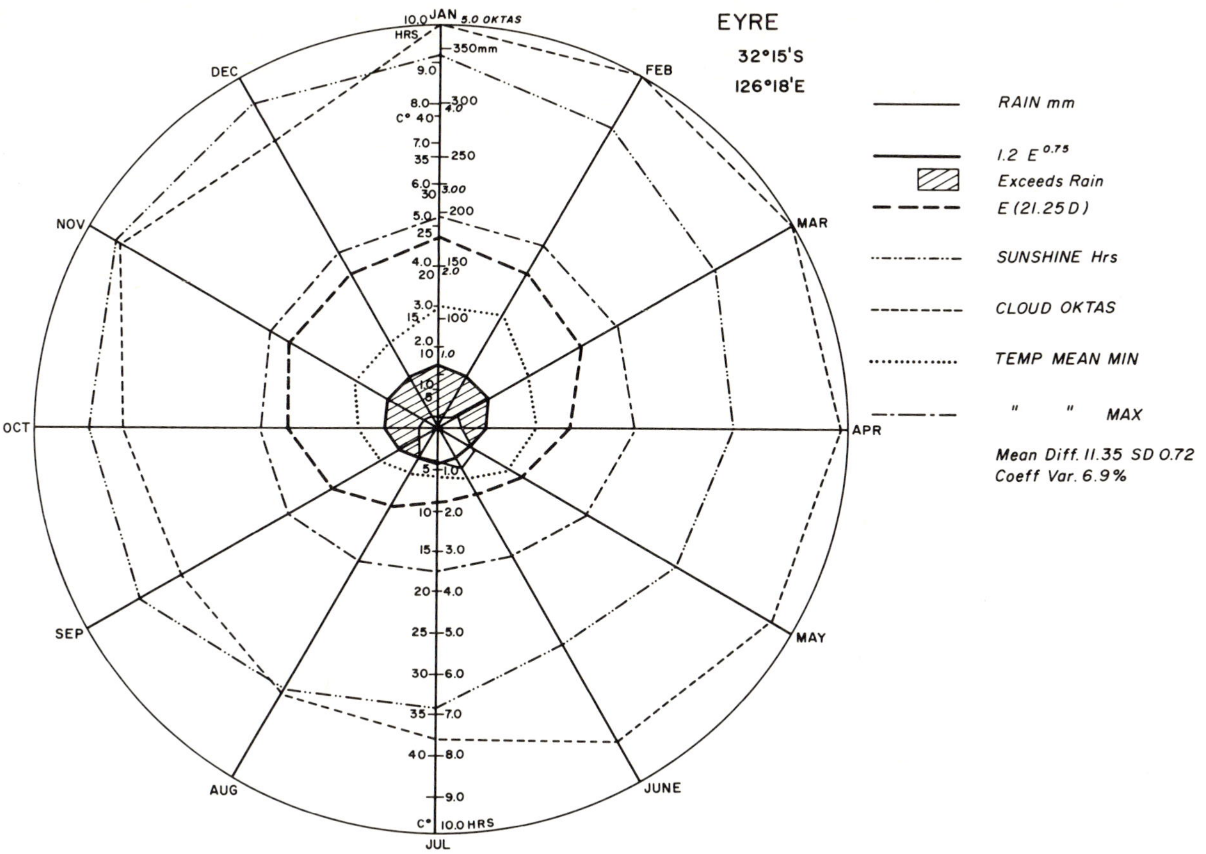
EYRE
32°15'S
126°18'E
RAIN mm
1.2 E^0.75
Exceeds Rain
E (21.25 D)
SUNSHINE Hrs
CLOUD OKTAS
TEMP MEAN MIN
" " MAX
Mean Diff. 11.35 SD 0.72
Coeff Var. 6.9%
JAN
FEB
MAR
APR
MAY
JUNE
JUL
AUG
SEP
OCT
NOV
DEC
5.0 OKTAS
350mm
10.0 HRS
C°

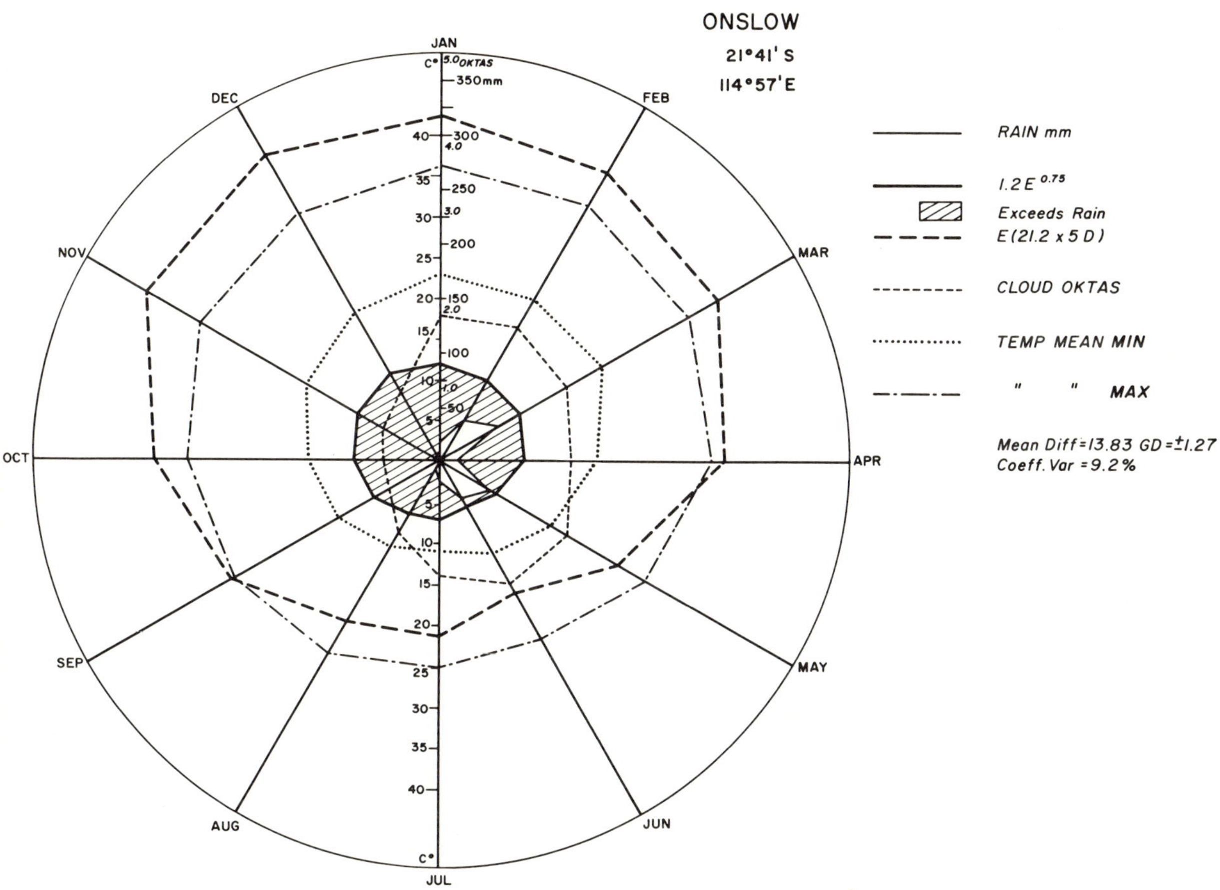
ONSLOW
21°41'S
114°57'E
RAIN mm
1.2 E^0.75
Exceeds Rain
E (21.2 x 5 D)
CLOUD OKTAS
TEMP MEAN MIN
" " MAX
Mean Diff = 13.83 GD = ±1.27
Coeff. Var = 9.2%
JAN
FEB
MAR
APR
MAY
JUN
JUL
AUG
SEP
OCT
NOV
DEC
5.0 OKTAS
350mm
C°

these materials to unpleasant or toxic levels at any site is due to inadequate ventilation at ground level and is associated with either high buildings clustered closely together or with a containing range (or ranges) of hills such as at Los Angeles, Santiago, and Adelaide, or with the development of temperature inversion. A photochemical smog is produced by the bright sunlight on the accumulated pollutants. It could consequently be a hazard in a desert city.

From first principles it would seem that an increase in the surface roughness, as caused by an increase in the height of the buildings, must decrease the extent of ventilation at ground level.[16] A low profile city with wide streets would be expected to allow better ventilation, and hence better dispersal of air-borne pollutants. Literally all the desert coast of the Bight and the northwest coast as far north as Onslow and beyond is free from containing ranges of hills, while the high insolation of the desert allows optimum use of solar energy and so reduces the likelihood of a temperature inversion.

Better use of solar energy

At the present time, equipment has been developed to allow the use of solar energy for domestic hot water heating, for food processing, and for space heating for offices, shops, and houses. For this production of low-grade heat, the solar collectors are large; but building them into the roof construction both reduces total cost and is more aesthetically desirable. The low land cost of the desert sites is important in making the greatest use of solar heat, for a low profile city with mainly single- or two-floor buildings provides the most effective balance between floor area of the building and the area of solar collector needed to provide the heat required.

Since the solar energy use for heating buildings and homes is merely the sun's energy that would be falling on that part of the earth's surface anyway, the supply of this part of the energy needs of the city by solar power would not contribute at all to the development of the heat domes and consequent inversions that characterise all our present metropolises. The use of solar energy in these ways would consequently also help minimize smog.

Effective ventilation

It is also possible in a single-floor building to make better use of natural wind action for ventilation. The addition of sliding glass walls would facilitate the advantages of breezeways throughout the house. By allowing the cooler night air that is characteristic of deserts to circulate throughout the house, one could make the building a unit for storing coolness. Houses designed on such lines have been used in Accra in West Africa for ten years, but they do not seem to have been widely used elsewhere. Furthermore, winds can penetrate better through wide streets. In coastal localities, sea breezes cool the land and reduce the power load for the domestic cooling.

Infrastructure

Because people fear an increase in the cost of services such as roads and power, water, and sewage lines, they sometimes criticize low density housing. However, the practical and financial advantages of this plan are numerous. In estimating the cost of roads and services, it is essential to consider not only the initial cost, but also the costs of maintenance.

Adjustments to the pattern of road construction used in existing urban areas may provide considerable savings. For example, it is not necessary for a wide suburban road in a city in the desert to be wholly covered with bitumen. If only the central strip (about two lanes wide) were bitumen and the rest were left as soil, the cost would be minimized and other advantages would follow. It would, for example, reduce the advective heat load which is inevitably higher from a bitumen road than from a vegetated verge. In addition, there would be no need to make curbs or to construct storm water drains. Provided the roadway were wide enough, rain falling on the bitumen strip would be absorbed by the soil at the roadside and would encourage plants to grow along the road's edge. The space between the bitumen road and the footpath would support trees and shrubs, selected from amongst the attractive drought-tolerant native species. Beneath the surface of the strip of land between the sealed roadway and the fences of the gardens would be various supply lines—gas or electricity, water, and sewage.

As a result, the city could avoid the high cost of periodically breaking through and then repairing concrete roadways to alter or install supply lines. There also would not be the disturbance and restriction of transportation caused by the frequent 'road-up' signs that characterize life in most suburbs. The use of a wide strip of soil between the bitumen roadway and the footpath to soak up storm water is possible only in desert areas; in places with higher rainfalls, the total of rain plus runoff, would often exceed the infiltration capacity of the soil strip and turn the verge into a mud bath unsuitable for walking or parking.

General community needs

There are other advantages to smaller communities. They can supply most of the social, educational, and medical needs of most of their people. Both educationally and medically, however, they cannot supply all the opportunities and advantages; but where groups of cities in the 500 000 range are developed within easy reach of each other, all needs could be met. The success of the bush nursing and flying doctor services that were pioneered in Australia and that offered medical services to the small and scattered communities of the Australian arid areas clearly indicates what can be achieved.

One of the problems that afflicts every great metropolis in the world is the provision of good, reasonably priced housing for families whose sizes constantly change. The low cost of land for a city on the desert provides a solution. Houses with breezeways connecting the constituent parts help owners add and subtract additional rooms with minimal trouble. If the additional rooms were designed to be portable, they could be detached from the family home when the children move away and added onto another home. Such units, mainly of timber, do not need to have the stark unlovely lines of the asbestos-sided transportable homes that characterize Australia's iron-mining towns of the northwestern coast. They could be designed as buildings of beauty. With low elevation they would blend well into the bushland or into a garden filled with native Australian shrubs. With wide streets specified for the desert city, movement of the transportable room from one large block to another would present no difficulty.

Recreation

In these climates, the seacoast also has its greatest attraction as a recreational resource. On weekends the population of Brisbane, Sydney, Adelaide and Perth pour out to the beaches, and the traffic congestion on the main roads out of the city centres becomes intense. Webb says that in Western Australia where the population expansion is tending to move into a strip running north and south of Perth, people most frequently go to the beach that is nearest to them.[17] If the Perth pattern were repeated and population units were strung out along the coast, it would be easier to get to this prime recreational resource. There would be less wear and tear on nerves and vehicles and less cost in fuel than in more densely populated city centres. This problem is almost inevitable in a high-rise residential development.

How far apart the units of population need to be developed in order to maintain a sense of community within each of them is not yet clear. Something between 35 and 50 miles may be all that is necessary. If cities were to have 500 000 inhabitants and were 50 miles apart, the desert coasts of southern and north-western Australia could each carry 5 million people—10 million in all. There would be spacious living with a low demand for the energy of fossil fuels, a minimum pollution of atmosphere and sea coast, and no loss of good agricultural land.

CONCLUSION

In planning city development in arid zones, the aim should be to take advantage of both the biophysical and economic environments. The former include high insolation, large diurnal variation in temperature, frequent low night temperatures and low rainfall. The main economic advantage is the low cost of land. Making use of this advantage to develop a city of mainly one- or two-storey buildings allows maximum use of solar heating and thereby reduces the energy demand for fossil fuels. Wide streets with minimum bituminous surface allow optimum ventilation, remove the need for storm water drains and allow easy relocation or repair of service pipes and cables. Large blocks and wide streets facilitate the use of transportable parts of houses and permit consequent flexibility in the size of a house needed by a family at different stages of its development. There is need for fresh thought on house design for arid regions, particularly with regard to roof construction and to provision of courtyards and breezeways to take advantage of the climate.

NOTES

[1] A. K. Biswas, 'Loss of Productive Soil', paper given at a meeting of the International Federation of Institutes of Advanced Study (I.F.I.A.S.) Steering Committee for the Save our Soil Project, Solna, Sweden, 1977.

[2] R. A. Perry, 'Productivity of Arid Australia', in *Arid Lands in Transition*, ed. H. E. Dregne (Washington, DC: American Association for Advancement of Science, 1970), pp. 303–316.

[3] J. Simmonds, 'The Prospects for Nuclear Power', in *Options for Choice—Where Shall We Build Our New Cities*, ed. E. G. Hallsworth (Melbourne, Australia: Commonwealth Scientific and Industrial Research Organization, 1978).

[4] J. W. Holmes, 'The Availability and Cost of Fresh Water', in *Options for Choice*.

[5] D. H. K. Lee and A. Henschel, *Evaluation of the Environment in Shelters*, (Cincinnati, Ohio: US Department of Health, Education and Welfare, Division of Occupational Health, 1963).

[6] C. E. Hounam, 'Climate and Air-conditioning Requirements in Sparsely Populated Areas of Australia', Australian Bureau of Meteorology Working Paper 30/78 (Melbourne, Australia, 1969).

[7] H. A. Nix, 'Land Use Planning for Commercial Agriculture', in *Land Use Planning in North Queensland* (Innisfail, Australia: Australian Institute of Agricultural Science, 1974).

[8] D. J. Close, R. W. Dunkle and K. A. Robeson, 'Design and Performance of a Thermal Air-conditioning System', in *Mechanical and Chemical Engineering Transactions*, MC 4, no. 1 (May 1968), pp. 45–54.

[9] The population of metropolitan London has fallen by 700 000 between the last two censuses, ie, in the last 10 years.

[10] E. G. Hallsworth, 'Making the Most of Australia', *Journal of the Australian Institute of Agricultural Science*, 42 (1976), 167–175.

[11] L. M. Bornstein and H. G. Bornstein, 'The Pace of Life', *Nature*, 299 (1976), 557–558.

[12] I. E. Galbally, 'Ozone and Oxidants in the Surface Air near Melbourne', in *Proceedings of International Clean Air Conference* (Melbourne, 1972); idem, 'Nitrogen Oxides in the Air at Aspendale and Other Places in Victoria', *Clean Air*, 9 (1977), 12–15.

[13] C. B. Townend, 'The Economics of Waste-water Treatment', in *Proceedings of the Institution of Civil Engineers*, 15 (1960), pp. 209–230.

[14] A. R. Aston, R. J. Millington, and J. D. Kalma, 'Nutrients in the Sydney Area', in *The City as a Life System*, Proceedings of the Ecological Society of Australia, ed. H. A. Nix, 7 (1972), pp. 177–191.

[15] R. A. Lamontagne, J. W. Sinnerto, and J. Linnenbom, 'Non-equilibrium of CO and CH_4 at the Air-sea Interface', *Journal of Geophysical Research* 76 (1971), 5117.

[16] Bornstein and Bornstein.

[17] W. Webb, 'Report to Stirling City Council on Scarborough Beach Development' (Perth: Department of Geography, University of Western Australia, 1973).

SUGGESTED READING

Aston, A. R., R. J. Millington, and J. D. Kalma. 'Nutrients in the Sydney Area'. In *The City as a Life System*. Proceedings of the Ecological Society of Australia. Ed., H. A. Nix. Vol. 7 (1972). Pp. 177–191.

Bornstein, L. M., and H. G. Bornstein. 'The Pace of Life'. *Nature*, 299 (1976), 557–558.

Brunt, D. *Physical and Dynamical Meteorology*. Cambridge: Cambridge University Press, 1934.

Close, D. J., R. W. Dunkle, and K. A. Robeson. 'Design and Performance of a Thermal Storage Air Conditioning System'. In *Mechanical and Chemical Engineering Transactions*, MC4, no. 1 (1968), 45–54.

Holmes, J. W. 'The Availability and Cost of Fresh Water'. In *Options for Choice—Where Shall We Build Our New Cities?* Ed. E. G. Hallsworth. Melbourne, Australia: Commonwealth Scientific and Industrial Research Organization, 1978.

Lamontagne, R. A., J. W. Sinerto, and V. J. Linnenbom. 'Non-Equilibrium of CO and CH_4 at the Air-sea Interface'. *Journal of Geophysical Research*, 76 (1971), 5117.

Lee, D. H. K., and A. Henschel. *Evaluation of the Environment in Shelters*. Cincinnati, Ohio: US Department of Health, Education, and Welfare, 1963.

Simmonds, J. 'The Prospects for Nuclear Power'. *In Options for Choice. Where Shall We Build Our New Cities?* Ed. E. G. Hallsworth. Melbourne, Australia: Commonwealth Scientific and Industrial Research Organization, 1978.

Webb, W. 'Report to Stirling City Council on Scarborough Beach Development'. Perth: Department of Geography, University of Western Australia, 1973.

5 City building in the hot climatic zones

A. N. RIMSHA

City building is being improved constantly at a rate depending on the degree of a society's development. The available means of production and the material wealth are determining factors, as are the geographical and climatic environment in which man lives. Cities, during the course of time, have changed in accordance with the society's requirements. The town which formed as a result of the structure of community life had an effect in turn on the outlook of its inhabitants. Together with these social and related functional characteristics, the most important force in city formation is the geographical environment and the climate. The geographical environment, as the city's natural basis, constantly affects the organization of a settlement, accelerating or retarding its development.

A thousand years ago the famous Ibn Sina in the *Kanon Vrachebnoi Nauki* (Canon of medical science) paid a great deal of attention to the problems related to the local climatic and natural characteristics which must be considered when choosing a site for the building of a city:

> The one who selects for himself a residence, must know what the soil is like, the extent to which it [the site] is elevated or low-lying, whether it [the soil] is porous or non-porous, what the water is like, what the quality of the water is like, whether it [the site] is exposed . . . and whether it is high or low. He must know whether the chosen site is accessible to winds or whether it is in a hollow, what the winds are like—are they salubrious or cold, and also what seas, swamps, mountains, and mines are close by . . .[1]

Ibn Sina also devoted a good deal of attention to the environment inside the dwelling. He considered the insulation and ventilation of the premises to be absolutely essential, and he recommended that 'windows and doors should face the east and the north, that the east winds should penetrate into the building, and that the sun should be able to penetrate because it is the sun that sanitizes the air'. The teachings of Ibn Sina on the sanitation of the dwelling had a substantial influence on the formation of high density building in the cities of the eastern Soviet Union. The centuries-old architectural and town building practices of Soviet central Asia, in giving us examples of striking coincidences with the theoretical propositions of Ibn Sina, show that Sina's ideas are not totally original. For example, in the dwellings of Uzbekistan, the natural climatic conditions were considered wisely, and the types of dwelling were designed to differ from each other depending on the characteristcs of the local climate.

TYPES OF CLIMATE

Town builders know well the negative properties of a hot climate which affect man's life. But so far, scientifically-based norms and rules for town building in regions with a hot climate are absent from the literature on theory and practice. Moreover, the variety of natural climatic conditions on the earth requires, in each specific case, the establishment of rules for city building. Fortunately, the variety of climatic conditions observed on the earth can be reduced to a comparatively small number of types. Until recently the most widespread were the classifications based not on one factor but several natural conditions.

A very complete classification based on climate and landscape was given by the Soviet scientist L. S. Berg, who distinguished twelve types of climate: 1 humid tropical forest; 2 tropical forest steppe (savannah); 3 tropical desert; 4 subtropical forest; 5 mediterranean; 6 nontropical desert; 7 steppe; 8 monsoon regions of moderate latitude; 9 deciduous forest of moderate latitude; 10 taiga; 11 tundra and 12 perpetually cold regions.[2] The classification of the tropical climate by the English scientist G. A. Atkinson is also well known. It is based on the temperature and humidity conditions that are predominant over the whole year. He distinguishes six types of climate: hot dry; warm humid; mountain; maritime desert; monsoon (savannah) and oceanic island.[3]

For the purpose of city construction, however, planners can focus on the real climatological boundary between hot and cold, the +20°C isotherm. It tends to shift towards the poles on land and towards the equator over the oceans. The hot climate for town building spreads over all the territories of the earth which have an average annual temperature of +20°C or higher. For construction under the conditions of a hot climate, planners can usually consider the climate as either dry or humid.

A high air temperature and large diurnal temperature fluctuations are characteristic of a hot, dry climate. During the periods of dust storms and during periods of more moderate, dry winds, scattered radiation increases. A weak relative humidity with insignificant precipitation (250mm) is observed. To this type of climate one can assign the inner continental regions of Africa north of the equator, the Near East, and Australia, all of which are relatively poor in vegetation.

The characteristics of a hot, humid climate are a high relative humidity (up to 100 per cent), a large amount of precipitation (more than 500 mm a year), and low barometric pressure. The average temperature is approximately 30°C. The diurnal temperature variation is not large (5–8°C). The considerable cloudiness produces diffused solar radiation. This climate is characteristic of regions of the equatorial belt, and the land is rich in luxuriant vegetation. The hot, humid climatic regions occupy one-fourth of the world's land. These are the most populated territories of the intratropical latitudes, the majority of which are located close to bodies of water. The population density in these regions is considerably higher than in the territories featuring a hot, dry climate.

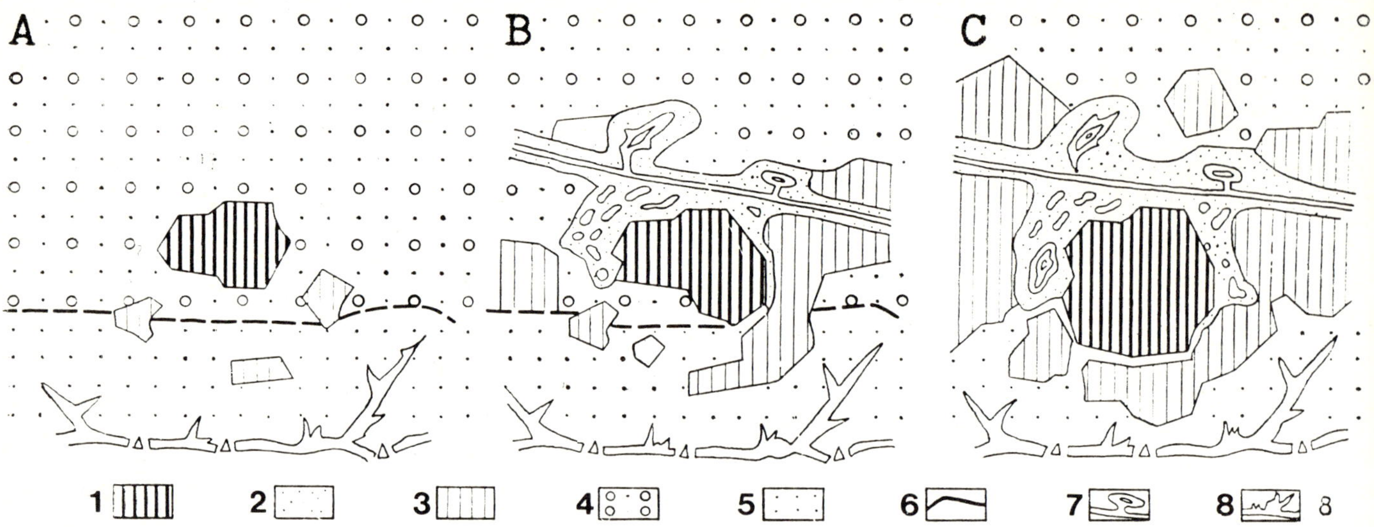

5.1 *Changes in natural surroundings of Ashkhabad, Russia, as a result of irrigation: A the town before the cutting of the Kara-Kum canal; B its present state and prospective future development; C the assimilation of the territory surrounding the town in the distant future; 1 high density construction; 2 green belt; 3 irrigated land; 4 desert; 5 developable territory surrounding the town; 6 boundary between the desert and the land surrounding the town; 7 Kara-Kum canal and reservoirs; 8 mountainous land*

CLIMATE AND URBAN DESIGN

Reducing impact of sun

The microclimate changes with the character of any high density construction. The planned characteristics of high density construction can also affect the microclimate of the surrounding territory. In a hot climate, to reduce the influence of its negative factors, special requirements should be established to ensure the correct orientation of dwellings and community buildings, so that they face (ie, have their broadest side face) principally to the north or south with a possible deviation from the longitudinal axis of 15–17°.

The most unfavourable orientation is in a westerly direction because the slanting western rays of the sun penetrate deeply into the premises and heat them up considerably in the course of five or six hours. A southern or northern orientation is more acceptable because at noon the sun is high in the sky and its rays do not penetrate into the depths of the premises. This orientation of the buildings, especially the dwellings, thus makes it possible to limit the direct solar radiation and also to improve the ventilation of the premises without decreasing natural lighting.

An important special feature of hot localities is the increased brightness of the sky. In these areas, the brightness of the light during the day is three to four times greater than in a temperate belt. Considering the character of the increased brightness of the sky in a hot climatic region, planners should provide for the construction of deeper premises and the reduction of the area of light openings, the arrangement of them along the elevated parts of the walls bounding the premises, and the equipping of them with sun shields.

Reducing impact of winds

The power of the natural climatic factors in different geographical regions of the earth demands special approaches to and evaluations of the qualities of the external environment of the land adjoining the town. In hot climatic zones planners must not try to perfect the environment; rather they should transform it in order to create a qualitatively new environment.

In a hot, dry climate the problem of the creation of a favourable microclimate depends to a great extent on the measures taken to isolate the city from the destructive influence of the desert by the planting and irrigation of trees and shrubs on the land adjoining the city. In the hot, humid zone, the struggle is with calms. Planners should create corridors of forest vegetation in relation to the prevailing wind conditions and try to attract fresh winds from the cool heights of the surrounding hills.

An interesting example of the artificial transformation of nature to create a more favourable, more healthful environment is the transformation of the desert land around Ashkhabad (USSR). Here planners not only improved the surrounding environment of the town, but took decisive measures to radically transform it. Figure **5.1** shows the transformation of the desert surrounding the town of Ashkhabad as a result of the digging of the Kara-Kum canal. Digging this canal made it possible to develop irrigated agriculture and to plant greenery in the territory. With this planting, the dustiness of the air decreased, and its humidity rose.

In the hot, dry regions, the wind (when the air temperature is high) is a source of additional heat. The hot and dry winds lapping around the buildings transmit heat to the walls and to the inside premises. A sandy soil is blown about readily by the wind. This mobile sand often covers oases, productive lands, settlements, etc. Sand dust acts perniciously on vegetation, enveloping its foliage with a layer of dust which reduces considerably the healthful properties of the vegetation. The introduction of water at Ashkhabad reduced this problem.

During the planning of populated places in arid regions, planners should consider:

1 The planned structure of the town must create a space that is isolated and insulated from the desert, ensuring within its confines both coolness and shade.
2 In so far as an exceptional lack of water is characteristic of these regions, it is necessary to search for water resources. The water supply can improve the climate and irrigate the territory.
3 A sandy land is not very suitable for planting. Its agrotechnical processing is required, or fertile soil must be brought in.
4 Because of the weakness of a sandy soil and its drifting which causes considerable dust in the air, the consolidation of the soil is necessary.

Additional considerations

Recently a new science, constructional climatology, has become developed widely. It works out indices for meteorologi cal influence on city building. During the more detailed scientific consideration of the climatic factors influencing urbanized localities located in hot climatic zones, it is important to consider also such factors as the city's latitude and longitude, its elevation above sea level, its average temperatures, the amplitude of temperature fluctuations there, etc.

With the completion of the city building tasks in which planners consider the influence of the physical and special climatic features of the area, it is important to foresee and provide for the town's influence on the climate and the possibility of improving the microclimate of the surrounding region. In other words, during the planning of a new town, it is necessary to visualize the changed microclimatic conditions which will affect the further planning and construction of towns. This possibility of changing the climate is especially important for town building practices in the developing countries of Asia and Africa and, in certain instances, in the regions of the 'fourth group' of the constructional-climatic belts of the USSR.

Urban configuration

Analyzing the influence of the various characteristics of a hot climate on town building in specific geographical regions necessarily leads to a search for the optimum combination of

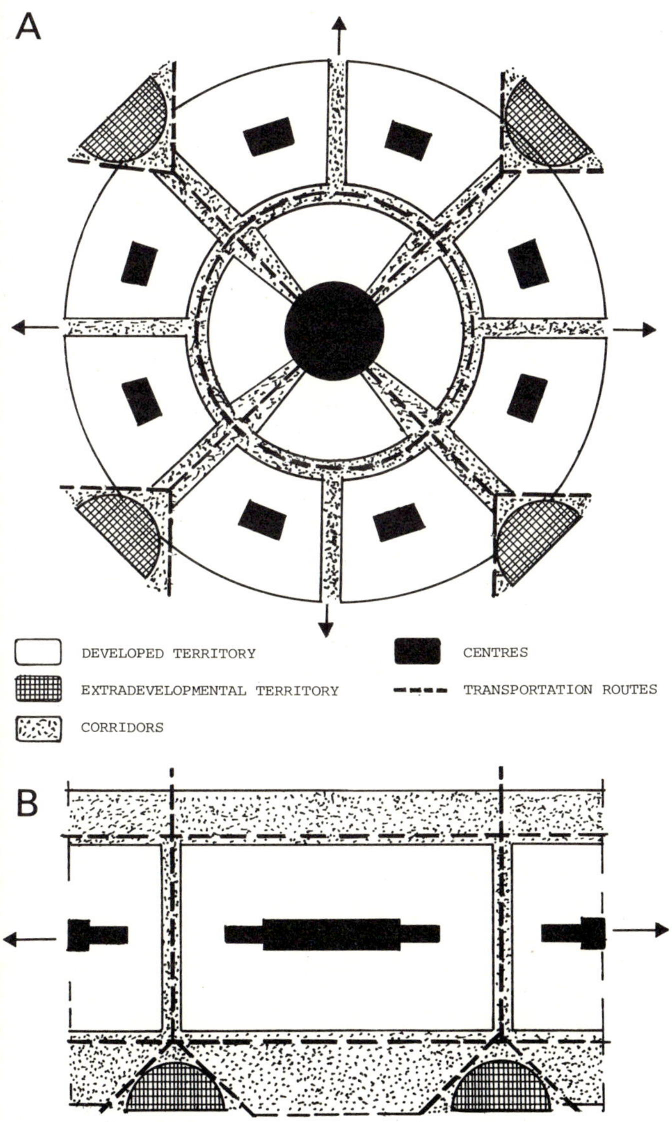

5.2 *Configurations of city planned in relation to climate: A closed; B open*

the various planning variables relating to the specific features of the climate. In general outline, such combinations, it seems to me, could be reduced to two theoretical kinds, closed and open systems, as depicted in **5.2**.

For hot, humid climatic zones, a belt or open system is incomparably preferable to a closed system because it opens up strong possibilities for combating or preventing calms and the possibility of the ventilation of the town environment. But the closed system can be used successfully for settlements located in hot, dry climatic zones where the problems of combating dust storms and excessive solar radiation necessitate the closed nature of the high-density construction.

The climatic and geographical conditions of the locality also exert an influence on the character of the construction of the different functional zones of a city, on their size, and on their configuration. These conditions, if studied, point to an optimum arrangement of an interaction among the zones. Observations carried out for many years on the functioning of towns under hot climatic conditions led Soviet planning specialists to the following conclusion: for towns where the greater part of the year consists of warm days, those urban complexes in which the natural space surrounding a town is in turn surrounded by walls are usual.[4] This special feature of such towns is a reflection of the national traditions that arose historically in localities with a hot climate. Today, the planning of the functional zones of a town in similar locales should be done in line with this practice.

The consideration of the planning principles relevant to the influence of the prevailing winds also is important in city building in hot localities. Moreover, one should bear in mind that in a hot, calm environment, the warm—and consequently lighter—air rises. This forms a certain difference in air pressure, as a result of which the surrounding cooler air flows into the places occupied with the more rarefied air. A local convectional exchange of air thereby occurs between the insulated and shaded areas in the town's residential zone. Thus, the direction and velocity of the air currents in the densely built-up environment are the result of the interaction of dynamic currents of air whose velocity and direction are caused by the aerodynamic properties of the densely built-up area itself.

THE CASE OF KABUL

My work on the general plan for the high density construction of the city of Kabul, a project which entailed the planning of both the high-density construction of the city's centre and the construction of two microregions consisting of a series of houses of large panel construction, has been introduced onto the national scene as an experiment in the field of contemporary city building in hot climatic zones.[5] Because of the nature of the project, it was possible to introduce recent scientific findings into the design of the high-density construction. These findings included the following: the planning of the entire city based on the principle of its microclimatic division into districts and taking into account the historical and socio-economic factors responsible for the development of the city; a full consideration of the influence of the natural climatic factors on the planning of the high-density construction of the centre of different microregions, of the individual buildings and their structures, and of the infrastructure; and the use of national and historical building traditions wherever possible.

The fairly intensive reconstruction of the city now under way and also the rapid mass construction of houses there testify to the importance of the consideration of the climatic factors which influence the general planning of a city. A characteristic feature of Kabul is that the greatest burden of these climatic factors is carried by the central part of the city where, because of the high density and the large number of horizontal areas for transportation routes, there are higher temperatures, while the opportunities for ventilation are impeded. Considering the special features of the climate, three principal axes from the center were developed: 1 the banks of the River Kabul; 2 Maivand Street; and 3 the highway linking the governmental facilities with the airport. By planning development along these three main axes as if along ventilation canals, it was possible to attract cool winds from the snowy heights of Pagman and to direct this fresh air into the web of the town's street network.

Considering the aforementioned climatic peculiarities, the planning of separate complexes in the centre of Kabul was undertaken following the principle of closed construction, even though each one of them must adjoin one of the three ventilation routes. Provision is also made for constant runoff into the Kabul River. The network of canals, artificial reservoirs, and cascades decreases noticeably the thermal burden of the environment.

Protection of the town's centre from dust storms is provided by a necklace of pictureque hillocks from whose summits the characteristic panoramas of the city open up. An extremely important element in the planning of the centre of the city of Kabul, considering its climate, is the use of subterranean space to provide comfortable conditions for pedestrian and vehicular traffic and for parking facilities. This construction also extends the supply of expendable resources and reduces the length of trips to service facilities. A special feature of the planning of the high density sections of Kabul was the consideration of principles derived from the historical national traditions of city building in such a landscape and climate.

CITY BUILDING ON THE GOLODNOI STEPPE

The principles for the planning and building of towns and agricultural settlements in arid regions are reflected also in the experience of city building in the territories adjoining the Kara-Kum Desert known as the 'Golodnoi Steppe' (fourth climatic zone of the Soviet Union). This territory is characterized by the following traits: environs featuring an enormous area of deserts from which winds constantly encroach without hindrance during both the hot and the cold times of the year; groundwater and soil containing a high percentage of mineral salts which are especially detrimental to concrete; groundwater amplitude fluctuating up to 1 m; water insufficient for irrigation and leading to the poor growth of deciduous trees; ground subsidence with an active seismicity of seven to eight marks; insignificant precipitation annually averaging approximately 100 mm of which 54 per cent falls in the spring; a summer shade temperature of +40°C falling to −20°C in the winter.

Given these complex microclimatic conditions, the experience of city builders in constructing a new town in this territory is interesting. The city is planned on land that was free from high-density construction. The measures they took, which included some high-density construction, made it possible to eliminate the natural phenomena which hinder the construction and sound organization of site and, to a definite extent, made it possible to create a new natural environment. They incurred the expense of various ameliorative measures such as providing the artificial irrigation required by the high thermal burden. For territories located in deserts with a high solar radiation and a low air humidity, the microclimatical effect of the artificial humidification of the environment by the provision of open expanses of water will be quite important. The prevalent conditions of a high saline water-table and a nonuniform distribution of the surface flow predetermined the need for a developed drainage system for the land: ie, the coordinated removal of surface, infiltrated, and drainage waters through engineered networks linked with the water supply, sewage system, and heating and cooling systems, etc. Experience has developed several possible variants of the basic irrigation system.

Variant I: a gravitational flow system with water supply from the water taken in along the system's pathways. With this variant it is essential to create an artificial contouring. This will require a large amount of earthwork, difficult in desert conditions. In such a system the water travels along pressured arterial water conduits up to the level of elevated guide marks. The supply and distribution irrigation networks are open canals and reservoirs which receive and simultaneously remove surface waters, possibly shedding some water to the residential microregions. The artificial contouring can be doubly sloped or singly sloped. The merit of this variant is the possibility of following it in the absence of building materials and the simplicity of the arrangement of the supply and distribution networks and the simplicity of the technique of watering. This variant also fits in with the national traditions of the population. The drawbacks of this variant are the necessity of a large volume of earthwork and the possible destabilization of the sanitary conditions of the city, if pollution of the open reservoirs occurs.

Variant II: a pressure system with watering through water sprinklers. With this variant the method of stationary sprinkling is completely dominant. This method is used widely because it increases the microclimatic effect of irrigation and decreases the volume of water consumption. The drawbacks of the method include its limited effectiveness for the sprinkling of large masses of vegetation and the expense of reliable automatic machinery.

Variant III: a combination pressure and gravitational system. This variant provides for the complex use of three kinds of irrigation: gravitational, sprinkler, and subsoil.

Calculations show that Variant II is the most advantageous from the point of view of capital investment and operational expenses. Variant III is just as economical. In light of the expenditures for Variant I per ha of territory, the most expensive economically is Variant I.

NATIONAL TRADITIONS OF TOWN BUILDING

After investigating the problems of town building in a hot climate at different periods of an area's historical development, experts have concluded that the bases of the national traditions are the natural and climatic properties of the geographical environments.[6] The basic conditions comprising the national traditions of town building also include the specific socioeconomic conditions in which the national traditions develop and the level of the technical development of the society. In the development of a national tradition, significant roles are also played by the demographic composition of the population and the aesthetic ideals of the people, including their original tastes, notions, and criteria in construction aesthetics.

Thus, the little peristyle courtyard arose as a result of the experience of many centuries in combating the natural factors which act negatively on the human body. The atrium surrounded by a gallery creates deep shade, cooling the walls and the living quarters arranged behind them, as **5.3** shows. The presence in the centre of the little court of a fountain surrounded with vegetation gradually during the course of a hot day creates coolness and humidifies the space surrounding the courtyard. International experience with low-rise buildings in countries with a hot climate produces many examples where the summer living quarters are similarly organized on an open space or little courtyard adjoining the house. Such

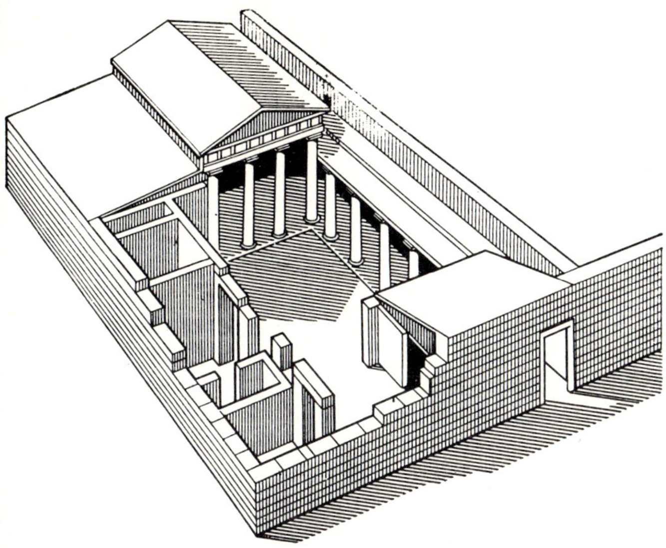

5.3 *Fourth century BC Greek dwelling featuring peristyle courtyard*

houses with little courtyards have become widespread in Spain, Italy, the countries of Latin America, etc. The choice of the type of summer living quarters is determined largely by local climatic features. The national building tradition in Uzbekistan demonstrates this. Three types of houses, Ferganian, Khibinian and Bukharian are found there.

On the Ferganian Plain the method of planning a house in which the inhabited rooms are connected by a terrace is widespread. The windows of the inhabited rooms and the terrace face the southeast and are shielded by a pergola entwined with a grape vine. In the summer this canopy of thick vegetation drastically reduces the effect of solar radiation, and the coolness of the planted space creates convection currents. In the winter the opposite is true; the sun's rays facilitate a certain warming of the terrace and the inhabited rooms. In Khorezm, cool winds which come from the snowy heights blow at night. In response to this, the region's architecture adapts by creating special wind catchers which bring refreshing currents of air into the Khibinian house, increasing the exchange of air and its coolness.

In Bukhara, on the other hand, the great drawback is the hot, dusty winds that announce the desert's proximity. Therefore, here the entire structure of the house is built to insulate the dwelling from the hot air of the desert and to reduce the thermal load of the environment on the house. In response to this goal arises the closed character of the dwelling, as **5.4** shows. The basis of the plan of the house consists of summer

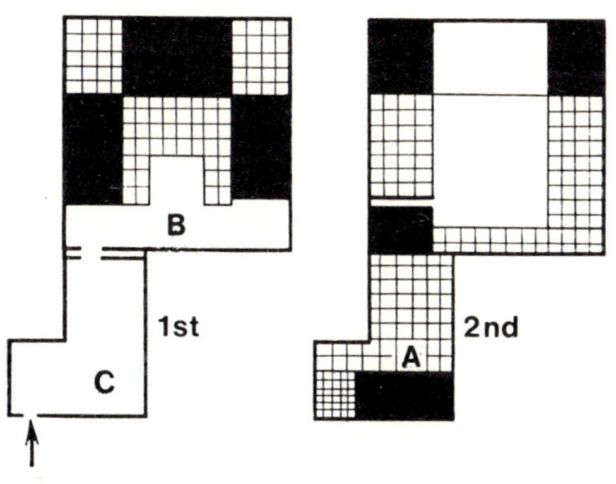

5.4 *Dwellings in city of Bukhara, Russia; featuring the winter dwelling quarters arranged at the sides of the summer quarters and the men's part of house A arranged on the second floor above the housekeeping area C*

and winter cells with a precise functional division of the house from its surrounding property. The inhabited part is separated from the street by a wall (*duval*) without any kind of windows. The narrow streets or alleys between the properties serve solely as a channel for traffic. Because of their restricted width and picturesque tracery, they are protected in the shade from the rays of the blazing sun. An important consideration in the planning of the area next to the dwelling (the enclosed little couryard) is its basic function as an addition to the summer cell. This, if well-designed, planted with trees and shrubs, and equipped with a reservoir or fountain, serves during the summer period (the most prolonged season) as the place where the family lives.

Dwellings in regions featuring arid climate should consist of winter quarters, equipped for protection from the cold of winter, and summer quarters, protected against sultriness, dusty winds, and storms. The optimum sizes of the inhabited rooms have been established by experience as the following: living room—16–18 m^2; parents' bedroom—10–12 m^2; second bedroom—8–10 m^2; third bedroom—7–9 m^2. The dwelling should also feature a separated sanitary complex, an anteroom with household cupboards, and a garage close to the inhabited premises or in the basement. Under the conditions of a hot, dry climate, the historically developed closed dwelling can be used widely by present-day town builders.

The formation of unique national architecture is also influenced considerably by the national economy and the landscape into which the building is inserted. A significant role in the determination of the characteristic traits of the national traditions is also played by such conditions as the level of available construction techniques and the character of the local building materials. The wooden architecture of the north; the white stone walls of ancient Moscow, the varied tints of Armenian tuff, the adobe residential construction of the east, etc, are characteristic only for a definite, specific locality. Finally, the domestic and aesthetic tenor of the people, which has arisen over the centuries and has determined the organization of daily life, has also affected the national traditions of city building.

DEVELOPING A MICROCLIMATIC MODEL

Examining in general the influence of a hot climate on town building, one can recommend that developing a method of planning should be based on the following: 1 an analysis of the factors which underlie the climatic basis of the town building plan (the microclimate, problems in construction physics, communal hygiene, physiological difficulties, etc); 2 a study of the contemporary science and practice of town building and of the history of the national architecture; 3 the preparation of a system for conducting natural microclimatic observations within the limits of an urbanized environment; and 4 an investigation of the problems caused by the various methods of high-density building in the microclimate of a locality.

As a result of an analysis of the natural and climatic conditions of a specific territory, it becomes possible to work out the preplanning materials dealing with microclimatic issues including a mapping of the microlimate into districts; a definition of the characteristics of bio-comfort; the establishment of the basic rules and norms for using sunshielding equipment and for combating calms, excessive humidity, and dust storms; the establishment of hygienic measures to neutralize the elements of a hot climate that act negatively on the human body and to minimize the harmful influence of the waste products of industry.

The first step in compiling a plan for the high-density construction of a sector of a city should be the investigation of the town's microclimatic model. An analysis of the influence of the extremes of a hot climate on city building makes it possible to define the boundaries of the microclimatic model of the city and to come up with recommendations for the planning of the high-density sector of the city. The final stage of the scientific study of the climatic characteristics of the urbanized locality, in my opinion, must be a climatic model characterized by the following: 1 the subdivision of the city as a whole into districts according to their microclimate and other planning elements; 2 the characteristics of bio-comfort satisfying the physiological needs of man in a given specific location; 3 a microclimatic description reflecting the degree of solar and reflected radiation, the temperature of the environment, the humidity, the character of the prevailing wind conditions, and pollution of the air (an important microclimatic consideration *vis-à-vis* the problem of communal hygiene, construction physics, etc); 4 the character of the earth's surface (hypsometry, scenery, landscape); 5 the influence of the national traditions of town building and the methods which meet the aesthetic tastes of the population; and 6 the character and density of the town centre which will influence the region's microclimate.

The purpose of the climatic model of a city is to create a practical structure which reproduces the specific town's environment in a schematized, geographically visual form. The model must be based on the observations and measurements derived from the data supplied by supporting meteorological observatories.

One of the possible scientific methods for designing a plan to facilitate optimum city building decisions for cities in a hot climate is to approach the problem using the cybernetic methods of systems analysis. The basis of such an approach is the formulation, in precisely and strictly determined terms and concepts, of the correlations between the alternative solutions to problems and the possible goals.

The correlation between the goals and the means for reaching them constitutes the model of the problem being solved. Moreover, this model is expressed with the aid of a sequential and logical chain of arguments which link the basic parts of the system being analyzed into a single complex. Ultimately, one can formalize a given problem and use mathematical methods to widen the circle of possible variants for urbanization and to seriously analyze the merits and defects of each of them.

The presence of a series of elements that are inherent to the problem being investigated makes urbanization a possible object of system analysis. These elements include the presence of a chain, the presence of constraints, the possibility of reaching the goal, and the formulation of criteria. A preliminary analysis of the possible states of the system being investigated with the aid of the construction of what is known as a 'tree of objectives' makes it possible to present a series of criteria in correlation with each other in a hierarchical structure as **5.5** and **5.6** show.

During the optimization of city building decisions, the aesthetic expressiveness of the city must be developed. The

5.5 *Possible 'Tree of Objectives'*

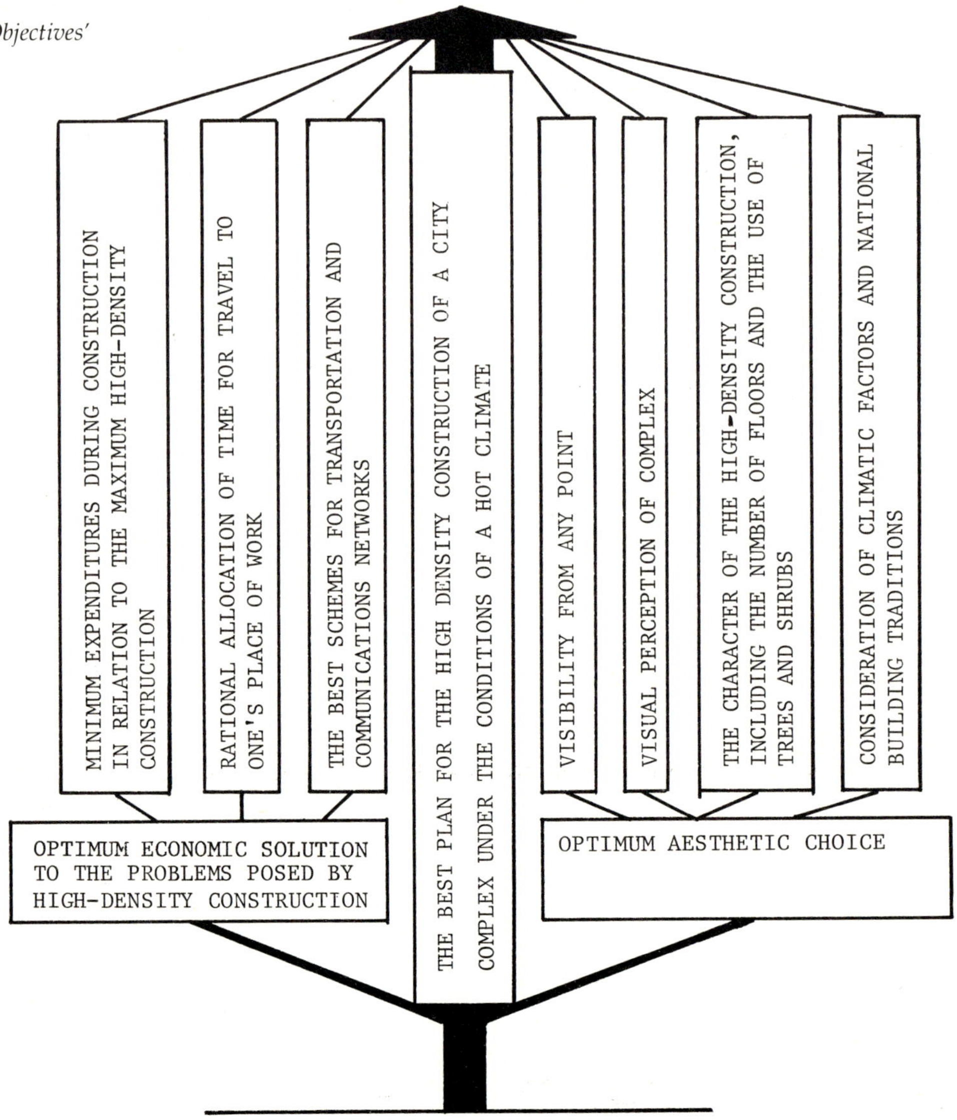

aesthetic side of the science of city building features a very complicated system of constraints which make possible a multitude of permissible solutions, each of which will be satisfactory to a certain extent. But these solutions will be distinguished from each other by the extent of satisfactoriness measured by definite criteria.

The most significant criteria in the arid region can include the character of terraced construction in a steep topography; the height of buildings (which depends on the geological basis); biological–psychological factors; the traditional aesthetic tastes of the people; the interdependence between the density of buildings and the climatic characteristics of the territory; and the extent and character of the regional infrastructure.

CONCLUDING RECOMMENDATIONS

General and physical

1 The amelioration of the thermal conditons in a town can be attained by the widespread planting of trees and shrubs and the watering of the territory. It can also be attained by a variety of methods of grouping buildings to ensure the essential ventilation.

2 As much large asphalt area as possible should be excluded. The provision of overhead cover along pedestrian routes should possibly be used.

3 The construction of buildings in a city should mainly be on columns (with the ground floor vacant) or in a ventilated cellar.

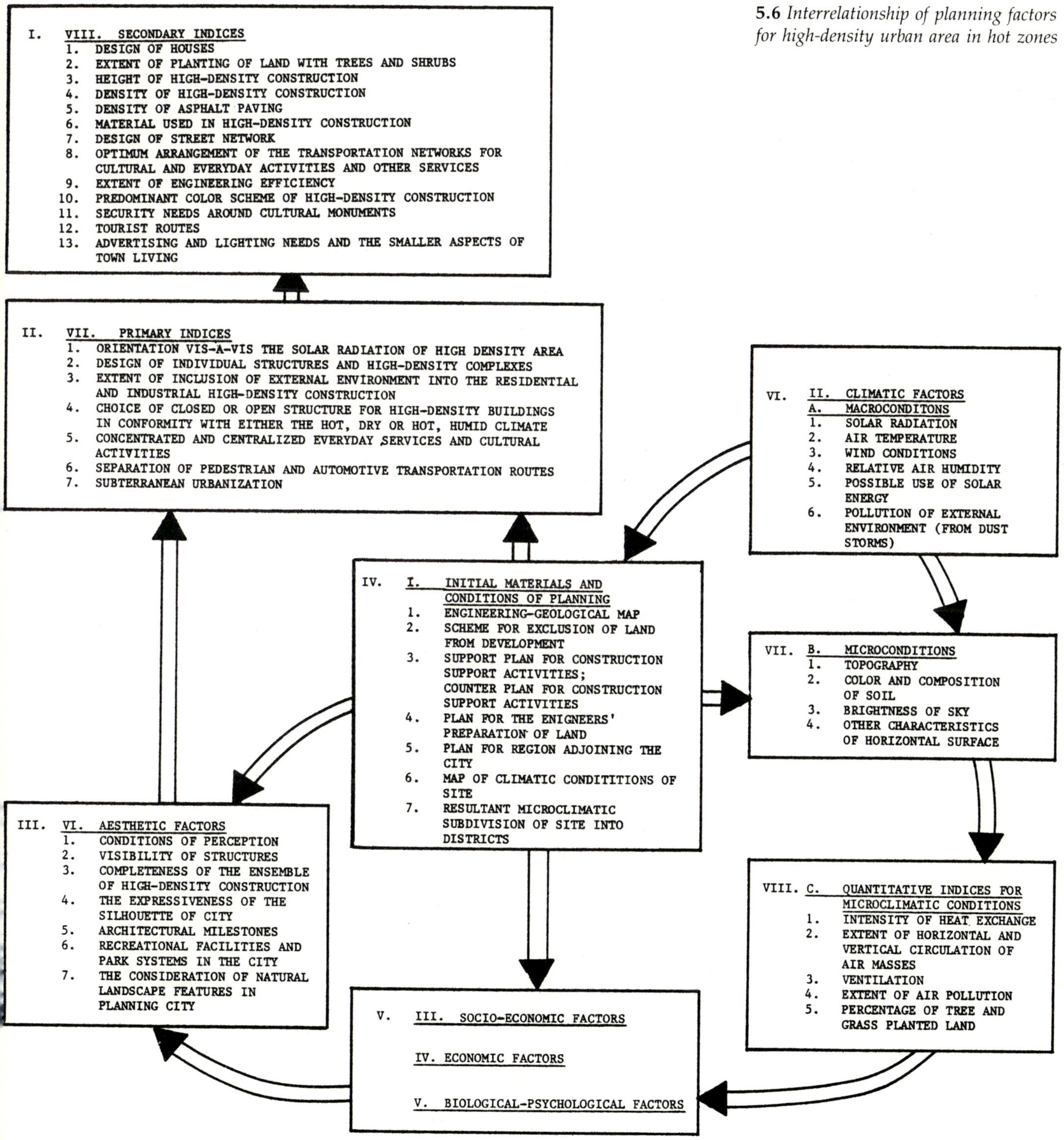

5.6 *Interrelationship of planning factors for high-density urban area in hot zones*

4 As much as possible, the underground space of the central part of the town should be used.
5 The increased solar radiation, the albedo of the surface, and the presence of calms require an abrupt reduction in the radii to cultural and domestic services.
6 The winds in the humid tropics dictate the necessity of orienting the directions of the principal streets of towns (especially pedestrian thoroughfares) parallel to the wind's direction. Under the conditions of a hot, dry climate, the possibility of dust storms compels the erection of barriers in the wind's path in the form of green plantings or the blind walls of buildings. In this case, the chief arteries of a city should be laid out perpendicular to the directions of the prevailing wind.
7 Irrigation should be used in the townspeople's parks and other recreational areas; artificial lakes, large open reservoirs, and a system of developed canals should be created.
8 During irrigation and watering, it is possible to form groundwaters which cause salination and the partial drowning of the territory. To prevent this state, it is essential to arrange a drainage system to prevent the lowering of the groundwater level by creating a constantly acting regulatory system for the groundwater at different times of the year.
9 Planners should create a protective green zone surrounding the populated places which will weaken the harmful influence of the desert, will safeguard the city from sand drifts, will help quench dust storms, and will moderate the force of the wind.
10 Planners should consider the possible use of water reservoirs as protective zones to weaken the harmful influence of the hot winds and to remove dust from them: ie, the use of water surfaces for the creation of favourable conditions facilitating the improvement of the microclimate of the populated place.
11 The planning of a network of streets and the configuration and orientation of the groups of dwellings must take into consideration the need to ameliorate the action of winds. The orientation of residential and service premises in the arid conditions is most favourable toward the south, permissible toward the north, and extremely unfavourable toward the east. The planning of high-density construction of populated places must provide for the most rational system of planting with trees and shrubs and watering of the city's land so as to create a favourable climate for dwelling and community and cultural buildings, productive businesses, and other places related to the daily domestic life, work, and recreation of the population.
12 Water and vegetation play a most important role in the extinction of the high temperature and in the creation of a favourable microclimate. As a consequence, they must become integral hygienic and architectural parts in the planning and high-density construction of residential settlements. The planting with trees and shrubs and watering must be done simultaneously with the construction of dwellings and community service buildings. Special attention must be paid to the system and methods of planting trees and shrubs along streets and in residential and community areas so as to ensure the shading of buildings and the land adjoining them from warming by the sun's rays; and not only the passive (irrigation ditches), but also the active use of water (water screens, fountains, water spraying equipment) should be used because the resulting increased evaporation absorbs heat and lowers the surrounding temperature.

Comfort-related

The experience of town building in regions with a hot climate has revealed that the biological state of man depends on age, sex, the surrounding environment, time of the year, etc. Taking into account the necessity of maintaining relative human comfort in hot climatic zones, planners must observe certain hygienic requirements when devising their plan. These include the following:

1 Avoid the orientation of one-sided apartments and terraces toward the west or southwest. In those cases where this cannot be avoided, the widespread use of sun-shading devices and shading by using the crowns of trees and climbing plants should be encouraged.
2 Gaps between buildings must be designed to ensure the possibility of the ventilation of that space and the penetration of the sun's rays into the lower floors for up to two hours each day.
3 In arid regions with sand storms, it is recommended to arrange buildings with their blind butt ends facing the wind direction.
4 In hot regions, the most acceptable design for a dwelling is the gallery type.
5 It is preferable to remove staircases and elevator installations from the basic core of the building; moreover, it is preferable to use perforated walls extensively to provide proper aeration.
6 Considering the elevated thermal burden of buildings, it is expedient to connect the complexes of community service structures with covered galleries which form a part of the system of pedestrian thoroughfares.
7 It is expedient to plan inhabited rooms to be isolated and linked with the gallery or anteroom.
8 For buildings of one-to-two storeys, experience recommends the following methods of protection of dwellings from direct solar radiation: use of walls and flat roofs of high thermal resistance; the planting of trees with developed crowns near the house on the eastern and western sides; the shielding of a flat roof from heating by the perpendicular rays of the noon sun by the arrangement on the roof of a horizontal trellis entwined with grapes or other climbing vegetation (this roof can also serve as a place for nocturnal rest).

A method for determining building height

In concluding my recommendations, I would like to deal with an important problem: the investigation and analytic determination of the dimensions and conditions of the height of the high-density construction on a complex hypsometric basis.

In connection with the scarcity of town land, especially in climates where the land must be irrigated, an urgent problem arises. The steep slopes of the topography must be used for terraced construction. It is known that this terraced construction on a steep topography makes it possible to use numerous interesting designs, and the rich geoplastic art makes it possible to create detached building formations. It is also known that the character of a broken topography influences the formation of the special features of a given microregion. The different conditions of a topography are capable of changing the direction and speed of the wind, of protecting the high-density areas from dust storms, and, because of the varied orientation of the slopes in relation to the sun, of regulating the temperature of the environment, thereby making possible the formation of convection currents.

Scientific investigations and verification of these factors when building at a high density the towns of Erevan, Bukhara, and Kislovodsk made it possible to create a graphic-analytical method of analysis of the perception of verticals in the construction of the town on a complex typo-geodesic foundation.

The essence of the method is the following. Let us assume that the observer is at the conventional sign of the earth 'K' and that 1.7 m is the height of the observer. If he directs his sight from a point 'N' to the barrier of his perception (a height on the existing high-density construction), point 'A', whose base rests on the accepted mark of the earth 'm', it is necessary to determine 'H', the optimum height of the high-density construction situated on the mark 'n' at a distance 'L' from the mark 'K'. 'l' (ell) is the distance between the observer and the barrier to perception. The horizontal plane is at the eye level of the observer, and the ray of perception will create two similar triangles which will make it possible to compile the relationship from the formula presented in **5.7**.

CONCLUSION

This chapter attempts to systematize a large number of positions drawn from the theory and practice of city building under the conditions of a hot climate. This systematization is directed to protect man from the environment surrounding him. In no way does the chapter pretend to be a complete examination of the problem being discussed because it was not possible, within the confines of a single chapter to analyze exhaustively all the multifaceted processes involved in the development of city building in hot climates.

NOTES

1 Ibn Sina, *Canon of Medical Science* (Uzbek, USSR: Academy of Science, 1954).
2 L. S. Berg, *Fundamentals of Climatology* (Leningrad: 1938) (in Russian).
3 G. Atkinson and B. Artch, 'Housing Construction in Countries With a Tropical Climate', paper delivered at First Congress of the International Council on Construction, Rotterdam, 1962.
4 Ernst Egli, *Geschichte des Städtebaues* (Climate and town districts) (Zurich: Erlenbach, 1951).
5 At present, the plan is being adjusted, taking into consideration a more urban base.
6 A. Rimsha, *Townplanning in Hot Climates* (Moscow: World Publishers, 1976); idem, *City and Hot Climate* (Moscow: Stroiizdat, 1975).

SUGGESTED READING

Aronin, D. *Klimat i arkhitektura* [Climate and architecture]. Moscow: Gosstroiizdat, 1959.

Atkinson, G., and B. Artch. *Zhilishchnoe stroitel'stvo v stranakh s tropicheskim klimatom*. Pervyi kongress Mezhdunarodnogo soveta po stroitel'stvu, Rotterdam [Housing construction in countries with a tropical climate. First congress of the International Council on Construction, Rotterdam]. Moscow: 1962.

Doksiadis, K. *Planirovka i razvitie gorodov-satellitov i novykh gorodov*. Materialy soveshchaniya ekspertov OON v Moskve [Planning and development of satellite towns and news towns. Materials of a meeting of experts of the UN in Moscow]. 1964.

Gradov, G. A. *Gorod i byt* [Town and mode of life]. Moscow: Stroiizdat, 1968.

Kililevich, L. N., and V. A. Kossakovskii. *Zhilishchnoe stroitel'stvo v usloviyakh zharkogo klimata za rubezhom* [Residential construction in the conditions of a hot climate abroad]. Moscow: Stroiizdat, 1965.

Lavrov, V. A. *Gorod i ego obshchestvenniy tsentr* [The town and its public center]. Moscow: Stroiizdat, 1964.

Perspektivy preobrazovaniya okruzhayushchei chelovaka sredy. Sb. trudov [Perspectives/prospects of the transformation of the environment surrounding man. Collected papers]. Moscow: Stroiizdat, 1973.

Rimsha, A. *Gorod i goryachii klimat* [The town and a hot climate]. Moscow: Stroiizdat, 1975 (monograph).

——. *Gradostroitel'stvo v usloviyakh zharkogo klimata* [Town building in the conditions of a hot climate]. Moscow: Stroiizdat, 1975 (monograph).

Ryazanov, V. A. *Sanitarnaya okhrana atmosfernogo vozdukha* [Sanitary protection of atmospheric air]. Moscow: Medgiz, 1954.

Sapozhnikova, S. A. *Mikroklimat i mestnyi klimat* [Microclimate and the local climate]. Moscow: Gidrometeoizdat, 1950.

Sbornik trudov nauchno-teknicheskogo izdatel'stva po grazhdanskomu stroitel'stvu i architekture [Collection of transactions of scientific-technological publishing house on civil construction and architecture]. Moscow: Publishing house of the Academy of Sciences of the Uzbek Soviet Socialist Republic, 1967.

Shelekhovskii, G. D. *Mikroklimat yuzhnykh gorodov* [Microclimate of southern towns]. Moscow: Academy of Architecture of the USSR, 1948.

Sina, Ibn. *Kanon vrachebnoi nauki* [Canon of medical science]. Uzbek, USSR: Academy of Science, 1954.

Voronina, V. L. *Opyt proektirovaniya zdanii v stranakh tropicheskogo klimata* [Experience/experiment in the planning of buildings in countries with a tropical climate]. Moscow: Stroiizdat, 1966.

Yaralov, Yu. S. *Natsional'noe i internatsional'noe v sovetskoi arkhitekture* [The national and the international in Soviet architecture]. Moscow: Stroiizdat, 1972.

5.7 *Formula for determining appropriate height of high-density construction in hot zones: N–point perceived from; Am–barrier to perception; h–height of said barrier; NC–ray of perception; kmn–geodesic marks; H–optimum height of high-density construction; L–distance to high-density construction*

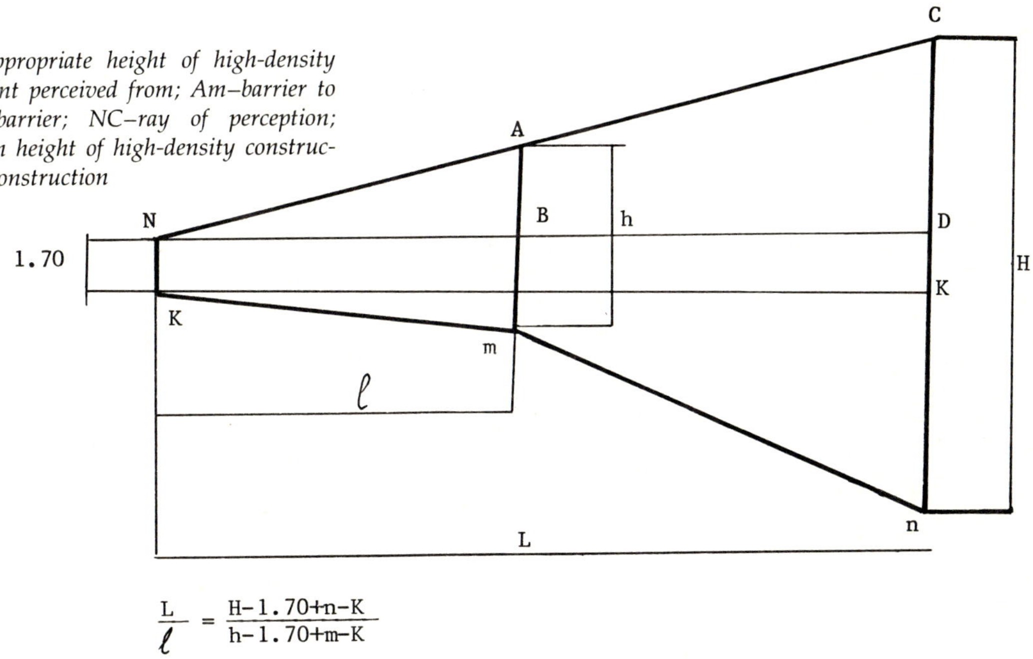

$$\frac{L}{\ell} = \frac{H-1.70+n-K}{h-1.70+m-K}$$

6 A systems approach to urban planning in arid zones

JOHN F. BROTCHIE

The systems approach in planning for arid areas provides a means for concentrating on unique features while capitalizing on universal community characteristics. For instance, transportation and communication problems may be handled by techniques developed for nonarid cities, but drainage, water supply, and recreational facilities require special consideration in those areas where water is a scarce resource. As a consequence of this scarcity, communities may also lack social and economic opportunities and amenities. Where mineral resources exist, there is at once a means of developing regional economies and of meeting social and environmental needs. These needs may include increased employment, improved housing and services, and a more pleasant environment. The equity of distribution of these benefits, diversity of the opportunities offered, and a means for the community to participate in decision making are further factors for consideration.

A systems approach to the planning problem first identifies the needs of the community, the means by which they may be satisfied (eg, the components of the urban system, such as residential development, industry, transportation and other services and facilities, recreation and environmental improvement) and the resources available for these purposes. This is the qualitative phase. In a second quantitative phase, the resources may be allocated to the components to best meet community needs (see **6.1**). Critical resources include water, capital, suitable land, energy, equipment, and human skills.

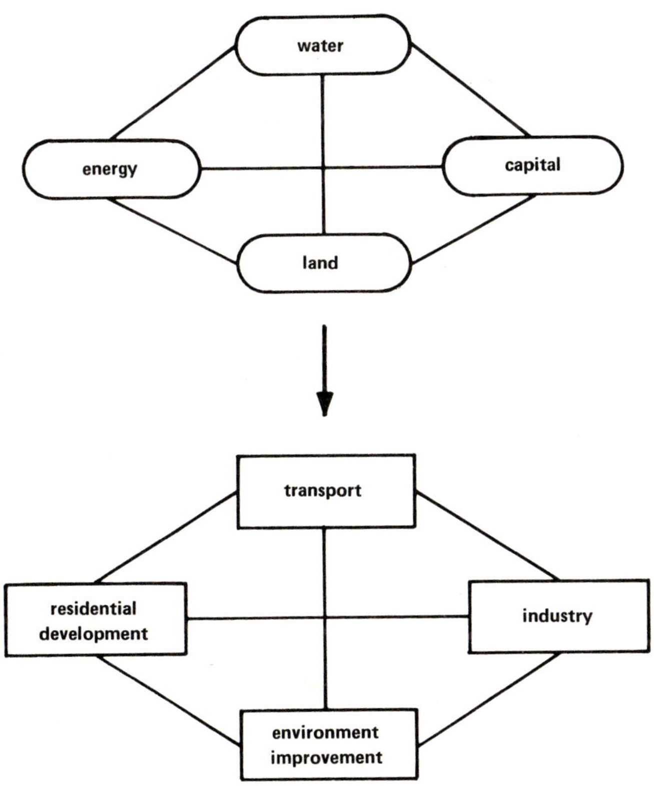

6.1 *Allocation of resources to urban activities in an arid region*

INTRODUCTION

Identifying the means of using resources to meet community needs can be a creative and innovative process which can use the unique features of the region as well as the more conventional components. Water, for example, is a prime consideration. The allocation of capital to its development and the allocation of the water among industry, residential uses, and environmental improvement must be considered. Improvements to the environment may take both conventional and more imaginative forms. For instance, there could be:

1 Tree planting along the streets to provide shade, cooling by evapotranspiration, and aesthetic improvement.
2 Urban parks (including urban forests) as an extension of these street plantations.
3 A supply of water in pipes or open channels to cater for the vegetation and provide additional evaporative cooling.
4 Swimming pools, lakes, and ponds.
5 Market gardens and orchards in and around the city.

The costs can be small in relation to the environmental benefits derived. Evapotranspiration, for example, can absorb a substantial part of solar heat gain (eg, 15–30 per cent or more depending on the vegetation and the watering it receives).[1] In fact, the average evapotranspiration rate during the summer is approximately 6 in per month (or more) over the planted area.[2] This can absorb 40–50 per cent of total solar radiation falling on the area. The net effect will depend on the percentage of an area planted but can be sufficient to reduce the temperature into a more comfortable range. Aesthetics, shading, and filtering of particulates are further benefits.

A systems approach can also be applied to the development of drainage systems. In most arid urban areas, the little rainfall that does occur is largely lost. Urban development has the effect of superposing impervious surfaces—pavements and roofs—over the previously natural terrain. These impervious surfaces are then connected by an efficient network of lined drains and pipes. The result is that rainfall runoff from these surfaces is nearly 100 per cent compared with 10 to 15 per cent before development. Furthermore, the network of drains and pipes increases the speed of runoff so that in any given period a much larger area contributes to the flow at a critical point. The results can be a one order increase in runoff, flash flooding, and erosion of natural rivers and creeks which now require protection by grading and lining. Most of the rainfall which would have contributed to sustaining urban vegetation is wasted, and additional water must be piped in to replace it.

A systems approach considers urban development comprehensively, a total system with interactions between its components. Water from roofs may be run into soak pits or porous drains to increase the water level in the soil. Residential street networks may have low connectivity (eg, courts) to reduce traffic and hence decrease pavement width and length. Concrete kerbs and channels may be omitted to allow rainfall to run over the natural terrain and percolate into the soil. The net runoff may then be handled by the natural drainage system, with some storage where required (eg, in lakes and ponds).

Such a system can be incorporated into a network of urban parks and can be integrated into the schemes previously outlined. This then will provide a cooler, more pleasant environment. A major indirect benefit of these improvements to the environment is the attraction they provide for residential development as well as for industry.

Capital allocation among industry, housing, transportation, and infrastructure also requires a comprehensive approach. This involves analysis of the consequences of each allocation and the interactions among them. Investment in industry provides extra jobs and income and hence capital generation to finance other needs. Investment in low-income housing also requires special consideration. Use of local materials allows local labour, skills, and knowledge to be used and money to be retained and multiplied in the region. Use of imported materials means less local employment, a wastage of local knowledge and skills, and leakage of much investment from the region. Local materials such as brick, stone, or mud brick have the added advantage of providing a thermal reservoir or insulation and a more comfortable internal enviroment. Less maintenance and better appearance are further potential attributes.

Transportation provides access to urban opportunities and a tool for guiding urban development. A network of transportation corridors can be reserved before development forms an amorphous sprawl. The transportation system can be conveniently divided into tracks, vehicles, and controls for purposes of systematic planning and design. Where resources are limited, road transportation can provide flexibility at low capital cost. The size of vehicles may be treated as a planning decision: in the case of public transport, economies of scale of large vehicles such as buses or trains may be traded off against the increased frequency and fewer stops of smaller vehicles, such as multihire taxis, small buses and personal rapid transit. Coordination of smaller vehicles and larger ones at an urban level can provide the best of both worlds—the economies of express-line haul on arterial routes between zones and the greater frequency and speed of the small vehicle for distribution within a zone.

A key factor in both water and energy use in arid areas is residential density. With low densities, the use of water *per capita* may be very high where the supply is not constrained. Demand in Perth, Western Australia, is about 650 litres *per capita:* well over the norm for less arid Australian urban areas. Much of this water is used on home gardens; thus, increased density will reduce this demand substantially. Higher density or clustering of residential and other facilities can also shorten travel paths, particularly for shopping and recreation. In many cases it will also allow walking or bicycle riding to replace the use of the car. The provision of tree-lined paths facilitates this change. The costs of maintaining these paths and trees may be traded off against the savings in vehicles, energy, the extra infrastructure they require, and the other costs of low-density development. Furthermore, the use of native vegetation reduces external watering demands.

Higher densities facilitate provision of network services and community facilities. Water saved on garden watering can be used for a borne sewage network or recreational facilities. Reticulation of chilled water for home cooling, drinking, or both is also feasible at a sufficiently high density. Heat pumps can provide an efficient means of heating and cooling, and an integrated heating and cooling system is then also feasible. Higher densities can thus result in savings in basic infrastructure which can make additional services and facilities feasible. Even some building costs can be reduced by the economies of scale involved and by the sharing between home units of common wall and floor/ceiling elements. At one extreme of density, a neighbourhood might be housed in a single large condominium of units with integrated services including conditioned water and air, cable telecommunication, and recreational and social facilities. At the other extreme, all houses might be single and detached on separate large lots.

In trading off between the two, the monetary savings, convenience, and additional facilities with the integrated, high-density system must be weighed against the greater freedom of individual choice and action possible in the lower density solution. This trading off can be done at two levels: 1 the community level where overall ranges of density are selected and 2 at the individual level where personal choices within these ranges are made. At the upper level, quantitative data are needed if informed choices are to be made. Thus the desirable components of the urban system including those features special to arid regions may be identified and assembled. The resources required for each may then be estimated, and key decisions on the allocation of these resources can then be considered.

In this quantitative phase, the unit benefits and costs associated with each component can be determined. The total benefits and costs associated with each allocation of resources can then be evaluated and used as a measure of merit. The expression of this merit in terms of the allocation decisions provides a model for planning purposes. Let us now consider the construction of such a model.

MODELLING OF THE URBAN SYSTEM

Since the allocation of resources to the various components of urban development (**6.1**) is a quantitative process, appropriate methods are required as a basis for it. A model which describes the benefits and costs or goals of the system in terms of the decisions to be made can be used as a basis for this allocation. This distribution can be varied systematically until measures of benefit-less-cost increase to a maximum. The allocations which satisfy various goals—or weighted combinations of goals—can be found and used as a basis for planning decisions. Allocations which are viable under a range of futures can be selected.

A model oriented to this type of planning process is the urban planning model, TOPAZ (Technique for Optimal Placement of Activities in Zones).[3] TOPAZ is primarily an allocation model, alloting land use activities to zones or regions on the basis of a comprehensive set of benefits and costs which may be divided into two basic sets:

1 Those which define the merit of the *absolute location* of an activity.
2 Those which define the merit of its *relative location* with respect to other activities.

The various benefits and costs belong to one of these two classes. An awareness of this division can be of great help to the urban planner. The costs of absolute location can include the costs of providing the network services and environmental improvements for the area, as well as the site development and the building and operating costs. A set of submodels for SUCO (Service Utility Costs) has been developed to help plan, schedule and determine the cost of these network services.[4] Three submodels have been specially geared to the hydraulic network services: water supply, drainage, and sewage. The general SUCO model can be specially geared to other services including power, communications, chilled water, hot water, and irrigation systems. The submodels allow optimal staging of the development of these systems, including their upgrading over a number of time periods, to be selected. This is done by trading economies of scale against the discounted values of future costs.

The costs of relative location can include communication and transportation between the pairs of activities in zones. Several modes of transportation including the assignment of flows can be considered simultaneously. The relative locations can thus be expressed in terms of alternative paths; the characteristics of each path will determine the flow between them and the consequential costs, travel times, energy use, pollution levels, and congestion.

Improvements to the transportation system can be expressed as a change in one or more of these characteristics. Improvements handled in this way can include improved control systems for traffic, construction of a new freeway over all or part of the route, installation of a rapid transit system, provision of a new fleet of buses, or a change in the pricing system. Each of these changes will effect the route characteristics and hence the various performance attributes including flow costs and trip times.

The model considers the net sum of benefits-less-costs involved and maximizes this total, thereby trading off cost increases in one area, eg, servicing costs, against cost savings in another, eg, transportation. Each component of benefit and cost is separately weighted. The model may maximize benefits-less-costs for the community as a whole, thereby providing the best planning solution. Alternatively, it may include only those benefits-less-costs incurred by the decision maker. In this case, it simulates behavioural location decisions. In each instance, a change in one component of the system, such as the road network, can have impact throughout the planning area. By obtaining both the optimal solution for the community and the behavioural solution, the inducements necessary to implement the best solution may be assessed.

Thus the model can maximize utility for the community as a whole or for the particular segment of the community considered. This may be the public sector or various groups within the private sector. If each of these groups is considered separately, 1 areas of potential conflict can be examined and avoided, 2 the distribution of benefits derived from changes to the system can be studied, 3 equity of this distribution can be evaluated and improved, and 4 incentives needed for implementation can be assessed.

6.2 *Structure of urban planning model TOPAZ-URBAN; general allocation model, submodels giving relative and absolute benefits and costs especially pertinent to urban planning, and data calibrating it to a particular city*

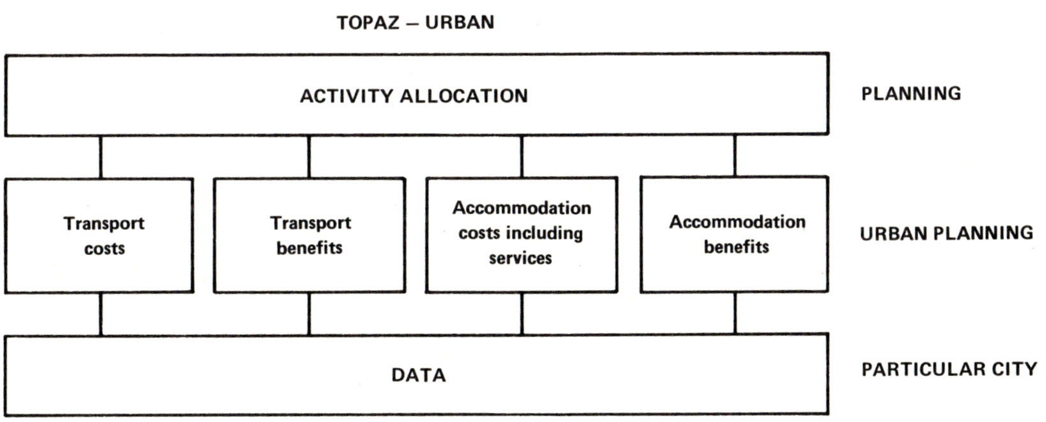

Using TOPAZ

The TOPAZ model is hierarchical like the city it describes. At one level, it allocates activities to zones and time periods. At another, it evaluates the subsystems which connect and support them. At the third level are the basic data which define them (**6.2**). At the first level, the model describes a general system. At the next, it is specialized to an urban system by the inclusion of urban subsystems. At the final level, it is specialized to a particular city by its data.

The submodels determine the unit benefits and costs for each component of the system. An information base includes both input data and the decisions made in the general model and other submodels. The solution process is thus a repetitive, sequential one with information flows among subsystems between iterations.

The transportation submodel is divided into several modules. In one, trip generation is considered. In another, the spatial pattern of flows is evaluated. Here, trip origins and destinations are related in a fully constrained gravity model.[5] Alternatively, a partially balanced model or a calibrated model may be used. In another module, the split of journeys between available modes is determined. Another unit assigns trips to the transport network. One considers the interaction between the trips and the network, for example the necessary cost of a network to meet the flow demands (or congestion effects when capacity is constrained). Another determines trip costs and travel times and their breakdown between activities and between zones. The cost coefficients for use in allocation are also determined.

A different submodel calculates the exhaust emissions from these traffic flows and, from meteorological information on wind velocities, computes their dispersal over the city and plots contours of pollution levels. A further submodel estimates the opportunities for access between activities using the transportation networks above. Accessibility is formulated in terms of the number of transportation opportunities and the travel times required to reach them. Since service system costs, trunk and headwork sizes, and stagings are also important, this information has been outlined. Finally, there is also a submodel which calculates land values and their changes. These submodels perform two functions, evaluating component benefits and costs required by the allocation model and also the total benefits and costs and showing breakdown between activities and between zones.

Necessary data

The data used in the model are those required for any comprehensive quantitative analysis of urban development.[6]

This information may be available elsewhere in one form or another—from a transportation study, from a land use study, from land transaction records, from census files, and from records of construction and operation of urban services. However, it will generally require some transformation to bring the information to the form and level of aggregation required.

The data required and the initial key decisions associated with them may be grouped into items defining:

1 The components of the urban system
   a activities or land uses—that is, number and types of activities and increments of each in specified time periods
   b region to be considered and number, locations, and sizes of potentially suitable zones within it
   c planning horizon and division of the time span into development periods.

2 The interactions and infrastructure
   a relationships between activities, or the matrix of flow between activities and the unit costs of each
   b relationships between zones—that is, the modes of transportation, path lengths, and travel times between the zones for each transportation mode and the unit costs of the paths
   c relationships between activities and zones or the set of services required to accommodate the activity in the zone, the unit costs of accommodating the activity including costs of service reticulation. (Each service cost may be separately included and weighted. The unit benefits of locating the activity in the zone, for example, or measures of environmental suitability or the costs of environmental improvement may also be included and weighted)
   d relationship with surrounding regions. For example, traffic flows to various recreational facilities or other (fixed) centres and also the trunks for the water supply and other network services including layouts and urban demands
   e interactions between time periods including the discount rate used for reducing the value of future transactions and future cost rises reflecting the depletion of resources
   f weighting placed on the various components of benefit and cost.

The data are an interface between the model and its users. This information facilitates a range of uses. The output from the model provides information on the resulting system and its performance. For instance, it provides 1 land use allocations, 2 traffic flows, 3 average journey times, 4 total benefits-less-costs, 5 accessibilities or opportunities, 6 air pollution levels resulting from traffic, 7 cost of land involved, and 8 costs and stagings of service systems including trunk and headwork sizes. The measures above are broken down to give their distribution between activities and/or zones.

Hierarchy of planning decisions

The model and its data encompass several levels in a hierarchy of planning decisions. At the first level are decisions concerning land use or the allocation of resources. At the second level are decisions on the infrastructure networks to service these uses. Below this level are the data which represent the following three levels of decision: 1 on further growth of the activities and on the spatial and time boundaries of the problem; 2 concerning social, economic, and ecological aspects; and 3 concerning community aspirations and goals and the weighting to be placed on them. Together the model and data form a five-level hierarchy in which the decision-making process becomes less automatic, and community participation increases at each level in the hierarchy as the bases of decision become more difficult to quantify. Thus the model handles the complexity of quantified components and interactions at one level, with human value judgement providing the weighting for trade-offs between factors at the other levels.

Weighting may be placed, for example, on environmental, economic, and social factors either explicitly or implicitly through value judgements on the relative merits of different total plans. From these latter overall value judgements, the weightings of factors may be determined by mathematical analysis. In either case, the process may be interactive; the consequences of the weighting may be evaluated by the model and may provide further information on which modifications may be based.[7] Some factors considered are illustrated in the examples.

Implementation

Implicit in this planning approach is the concept that there is an overall planning authority with legal or financial capacity to implement its plans using zoning powers, control over infrastructure, or taxes. By using the model in both its planning and behavioural modes (as previously discussed), the differences between the best solution for the community and the behavioural solution of individuals are found. The incentives necessary to bring the two together may then be determined. Similarly, areas of potential conflict among different groups can be analyzed and predicted and measures taken to avoid them. Where no authority has sole planning powers, conflicts are unlikely to be resolved without one or more sectional interests exercising power to achieve control. However, the model does provide information concerning the consequences of each alternative course so that sectional interests can see their own benefits and costs and those of the community as a whole (see later examples). An understanding of the way solutions overlap with the political system is also necessary, and perhaps an ultimate use of models is an aid to this understanding.

Studies

A number of urban studies, generally at a macroplanning level, have been carried out while using this modelling approach. Further studies for arid or semiarid regions are in progress. One study concerned the impact of various development options on land use and transportation in Tehran, Iran. Social, economic, and environmental consequences of each option were required as a basis for development decisions.[8] Two different model studies were made. One study used the prescriptive model, TOPAZ, to indicate optimal policies for long-term land release and infrastructure development along with behavioural consequences at the transportation decision level and evaluated the results including air pollution levels. The other model was a predictive model.[9] It was used in a larger study of the short- and medium-term behavioural responses to a number of policy packages. It also evaluated the results of these policies. The two model studies were complementary, and Tehran provided a unique opportunity to use the two models together in this way. The findings of the TOPAZ study (some of which appear to be typical also of other metropolises) indicated for the year 1977 that:

1 Significant savings (up to 10 per cent) in transportation costs and energy can be made by optimal location of new activities.
2 The density of these activities substantially affects both transportation and infrastructure costs.
3 Quality-of-life goals can be simultaneously satisfied by optimal location of new activities in environmentally suitable

zones (close to the city) at a negligible (eg, 2 per cent) increase in transport cost.

4 Policies to provide planned new cities in the region can be implemented without increasing overall transport costs (difference 1 per cent).
5 Consumer surplus benefits of proposed new transportation infrastructure are substantial (but so are costs).
6 Multiple centres of commercial activity are currently viable and will be even more so with future increases in transportation energy costs.
7 Variations in air pollution distribution with patterns and policies of development and with different wind conditions can be estimated.

Details are given elsewhere.[10]

Studies also have been made for various Australian cities using the TOPAZ model to evaluate and develop strategic or structural land use and transportation plans. The cities of Darwin, Gosford-Wyong, and Melbourne have been studied, and a recent study of Albury-Wodonga looked at transportation and land use options on a detailed structural level. A regional study in Western Australia is looking at the interactions and trade-offs between various competing land uses in a water catchment.

Other studies in New South Wales have used the model to determine the best groupings and locations of the various hospital facilities required.[11] The basis for this allocation has included patient and visitor access to each type of facility as prime considerations. Establishment and operational costs including staff organization and movement costs within the facilities are also considered. A decentralized, hierarchical system of health care facilities was found to be most appropriate. Mean access times have been almost halved by the facilities planned in this way. In another region, casualty centres are currently being organized using the same technique. Let us examine some of these studies.

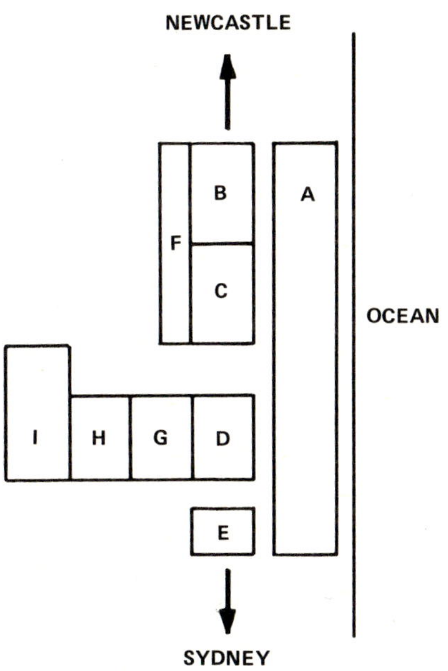

STRATEGY NO.	AREAS OF URBAN DEVELOPMENT	INCLUSION OF AIRPORT	LOCATION OF AIRPORT
1	A B C D E	YES OR NO	H
2	B C D E H I	NO	
3	B C D E F	YES OR NO	H
4	B C D E H	NO	
5	C D E I	YES OR NO	H
6	B C D E Part A F G	YES OR NO	H

GOSFORD-WYONG DISTRICT

The Gosford-Wyong district is situated on the coast of New South Wales, Australia, 50 miles to the north of Sydney, and to the south of Newcastle. Its present population of 100 000 people is expected to increase to 0.5 million by the end of the century to form a system city in the Wollongong-Sydney-Newcastle complex. The Planning and Environment Commission (New South Wales) has planned the growth of the area within the context of state development as a whole.[12] It used the TOPAZ model to evaluate six different strategies for development and to determine the optimal land use patterns and time stagings for each. The six alternatives are shown in **6.3**.

In these studies, planners considered the interactions with the larger cities on either side as well as the interactions between the activities and zones within the area. Two activities, residential and industrial, and 22 essentially homogeneous zone modules of approximately equal size were considered. Two modes of transportation were evaluated, and a travel time matrix between the zones was constructed for each from transportation data. This source of data also provided information on travel demands between and within the activities. The three time periods considered were each of ten years. The strategies were ranked in order of merit on the basis of private and public transportation cost and service costs. Some of the results are shown in table **6.1**. The provision of an additional international airport for the Sydney region was also considered in the studies. Supposedly, development of this area would be phased over the three time periods. A structural level study using TOPAZ was then made based on strategy number six. The study considered and evaluated different land use patterns and transportation plans. Another benefit considered within the model at the structural planning level was environmental suitability of the zones. Reports on these two levels of study are incorporated in the Gosford-Wyong Strategy and Structure Plans Studies published by the Planning and Enviroment Commission of New South Wales.

MURRAY VALLEY AREA

A report of the Western Australia Murray Valley regional study (of water catchment planning for an arid zone) is given by Bennett et al.[13] Control over land use in water catchments varies considerably, depending on the history, size, and characteristics of the catchment and the existing uses. The problem concerns the amount of control which should be exercised. The objective could be to maximize the merit or utility of the catchment to the community. This involves selecting an optimal set of activities for each zone of a catchment. It is unlikely that this optimal set will be achieved

Table 6.1 Gosford-Wyong (Australia): performance of strategies ranked in order of cost

Measure of performance	Private transportation only	Test series private and public transportation and servicing	
		transportation	servicing
Best strategy number	6	6	3
Second	4	4	6
Third	5	1	5
Fourth	1	3	2
Fifth	3	2	1
Sixth	2	5	4

Source: *Gosford–Wyong Urban Strategy*. Sydney, Australia: State Planning Authority of New South Wales, 1973.

without external controls because some of the costs and benefits do not accrue to the landowners. Catchments yield, among other things, water, salt, sediment, and aquarium fauna in the river system; timber, crop, animal and mineral products from the land; industrial sites, transportation links, land recreation and living areas for man's activities; and biological populations and scenic beauty as justification for conservation. Landowners are not paid for many of these products, and therefore legislative control of catchments is often required. The model used to represent catchment management decisions is a version of TOPAZ with additional constraints placing limits on components of the objective function. This model is now applied to the catchment of the Western Australia Murray River, an area of 2750 square miles adjacent to more arid areas including the metropolis of Perth, Western Australia.

The activities considered within the catchment are a national park, natural and plantation forestry, orchards, grazing alone and in combination with cropping, bauxite mining followed by protection or production forestry or farming and water storage. The catchment has been mapped into geomorphological units, and these provide suitable homogeneous zones for the model. Land use studies by the participating organizations provide a link between geomorphology, vegetation, and land use.

Precipitation across the catchment varies from 20in per annum in the headwaters in the agricultural areas to 50in near the escarpment of the Darling Range. Approximately 45 per cent of the catchment is already cleared, causing severe salinity problems (the present average is 1200 parts per million of total dissolved salts), and the river or its tributaries are expected to be used for metropolitan water supply for Perth in the second half of the 1980s. If this is so, then a carefully formulated catchment land use plan is urgently required.

The benefits and costs have been divided into a number of separate matrices in order to differentially weigh them and constrain them for different solutions. The matrices developed in this study are net present value of the land use activity (royalties, taxes, and other transfers are not included in the costs); water, salt and sediment yields of each zone under each land use activity, and conservation and recreational values.

It is necessary to use additional constraints for the allocation of land to mining activities in any time period. Also, due to inadequate representation of market supply and demand and the current poor prospects for orchard fruit sales overseas, orchards are constrained to their existing area. Constraints are also used to ensure the value of conservation (otherwise, recreation is maintained or increased over time) and to ensure water storage requirements.

There are three additional problems in describing conservation values. First of all, aggregation of a number of species with differing habitat requirements to produce one conservation value may result in solutions where many of one species are conserved. Meanwhile, several of the others become extinct. This can be overcome by having individual values for each, since relatively few species require special conservation in this catchment. Scarcity also dictates value: 1000 numbats are worth preserving, but 1 million numbats would probably be considered a plague. The model is being adjusted to allow conservation values to be a function of the size of the activity in these situations. Finally, conservation values may well depend on activities in other catchments or other states. These boundary problems can only be studied by sensitivity studies of the solution to different conservation values.

By changing the weight of components of the objective function, the viewpoints of the different parties concerned with land use can be represented. Table **6.2** represents the sign of the weighting coefficients for possible groups. It is unlikely that the parties are as single-minded in real life as they are represented in the table, but conflicts in viewpoint and plan can be explored by sensitivity tests of these weightings. The present formulation of the problem permits changes in both the weightings and the data specifying the catchment; therefore, as values change with time, new solutions can be generated and modified plans and policies obtained. The study has involved the collaboration of state authorities and two divisions of Commonwealth Scientific and Industrial Research Organization, Australia (CSIRO), and this summary is based closely on their initial brief report.[14] A futher comprehensive report is being prepared, and a paper has been presented.[15] An important finding of the model study is that salinity can be better managed and the net benefit to the community is almost doubled if two dams are used for catchment instead of one.[16]

Table 6.2 Possible weightings of the benefit/cost matrices by parties interested in catchment management

	Land owner	Water supply agency	Average citizen	Land use agency
Primary production	+	0	0	+
Interactions between users	+	0	0	+
Water yield	0	+	0	+
Salt yield	0	–	0	–
Sediment yield	0	–	0	–
Recreation	0	0	+	+
Conservation	0	0	+	+

Source: Bennett, D, et al., 'An Allocation Model for Catchment Land Use Planning'. In *Proceedings of Hydrology Symposium*. Perth: Institute of Engineers Australia, 1973.

PLANNING A NEW TOWN

The last two examples cover particular regions and actual studies, but it is also important to consider the planning of a typical new town in the arid northwest of Australia; Nevil Shute touched on the physical, economic, and social needs of the isolated Australian town in his book *A Town like Alice.*[17] Brealey et al. have surveyed a number of remote communities with respect to physical and social needs.[18] The approach considered here is based in part on these studies. The activities treated are living, shopping, recreation, and employment; external communication (travel, freight, transportation, and telecommunications) is another consideration. A key parameter is density.

At one extreme of density is the 'Desert Hilton' concept. Here the whole community would be accommodated in a large hotel where there are many services and facilities. At the other extreme, suburban residential densities within separate neighbourhood cells may be considered. In this case, an urban transportation system and its infrastructure are also required, thereby decreasing the resources available for a range of additional services, facilities, and environmental improvement that can be allocated. Transportation costs incurred by a location decision can equal initial establishment costs when summed over time and will increase even further in the future. Savings with high density can be substantial and can be traded for a range of other benefits also. The reduced access opportunities to central facilities resulting from low density decreases their demand and viability. Consequently, the diversity of goods and services and the amount of income retained in the town are also to be reduced, along with the range of employment opportunities, particularly for the young and women. Thus, population size and diversity is less, and the ability to find a suitable niche decreases. Intermediate and diverse densities may also be tried. (A critical parameter is the urban size for which a car is needed. This parameter may be increased by the provision of shaded walkways and cycle tracks).

For each of these alternatives, activities may be allocated to available spaces and development periods on the basis of minimizing community costs. (Alternative bases for solution are maximum convenience or maximum access opportunities.) Additional services may be added or deleted until the same total cost pertains to each alternative; this method allows evaluation to be on a qualitative level. The price tags of the further services and facilities may also be separately given so that tradeoffs may be made. These further services and facilities can include the following: better communication with the capital cities such as Perth; water recreational facilities; other recreational and entertainment facilities; shaded walkways and cycle tracks; and parks, gardens, orchards, and urban forests. Vegetation and a surface water system of creeks and lakes are also reasonable options if sufficient water is available. At one extreme, water may be merely released to form its own drainage and vegetation system; the price involved can be traded against other services and facilities. By listing such options and evaluating their consequences, one can make informed choices. For some parameters such as density, provision of a diverse range allows real choices to be made and feedback to be obtained. This, of course, helps the urban planner.

Each of the studies discussed involves communities of different sizes with different land uses and resources but uses the same basic model. Other studies take different approaches. For instance, the Albury-Wodonga study deals with specific transportation proposals. It examines where to plan new river crossings and highways and how to anticipate their impact on the newly planned urban centres. The region is not arid, but water is again a scarce resource. The hospital planning studies are even more detailed. They treat the grouping of different hospital tasks, their location throughout the metropolitan area, and even their layout on a site.

CONCLUSION

The systems approach can thus be used to consider the various options open to the community, the interaction among them, and their various consequences. It allows the impact of one component of the system (eg, a rapid transit system, an extension of the system, or even a change in its operational schedule) on the performance of the city as a whole to be determined and goes even farther to allow a set of changes to be made systematically to improve this performance subject to constraints. Thus the systems approach may be used to evaluate and improve the system and to provide information of use in planning decisions. This information is comprehensive and provides the planner with a rational basis for policy decision.

The systems approach and the models associated with it are now practical planning tools which have been used or are being used for urban planning in arid regions in Iran and Australia and in other nonarid areas around the world. The techniques provide a means of evaluating alternative options and a framework and context in which the overall problem may be identified and decisions rationally taken. These decisions may be about housing type and density and the facilities, services, and environmental improvements that can be afforded as a supplement or which are basic essentials for comfortable living and for attraction of residents in some arid areas.

In arid areas, water is a scarce resource requiring careful and optimal allocation by planners who take qualitative as well as quantitative factors into account. Density is a critical factor in these areas, affecting style of living, mode of travel, water use, and the resources available for environmental improvement. Increased densities may allow diversion of resources to additional services and an enhanced environment, including parks and gardens and shaded paths to make walking and cycling more attractive, and may make public transportation more viable in larger towns.

The tools discussed are primarily macroplanning aids to enable alternative land use and transportation patterns to be evaluated or evolved. Their present state of development is such that smaller scale components (eg, the inclusion of a new freeway or rapid transit system, an extension of the system, or even a change in fares or level of service) may also be treated. Similarly, new housing developments or commercial centres may be evaluated or located, and even hospital facilities may be grouped, located, and evaluated by planners using these techniques. Planners and decision makers are therefore looking increasingly to these methods to ensure that scarce resources, including water in arid regions, are effectively and efficiently used and that all resources are allocated to best meet planning goals.

NOTES

1 J. F. Brotchie, 'Some Systems Concepts for Urban Planning', *Royal Australian Planning Institute Journal*, 12, no. 2 (April 1974), 43–50.

2 R. S. Foster, 'Roots—Caring for City Trees', *Massachusetts Institute of Technology Review*, 79, no. 8 (1977), 28.

3 J. F. Brotchie and R. Sharpe, 'A General Land Use Allocation Model: Application to Australian Cities', in *Urban Development Models*, Land Use and Built Forms Conference Proceedings No. 3, ed. R. Baxter, M. Echenique and J. Ovens (Lancaster, England: The Construction Press, 1974), pp. 217–236.

4 G. Davis and J. F. Brotchie, *An Urban Services Planning Model and its Application to Melbourne*, CSIRO Division of Building Research Report, 1980.

5 A. L. Wilson, 'A Statistical Theory of Spatial Distribution Models', *Transportation Research*, 1 (1967), 253; and J. W. Dickey and R. Sharpe, 'Transportation and Urban and Regional Development Impacts', *High Speed Ground Transportation Journal*, 8, no. 2 (1974), 71–80.

6 J. F. Brotchie, J. W. Dickey, and R. Sharpe, *TOPAZ—General Planning Technique and its Applications at the Regional, Urban and Facility Planning Levels* (Berlin: Springer-Verlag, 1980). Lecture Notes in Economics and 'Mathematical Systems 180.

7 J. F. Brotchie, J. R. Roy, and R. Sharpe, 'Planning for and by the Community', *Proceedings of the Fifth Australian Building Research Congress* (Melbourne, 1975).

8 Tehran Development Council Secretariat, *Major Planning and Development Issues Affecting Tehran's Future* (Tehran, Iran, 1976).

9 Tehran Development Council Secretariat, *Land Use and Transport Model Pilot Study, Final Report* (Cambridge: Applied Research of Cambridge Ltd, 1977).

10 J. R. Roy and J. R. Crawford, *An Urban Land-use Transporation Study in a Developing Country*, CSIRO Division of Building Research Report, 1980.

11 *Planning Hospitals for the West of Sydney: A Preliminary Report* (Sydney: Health Commission of New South Wales, 1975).

12 *Gosford-Wyong Urban Strategy* (Sydney: State Planning Authority of New South Wales, 1973); and *Gosford-Wyong Structure Plan* (Sydney: Planning and Environment Commission, New South Wales, 1975).

13 D. Bennett, et al., 'An Allocation Model for Catchment Land Use Planning', in *Proceedings of Hydrology Symposium* (Perth: Institute of Engineers Australia, 1973), pp. 181–183: and D. Bennett, et al., 'Linear Programming as an Aid to Catchment Land Use Planning', in *Proceedings of Third National Conference of the Australian Society for Operations Research* (Adelaide, 1977).

14 Bennett et al., 'An Allocation Model'.

15 Bennett et al., 'Linear Programming'.

16 Ibid.

17 Nevil Shute, *A Town Like Alice*, reprint ed. (London: Heinemann, Octopus, 1976).

18 *Living Way Out* (Melbourne: Commonwealth Scientific and Industrial Research Organizaiton, Division of Building Research, 1972).

7 Integrated rural area planning: the Lachish region

YITZHAK ABT

Israel's experience in planning and building the Lachish District is related to some of the basic principles of integrated rural planning. Such planning is looked upon as an intermediate stage of planning, affording the possibility of coordination between national plans and local, rural modernization programmes within the concept of linked rural and urban development cycles.

I will discuss the emergence of the formerly barren Lachish District as a prosperous agricultural hinterland for its district centre, Kiryat Gat, in some detail. I will briefly describe the main features of the planning sequence involved in integrated planning, referring to the importance of feedback of implementation data to policy makers and planners, both national and local, in order to make the planning and progressive implementation a self-correcting process.

There is a growing realization that the tremendous contemporary advance in agricultural technology has not brought the rural communities in many parts of the world into the orbit of the market-oriented economy. This situation demonstrates the need for a broad set of policies to be followed by implementation in such spheres as ordered development of hitherto inadequately exploited resources, agrarian reform, rural consolidation, and agricultural settlement. In many cases, the situation also mandates the need to create or strengthen demand in the rural areas to stimulate the development of market-oriented agriculture.

Today many countries face the phenomenon of uncontrolled rural exodus to metropolitan areas that has aggravated urban problems with serious social and political repercussions while leaving the countryside impoverished in trained manpower and leadership. In many cases, the factors influencing population migration do not stem only from the economic advantages that the cities offer, but also occur because of the pressure exerted by an evergrowing rural population that can no longer subsist on farming. In other cases, failure to fulfill greater expectations for better services and recreation in rural areas induces the young and the mobile elements of the community to look for better amenities in the cities. The resultant social cost is often exorbitant.

A related problem confronting a number of countries is the need to buttress the economy of existing small towns in rural areas. Thousands of such towns exist in Africa, Asia, and Latin America. Often these towns are being drained of technical manpower because of lack of employment opportunities in the town where the workers have been trained. The result is that these towns become stagnant.

Integrated planning of rural areas mirrors a comprehensive concept of harnessing the total development potential of the farming sector and of concurrently modernizing rural communities. It is, in fact, at an intermediate level of planning that national plans and local programmes can be realistically adjusted and coordinated, while intersectorial factors to support rural modernization can likewise be introduced. This also involves an approach whereby the cycles of rural and urban development in defined areas can be effectively linked through spatial and locational planning to the benefit and dynamic growth of both systems. Therefore, the concept and the associated planning are of interest to both the developed and developing countries. It stands to reason, however, that neither the concept nor the associated planning can be indiscriminately applied since the basic objectives of integrated rural area planning differ from country to country and from one region to another in any one country.

The aim of this chapter is to focus attention on some of the ways in which integrated rural area development can be undertaken. The discussion is based on the experience Israeli planners gained during its development, as well as on my participation in development projects in some 20 countries in Asia, Latin America, and Africa.

PLANNING, IMPLEMENTATION, EVALUATION

The objective of rural modernization can be met, wholly or partially, through the application of comprehensive rural planning techniques. The planning process must be closely linked to experiences derived from implementation. This will permit a continuous feedback of data to policy makers and planners at both the area and the national levels. The data will, of course, reflect the results of the introduction of technological innovations. Such communication requires a decentralized institutional framework which would provide a possibility for harmonious relationships between the implementor, planner, and policy maker. Framework could be provided, for example, by an area development authority closely related to a national development agency or planning authority. The planning process can also be conveniently divided into interrelated stages designed to integrate the rural-urban development cycles in an area.

Formulation of objectives

The area objectives should state concrete income, employment, and production targets and spell out what is to be achieved through settlement or agrarian reform. Often the objectives are achievable in conjunction with larger capital investment projects such as irrigation, drainage, or land development.

At this stage the development team must be formed to collect agro-economic, physical, and sociological data pertinent to rural development and to survey the existing urban communities in the same area. This information can be divided into two main categories, information of national interest and information of local importance. Information is further categorized into three basic spheres, economic, physical, and sociological.

Success of integrated rural development programmes often depends on the components of physical planning which can be modified only at high expense once their implementation is under way. Hence, basic physical information requires accurate corroboration. The most practical way to record the bulk of such information is to map the data in such a manner that the policymaker is able to test his ideas on these maps before implementation. Overall planning calls for data on land tenure, topography, soils, climate, vegetation, land use, and irrigation potential.

Delineation of development area

The area chosen for development should be delineated as a functional unit. It should be sufficiently large to make a link between the agricultural, service, and industrial sectors practical, without, however, being unwieldy for planning and implementation purposes within a reasonable period of time. The boundaries of such an area should be sufficiently flexible to permit change as the scope of development widens.

Projection of development potential

One of the main aims of integrated rural development planning is to harness the development potential of the region. Part of the potential is inherent in the population and its productive capacity, which need to be directed into proper channels. Once agricultural development gains momentum in a given area, its results are exemplified by greater *per capita* productivity and the consequent release of manpower for other forms of employment and by greater quantities of upgraded farm produce needing new marketing outlets and processing possibilities.

The bulky nature of most agricultural products tends to warrant the establishment of processing industries in or near the production area. In an integrated plan of action, these two basic elements can be brought together within the region itself. Industrial processing and provision of diverse services to the farmers form the second tier of development, whereas agricultural development represents the first tier. As a result, people displaced from agriculture need not leave the region but can become involved in an overall betterment programme based on the output of an efficient farming community and the development of other resources in the region. A wide range of processing industries related to agriculture in the region can be accommodated.

Formulation of policy

The goal of this rural planning is formulation of a policy for integration of rural and urban cycles into a comprehensive and balanced strategy so that the improvement of one has a positive effect on the other and vice versa. This policy should be translated into a specific development programme for agriculture, industry, and services whereby production data from an agricultural plan will indicate the possibility of establishing industries and services within the urban development cycle. Urban data related to demand and employment will indicate the support that agriculture can expect for its development cycle from the urban sector.

Evaluation of plan and implementation

Detailed agro-economic and physical planning for the first stage development of the rural sector in the area now begins. The composition of the planning team for this work will depend, to a large extent, on the challenges of rural development for the area. The team coordinated by a project leader elaborates agro-economic plans, setting forth farm patterns and prototype farms, market mechanisms, crop credit programmes, and the investment for the agricultural development of the area.

Social planning involves a determination of the pattern of settlement and of service needs, taking into consideration the tradition of the farm community. A physical plan is an expression of both the agro-economic and social plans. Once these planning components are completed, an overall investment programme for the rural development of the area can be prepared and phased in according to the strategy suggested by the whole planning team. Similarly, detailed plans for specific industries and marketing installations related to agriculture need to be prepared. Results should be evaluated before the planning of future development.

RURAL PROGRAMMES

Israeli experience in rural area development and the experience of a number of countries undergoing agrarian reform suggest that the modern rural community might be organized according to a variety of coexisting patterns, all of them providing an adequate social and economic framework for the farmer.

The farmer should be free to choose the social and economic framework within which he lives and works. In Israel there is a variety of settlement patterns: The *moshav*, a cooperative of family farmers, usually comprised of a village of 80–100 families where each farmer tills his own fields and makes his own decisions; the *kibbutz*, a production and consumption collective managed by an elected secretariat and operated as a single unit; the *moshave shitufi* which combines collective production with private consumption; and the *moshava*, based on private land ownership and individual farming.

Production units and resources should be sufficiently large to provide acceptable income to farmers. The objective should be to provide each family with an opportunity to achieve, in certain stages, set income targets which central policy has established nationally or regionally. These targets are related to certain internal and external yardsticks. Since agricultural development is a gradual process and the receptiveness of the people to change is a variable, income targets can, in most cases, be achieved in three stages, each of which can serve as a benchmark in the process of rural modernization. These goals could be as follows:

1 Immediate income targets related to the cost of a balanced minimum food basket at rural consumer prices. In traditional agricultural communities, this target is often equivalent to the typically stated aspirations of farm families in the area. The goal can usually be achieved by concentrating action on improving the existing crop practices, particularly since, in absolute terms, the immediate goals are relatively modest and require few additional investments. The strategy to attain such targets usually centres around crop credit programmes linked to agricultural extension campaigns. Related to it is the introduction of improved inputs after trials on pilot and demonstration plots in the area. With the improvement of tillage practices and the introduction of better seeds or reproduction stocks, marketing arrangements will have to be improved to handle added production. The total effort must be geared to show the farmer the absolute profitability of his additional efforts.

2 Consolidated income targets equivalent to an income sufficient to provide all basic necessities and high enough to leave a margin of savings for further investment in farm and home improvements. In Israel, this target became the basis for the introduction of intensive irrigated agriculture on farm holdings capable of adopting modern agricultural techniques.

3 Parity income targets related to earnings of organized labour in industry. The target should be related to wages in a reasonably labour-intensive industry, and the production unit should have the resource potential and supporting services to keep pace with the future wage advances in industry. In Israel the size of farms planned on the basis of standard prototype farms proved to be quite adequate to attain parity income targets in many settlement areas of the country. With the introduction of vegetable and flower production for export in a number of farm areas, the original allocation of physical resources proved to be sufficiently flexible to cope with the new patterns of production required for the country's export drive.

Planning of contiguous land blocks free of housing and farm buildings would make future tenure reforms less costly and technically less complicated. In this way the size of plots could be adjusted in response to a possible decline in the active population engaged in agriculture without embarking on costly land consolidation programmes.

Production should be specialized around a few main crops or branches. Although operation may be sufficiently diversified to provide continuous employment for the farmer and members of his family, their activities should be centred around one or two main branches suited to the particular region. This is the only way to promote the adoption of technological improvements essential for raising yields and reducing unit production costs.

In effect, the process of planning and implementation described involves continuous adjustment of human responses to innovations, planned farm output, and planned urban development for the area until dynamic equilibrium is achieved. As the subsequent stages of rural and urban development are implemented, the socioeconomic level of the region spirals upward, retaining more of its people and offering them opportunities for a higher standard of living.

RURAL COMMUNITIES AND SERVICE CENTRES

An ideal modern rural community should be efficient and profitable, deriving the benefits accruing from advantages of scale on the one hand and promoting initiative by farmers on the other. The physical infrastructure of the community should be adequate. The community should be connected with the national communication network in order to facilitate transportation of products and to encourage exchange of goods at reasonable costs. To some extent, the infrastructure built for the community is a function of agro-economic and service planning.

Social amenities and rural community services should be similar to those enjoyed by urban communities, especially in matters of health, education, recreation, and municipal services. Their cost to the farmer should be compatible with his income, and their operation should be gradually taken over by the farmer and his community. The modern rural community should be capable of maintaining and paying for its economic and municipal services on the same basis as an urban community carrying similar social and economic obligations and granting the same privileges. Rural service centres can reduce the unit cost of services by virtue of their serving a larger geographical community.

Israel's experience in rural services planning is relatively recent. Over the past two decades, rural service centres have been established throughout the country to provide social and economic services for groups of six to eight villages (some 2000 persons). The distance separating service centres from the farmsteads was a fundamental consideration in the spatial planning of the centres.

The pattern known as the composite rural structure embodies the contemporary concept of a larger rural community. In this way, the economies of scale were attained in the establishment of schools, health clinics, recreational amenities, and larger scale economic services such as the provision of heavy farm machinery, marketing, and transportation services. By making rural centres both habitable and hospitable through the introduction of adequate housing facilities and other amenities, professional personnel to man the services can be attracted to these centres. The presence of doctors, nurses, teaching staff, and rural extension personnel undoubtedly contributes to the development of a dynamic rural community.

On the other hand, those services required by the household on a daily basis such as the kindergarten, the grocery, first aid, postal service, etc, were located in each village. Operation of rural service centres and of village services has permitted their upgrading by administrative measures. Outstanding are the efforts and the results obtained in education, vocational training, and preventative medicine.

At present, elected officers run most of the service centres in Israel and represent their centres on district councils. The village cooperative, the *moshav*, serves as the legal primary municipal unit of the community. The community, through its municipal and district cooperative institutions, can be identified as a bargaining group, especially in purchasing, marketing, and processing activities. Even before the community becomes a registered cooperative, it must be organized on a precooperative basis in order to become eligible for seasonal credits. This organization has become an accepted policy aimed at encouraging commercial agriculture and preparing the community for capital investments and other resource allocations for its future consolidation.

The ultimate goal of integrated rural area development planning is to create modern communities through transformation, consolidation, or settlement. Ideally speaking, planning of modern rural communities should be integrated into an overall national development plan so as to link macroplans with local planning which reflects local conditions. At the very least, local patterns of production should be related to the central national development goals.

URBAN CENTRES IN RURAL AREAS

The farmer has an active economic interest in having his products processed in order to enjoy part of the resultant added value. In recent years, farmers in Israel, through their village cooperatives and district councils, have established their own processing centres adjacent to the regional towns in the area. Dynamic employment and reasonable living conditions must be available for the sons of farmers within the bounds of the rural community in agriculture, services, and industry. It is at this juncture that development of the existing towns and establishment of towns and service centres in the rural area becomes vital. Stagnant rural towns should be linked to the agricultural development plan of the area and should benefit from the introduction of processing industries which would provide impetus to further economic growth.

Israel's experience with the establishment of new towns whose economic development was closely coordinated with that of the hinterland has shown that agricultural processing industries became the forerunners of further urban development cycles by involving the growth of nonagricultural industries. At this stage of development, the rural area towns were able to absorb the sons and daughters of farmers in gainful employment and thus prevent their migration from the region.

The modern rural community should provide a suitable framework for the establishment of harmonious relationships between the team of technicians working for the area's development—the promoter group—and the community of farmers—the producer group. A development breakthough characteristic of the modern rural community can be said to have been achieved when the producer group assumes a major role in promoting technology. It is at this stage that the community begins to display initiative and assumes an active role in decision making. This drive for modernization should be spearheaded by an elected rural leadership capable of representing the interests of the community before national and regional agencies.

THE LACHISH REGION

Background

Integrated rural area development in Israel became one of the cornerstones of the country's development policy from 1954 onward when policy makers and planners had to look for new ways to meet the challenge of absorbing and integrating masses of immigrants, mainly from less developed countries and traditionally patriarchal societies. At that time, it was necessary to absorb a substantial part of the mass immigration in agriculture, due to the absence of adequate employment opportunities in services and industries. Availability of water resources, on one hand, and the possibility of settling the comparatively large and sparsely populated tracts of land in the north and the south of the country, on the other, brought into focus the necessity and benefits of integrated functional area development. A major contributory factor for embarking on the programme was the clearly stated government objective of population dispersal, which, among other things, would avoid population pressure in the existing cities as far as possible.

The Lachish Regional Development Project, initiated in 1954, became a turning point in Israel's rural development. The region, located midway between Tel Aviv and Beer Sheva at the gateway to the Negev desert region, was a very sparsely settled area covering 100 000 hectares (ha) wedged in between the Gaza Strip and the Hebron Hills. With the completion of the main irrigation pipeline from the Tel Aviv area to the south, it became possible to embark on a comprehensive development programme to settle the area.

Plan

An integrated area plan was evolved for the region involving the establishment and consolidation of 3900 farm holdings in 56 settlements including 32 *moshavim*, 16 *kibbutzim*, 6 rural centres and a regional town.[1] Most of the new settlements, 26 in all, were established in the central and eastern districts.

Three ecological districts were identified in the region. After evaluating the needs of the national economy and the goals of national agriculture, prototype farms were planned in relation to the potential of soils, unit price of water, and market possibilities and limitations. Three major farm types emerged: dairy citrus, along the coastal belt, ie, in the vicinity of Ashkelon, 7 acres, each with an annual quota of 19 000 m^3 of water; field crop, east of the coastal strip in the heart of the region, 10 acres, each with an annual quota of 15 000 m^3; and extension grain-and-ranching, 50 acres, each with an annual quota of 9000 m^3 of water in limited irrigated areas of the more fertile valley of the district.

Implementation

Today the Lachish region is a major agricultural production area producing crops at a value of $80 million of which $30 million are exported in the form of flowers, vegetables, cotton, citrus, avocados, and groundnuts (table **7.1**). The changing agricultural pattern in the region in the last decade is most evident in the shift of farm activities toward export oriented production during the autumn-winter season (from October to May). Vegetable growing and floriculture for export to Europe has become a dominant activity. During the 1975 season some 3000 tons of onions, 1500 tons of watermelons, 2000 tons of sweet melons, 300 tons of celery, and 400 000 heads of lettuce were shipped to Europe. During the same season 36 million roses, 3 million carnations, and 9 million gladioli from the region were flown to the markets of Western Europe.[2]

The use of plastic covers for growing winter vegetables in hothouses for flower production and for weed control has enriched and diversified the seasonal occupational cycle of farmers. Thus, growing peppers, tomatoes, cucumbers, lettuce, etc, in winter through the use of 'plastic technology' contributes considerably to keeping the farmers occupied throughout the year.

At the same time, the industrial crops that were originally introduced 20 years ago have remained an important source of income. Cotton is grown at present over an area of 3200 ha with an average yield of 4.7 tons per ha during the 1975 season. Sugar beets are grown over an area of 1200 ha in the region, producing over 9000 tons of processed sugar.[3] Both cotton and sugar beets are processed in the regional town of Kiryat Gat. Over 20 000 ha are presently under sprinkler and drip irrigation, and another 20 000 ha are under dry land grain production (table **7.2**). Yields and labour productivity rose substantially with spectacular improvement in agricultural output in relation to unit water use.

In each settlement individual and group extension services were provided by experienced instructors from established villages who volunteered their services and lived during the

Table 7.1 Yield per hectare in the Lachish region (Israel)

Crops	Yield (1957)	Yield (1975)
Cotton	2 tons	4.7 tons
Sugar beet	40 tons	60 tons
Groundnuts	2.5 tons	3.5 tons
Tomatoes	40 tons	70 tons
Onions	15 tons	35 tons
Peppers	10 tons	20 tons
Potatoes	12 tons	20 tons
Lettuce	unknown	40 000 heads
Celery	unknown	30 tons
Wheat (under irrigation)	2.5 tons	7.0 tons
Sorghum (dry farming)	2 tons	3 tons

Data from Ministry of Agriculture, Lachish District Extension Service, March 1976.

Table 7.2 Agricultural land use*

Land use	Area in hectares
Field crops under irrigation	13 200
Fruit orchards under irrigation	6 300
Field crops and orchards under dry-farming systems with some auxiliary irrigation	17 900
Nonirrigated improved pasture and range lands	11 400
Forestation	25 000
Total	73 800

The cultivated area of the region, some 37 000 ha, represents about 9 per cent of the total arable land of the country of which 19 500 ha are under full irrigation requiring 76 million m^3 of water annually.

*Data from District Agricultural Extension Office, Ministry of Agriculture, Kiryat Gat, March 1976

first few years in the new villages. Planning of each village followed the specific types suited to the physical conditions of each ecological district. New parcellation systems were evolved, based on the principle that the lands of the farm holdings should form a continuous land block, free of houses or farm buildings. Holdings comprise between 0.5–1.0 ha adjacent to the house and another 3.5–4 ha in one or two plots which form the extended land block. In this way, cooperative mechanization, aerial spraying, and other services could be organized to derive the advantages of scale without affecting the individual farmer's ownership.

While some of the second generation farmers, attracted by modernization of agriculture, would have liked to have established their own farms, others sought local nonagricultural employment. Obviously, some of them found such employment in the towns' growing industry.

New Kiryat Gat, the regional town, owes its development to the progress of the surrounding agricultural hinterland. Populated by about 26 000 people, it has a processing industry based on cotton, sugar beets, eggs, citrus and flower packing, and groundnuts. It has cold storage facilities for perishable fruit and vegetables. The town today has 5500 active employees; 45 per cent work in industries, with 3000 persons employed in a vigorous modern textile complex producing 3 million m of high quality cloth and a wide range of textile consumer goods.[4] A packing house for citrus and silo installations for 30 000 tons of grain are under construction. There are three banks in town, with branches in some of the rural centres. It is estimated that Kiryat Gat handles about 40 per cent of business done in the district.

There have been significant changes, not all of them planned, in the administration and economy of the region. In the administrative and planning sphere, for example, responsibility has been shifting from the settlement authority, originally charged with the settlement and consolidation of thousands of newcomers to the area, to the district development company, owned entirely by the farmers. The company cultivates the dry lands belonging to the villages by using agricultural machinery hired from individual farmers; it also promotes and plans the establishment of grading, packing, and processing plants to meet the needs arising out of the changing crop pattern of the region. In the initial stage of the regional development, some 150 professionals and technicians were employed to implement the project; now there is an agricultural extension service operating in the region. Some 50 agronomists, 5 veterinarians, and technicians are employed at an annual cost of $300 000 but this is less than 1 per cent of net output value of the region.

CONCLUSION

This chapter has given me a unique opportunity to relate the present activities of Israel's Technical International Agricultural and Rural Cooperation, which I have the honour to head, in connection with the substantial achievements of the Lachish regional development experience. I lived and worked in the Lachish region both as a new immigrant and as a recruited farm planner some 20 years ago. This has enabled me to assess, on a continuing basis, some very interesting and important aspects of the development experience. Some of the lessons learned have been successfully applied in development projects abroad. Fortunately, when we worked on our cooperative projects abroad, there was an early realization that successful experiences in Lachish cannot be transferred or copied *per se* in other countries; yet principles can be adapted to specific economic, social, and ecological conditions, provided there is a consensus on the need to develop rural regions to generate employment and income opportunities.

The Lachish development experience, therefore, has become of particular relevance to many policy makers and professionals of countries undertaking agrarian reform programmes aimed at reconstructing and reorganizing peasant communities. In Lachish, traditional, often family or clan groups, have grown into modern, highly productive communities in less than 12–15 years, presenting a spectrum of planning experiences, technology transfer policies, and development procedures which are of interest to those planners in developing countries who are grappling with the challenge of promoting communities for active participation in a market-oriented economy. Planning criteria originally employed in the Lachish development experience have been adapted to projects under very different ecological, economic, social, and political settings. Israeli project experience abroad over the last 15 years indicates that universal criteria seem to be in the making.

The development cycle of a region or a nation in which the agricultural sector is dominant (ie, other sectors are as yet underdeveloped) can best be set in motion by first improving rural life and agricultural productivity. Rural development implies the improvement of incomes and their distribution, better employment opportunity in the area for the active working population, and a parallel improvement of social and educational services. The investment required to modernize rural communities, to improve peasant incomes, and to generate new jobs in the area is less costly than the social and economic cost of uprooting traditional rural communities and settling them in the big cities where nonagricultural job opportunities have to be created.

The planning of production units and resource allocations should be aimed at providing farmers with an acceptable income opportunity. The objective should be to provide each family with an opportunity to achieve, in certain stages, net income targets related to internal and external yardsticks. The phasing in of income goals can serve as landmarks in the process of rural modernization. My experience has indicated that these targets could be set and reached in the following stages:

1 Immediate family income targets related to the cost of a minimal food basket at rural consumer prices. This target is often equivalent to the typical aspirations of subsistence peasants whose sole aim is to have a sufficient food stock in reserve in anticipation of adverse climatic or national conditions, and it is being achieved in Israeli technical cooperation programmes by concentrating action on improving existing crop practices and supplying critical inputs through the organization of seasonal crop credit programmes. There have been rather satisfactory results in over 12 projects in the last decade. The directed credit system was an adaptation of the credit-*cum*-extension system that was developed by the Lachish development team.

2 Consolidated family income targets equivalent to an income sufficient to provide all basic necessities and services and high enough to leave a margin of savings for further investments in farm and home improvements. This goal is being achieved in a substantial number of projects through the introduction of intensified agriculture (eg, multi-cropping where irrigation facilities exist or intensified livestock production where a market demand prevails). In absolute terms, the income is only a percentage of incomes earned by industrial labour in the cities; yet in qualitative terms the target often offers a living standard sufficiently attractive to reduce the rural exodus to metropolitan areas.

3 Parity income targets related to the earnings of organized labour in the industrial centres. Usually in Israel's cooperative projects abroad, the yardstick is related to wages in a

reasonably labour-intensive industry where the production unit should have the resource potential and supporting services to keep pace with future wage advances in outside industry. In this category, a more sophisticated regional development approach is required, due to the attendant greater labour mobility. Rural-urban integration systems for delineated areas are planned. Agro-industrial projects become an important component of such plans and their implementation.

During the last decade, five such programmes were planned, of which four are presently being implemented. The incentive system for the introduction of agricultural industries used in the Lachish region was partially adopted in three out of the four projects mentioned, all in Latin America.[5]

The Lachish experience of physical rural planning was noted for its flexibility and tendency to maximize the economic advantages of scale through parcelling commercial holdings on contiguous land blocks, free of housing and farm buildings. These characteristics have been adapted, to a large extent, in many countries grappling with the problems of agrarian reform. In most cases, projects abroad are oriented to establish viable family farms with the possibility of adjusting plot sizes in response to new technologies, particularly farm mechanization, and to decrease the active population working in agriculture without the necessity of embarking on further costly land consolidation programmes. The Lachish experience of intensive farmyard investments where the farmer's initiative has been so successfully mobilized has made it a mainstay of Israel's technical cooperation projects. The composite rural structure as planned and implemented in the Lachish region has become the basis for developing the concept of creating viable larger rural communities abroad. The economy of scale for social, economic, and selected municipal services is encouraged through the construction of rural service centres where schools, health clinics, recreational amenities, and large-scale economic activities are located. In Israel's cooperation projects abroad, doctors, nurses, teaching staff, and rural extension personnel are right in the rural centres.

With regard to settlement patterns, the Lachish experience has clearly shown that it is inadvisable to impose settlement systems; the farmers themselves should be free to choose the ultimate social and economic framework within which they intend to live and work. Perhaps it is this understanding which led to such a demand for Israel's technical cooperation abroad, particularly in Latin America where the system for organizing peasant communities in cooperative endeavours is such a critical subject for discussion among policy makers and technicians.

Lachish experience, while demonstrating the importance of adopting policies and plans compatible with cultural values of the communities served, once and for all disproved the classic belief that cultural constraints are insurmountable when trying to develop a project. Lachish has taught us that the total effort must be geared to show the farmer the absolute profitability of his additional efforts. For Israeli experts, this became the main motivation for true development, irrespective of culture, religion, or a specific value system. Thus, we have registered some interesting success stories both from the economic and social points of view in projects in southeast Asia, Africa, and Latin American countries.[6]

With regard to the transfer of technology, the Lachish experience of specialized production centred around a limited selection of crops or branches was found to be very relevant for projects abroad. Although farm operations are planned so as to provide continuous employment throughout the year for the farmer and members of his family, the farm's activities are nevertheless centred around one or two main branches suited to the particular region. In this way, Israel's technical cooperation was able to stimulate the beginnings of a more specialized and more intensive extension service linked to seasonal credit programmes.

The Lachish experience showed the importance of several major spheres of investment. First, investment must be staggered in a number of developmental phases and in relation to income targets. Second, investments in the supporting system should be encouraged because they make primary investments incurred in resource development so much more profitable. Third, investments per family unit in rural areas are cheaper than investments needed to settle rural families which have migrated to the cities. Fourth, an investment/income ratio of 4:1 for a rural development project is quite satisfactory. Fifth, investments in rural housing of up to double the annual planned family income is both advisable and economically feasible. Finally, we now have proof from three major project evaluations that for every family settled, two or three additional permanent jobs are created.

The Lachish experience in implementation processes has been a major contributory factor to successful project development abroad. First, the interdisciplinary organization of the development team was a pioneering innovation for rural development efforts in some six Latin American countries in the mid-1960s. The idea that such a team should work and live in the project it is serving and that attractive living facilities should be provided by government to keep the team in the region is today accepted in most projects involving Israeli technical cooperation.[7] Second, both planners and implementors are integrated in the same development apparatus, operating as an executing agency. Third, the importance of farmer participation in decision-making and implementation, usually through a cooperative representative body, has demonstrated its value. In recent years, this has been a mainstay of Israeli technical advice due, in no small measure, to the realization that in the absence of such participation, the project has distinctly paternalistic overtones. Fourth, just as planning is linked to implementation, so applied research, too, is linked to the developing process.

The main application of the Lachish rural regional development experience lies in the very concept of integrated rural area development, geared to the creation of modern communities through the transformation and consolidation of settlements. In Israel's dialogue with developing countries, considerable emphasis is placed on the need to plan areas as a cross-function between macro-(national) and micro-(rural) community plans so that, at the very least, production patterns in a given area are compatible with national development goals. The 'feedback' mechanism that worked so well in the Lachish experience is being adopted now by many countries as a way of refining national plans adapted to the reality of the field and the local farm community. The experience of developing the Lachish region indicates the vast economic benefits that can be attained for the development of arid and semiarid regions, if applied research is undertaken at the same time as the developing process.

Originally, the agricultural plan for the Lachish region was conceived for the production of irrigated summer crops; only in later years through a significant input in applied research was it possible to fully exploit the advantage of aridity and sunlight for production of *winter* crops. The use of plastic ground covers and greenhouses (an agro-plastic revolution) and the introduction of drip irrigation (the blue revolution) enabled the Lachish region to become an important area for the export of winter vegetables and flowers. It is on the basis of this experience that Israel today is cooperating with five countries in Latin America in a specific programme of applied research and transfer of technology for development projects in arid and semiarid regions.

NOTES

[1] *The Composite Rural Structure* (Jerusalem: Agricultural Settlement Department, 1960).

[2] Data from Lachish District Extension Service, March 1976.

[3] Ibid.

[4] Data from Kiryat Gat town council offices, January, 1977.

[5] These projects are in the Dominican Republic, Nicaragua, Guatemala.

[6] Y. Abt, 'Types of Agricultural Projects Used in Developing Countries', in *Rural Development in a Changing World*. ed. Raanan Weitz. (Cambridge, MIT Press, 1971).

[7] *Rural-Urban Integration: An Approach for Developing Countries Based on Israeli Experience* (Tel Aviv: Institute for Planning and Development and CIAC, 1969).

SUGGESTED READING

Abt, Y. *Institutional Approach to Agrarian Reform and Related Tax Problems*. Norfolk, Va: Lincoln Foundation, 1966.

'Types of Agricultural Projects Used in Developing Countries'. In *Rural Development in a Changing World*. ed. Raanan Weitz. Cambridge, Mass: MIT Press, 1971.

Clark, C., and M. Haswell. *The Economics of Subsistence Agriculture*. New York: Macmillan, 1966.

The Composite Rural Structure. Jerusalem: Agricultural Settlement Department, 1960.

Schultz, Theodore W. *Transforming Traditional Agriculture*. New Haven, Conn: Yale University Press, 1964.

PART TWO

ENERGY AND WATER RESOURCES

8 Land use and energy conservation in an arid-zone city: Tucson, Arizona*

PETER MAHONY

The relationship of energy consumption to land use and community development patterns is a complex one, generally understood in a qualitative manner. Such factors as income, family background, and other cultural and socioeconomic characteristics often cloud the basic relationship between development patterns and energy consumption. To separate these is quantitatively problematic. Those studies available, conducted for specific localities, only provide insight on historical data and reinforce the mitigating role of socioeconomic factors. It is necessary to determine which urban design and planning factors could play a decisive role in reducing energy consumption in future development.

In order to obtain sufficiently generalized data for the nation, researchers adopted a case study approach and selected several planned population units (PPU's) which were exemplary of future regional urbanization patterns. One of these was Tucson, Arizona, exemplary of the arid city. The selection was based on the probability that new development in the United States will be in large neighbourhoods with growing developmental areas and not in new towns or areas with small incremental growth. Here people could use potential energy system benefits. In the study itself, investigators analyzed only existing technology, for ample savings could be realized without looking at future technologies.

For evaluating alternative development and transportation configurations, land use mixes and technologies, an approach of comparing two projected futures for the year 1995 was adopted. *Laissez-faire* unplanned growth was contrasted with energy conserving planned growth. This chapter is based on studies leading to a report by the researchers, *Reading the Energy Meter on Development*.[1]

*The Federal Energy Administration (FEA) with technical planning assistance from the Energy Research and Development Administration (ERDA) sponsored studies of the characteristics of growth in energy consumption as it related to raw land development and renewal so that policy guidelines could be established for minimizing this energy growth.

RESEARCH FINDINGS

The per cent growth in energy consumption for the years 1975–95 was compared with the growth in each land use sector (residential, commercial, institutional, industrial, and transportation) for each scenario as it was affected by the basic planning measures of land use mix, spatial configuration, transportation change, and the relationship of land use mix and design to the use of energy conserving technology. This was measured as energy elasticity:

$$\frac{\text{\% energy consumption growth or decline}}{\text{\% development increase or decrease}}$$

The resultant ratio indicates that particular energy conserving design's rating:

A ratio larger than one is unfavourable because it indicates energy consumption is growing *faster* than development.

A ratio of unity indicates that energy is growing at the *same* rate as development.

A ratio smaller than unity is favourable because energy is growing at a *slower* rate than development.

A negative ratio indicates that energy use is *declining* despite development increase; this change is the most favourable of all.

The energy elasticity ratio is desirable because it is a comparative tool, and it can dissect the effectiveness of individual energy conserving design measures. Those considered most important are the design measures grouped according to three energy saving sectors: land use mix, spatial arrangement affecting transportation requirements, and energy system technology. They interact with the four land use categories separately, all land uses together, and transportation.

Table **8.1** of the Tucson case study shows that the effect of innovative energy systems in conjunction with good planning measures can conservatively produce a declining rate of energy use approaching 60 per cent for each additional ft^2 or dwelling unit (du) of land use. Transportation planning

Table 8.1 Ratio of energy increase to development growth (Tucson, Arizona)

	Unplanned	Planned			
		Due to land use mix	Due to spatial arrangement	Due to technology	Reference units
Residential	1.37	.593	—	(−) .597	Dwelling units (du)
Commercial	1	.638	—		Square feet
Institutional	1	1	—	(−) .525 (Commercial, Institutional, Industrial)	Square feet
Industrial	1	1	—		Square feet
Overall consumption less transportation	2.45	.96	—	(−) .578	Population
Transportation	1	—	(−) .2		Passenger miles

measures which reduce travel can result in a declining rate of energy consumption of 20 per cent for each additional increment of passenger miles travelled. These are the two most significant measures followed by residential and commercial land use mix design actions related primarily to configuration and density; they can result in 60 per cent slower rates of energy consumption when compared to the rate of development. Given the high score of innovative energy systems with necessary urban design measures of density and configurations, these savings are further classified according to type. Highest savings of up to 25 per cent of BTU consumption are achieved by the use of selective energy and solid waste heat recovery systems. With the use of total energy and solar systems, savings of up to 20 per cent are achieved. Through central heating and cooling plants, savings of up to 15 per cent of the specific land use category that benefits from the energy system are possible.

On the simplest level, in order to conserve energy, future planning should be directed toward reducing transportation requirements, increasing the energy consumption efficiency of building configurations, and promoting shared energy-utility systems among various land uses to maximize mutually compatible heating, cooling, electrical, and waste product disposal needs. These have been generalized in **8.1** and **8.2**, which show a generic urban and suburban context, respectively. The design measures themselves can be reduced to a list grouped according to the specific objectives to be achieved and the general energy savings sector.

8.1 *Generalized urban context: energy conserving measures*

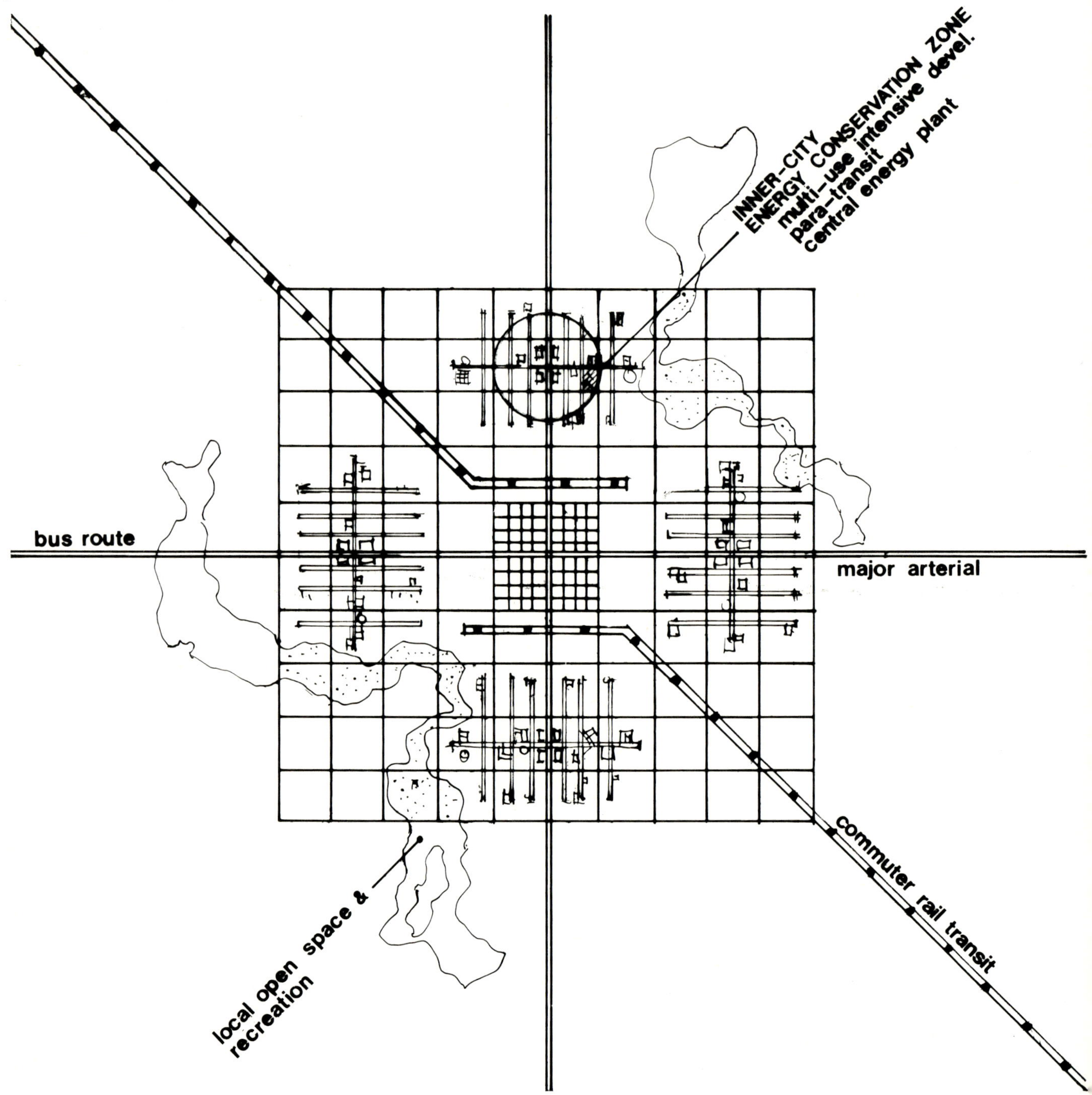

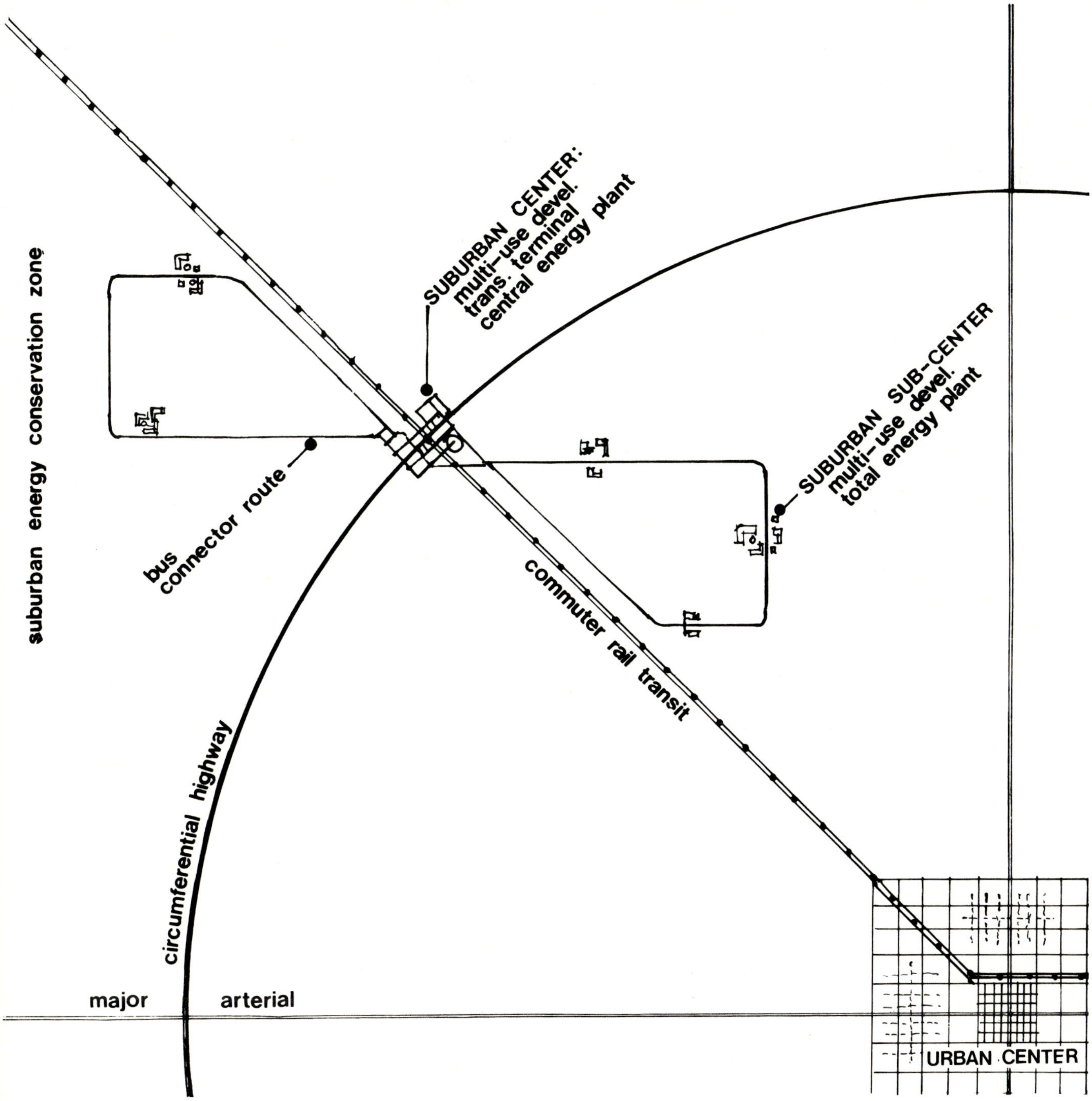

8.2 *Generalized suburban context: energy conserving measures*

Home-to-work travel reduction spatial arrangement:

1 Development of new multifamily housing units adjacent to places of existing employment and services.
2 Improvement of the proximity of housing at related income levels to nearby employment opportunities.

Land use densification and revitalization/land use mixture:

3 Redevelopment and increased densification of older housing communities.
4 Revitalization of existing town centres and an increase of multi-use aspects.
5 Intensification of industrial land uses within prescribed areas.

Local auto travel reduction/spatial arrangement:

6 Reduction of quantity of urban streets devoted to automobile use.
7 Initiation of transportation modes alternative to individual automobile travel.

Energy system integration into mixed land uses/energy technology:

8 Introduction of energy-efficient utility systems and infrastructure.
9 Development of new multi-use complexes of residential and business functions.
10 Integration of residential, commercial, and industrial energy uses.

Regional growth management to achieve trip reduction/spatial arrangement:

11 Preservation of open space and the containment of sprawl.
12 Revitalization of existing recreational facilities in proximity to residential units.
13 Development of new local recreation facilities available to residential units.

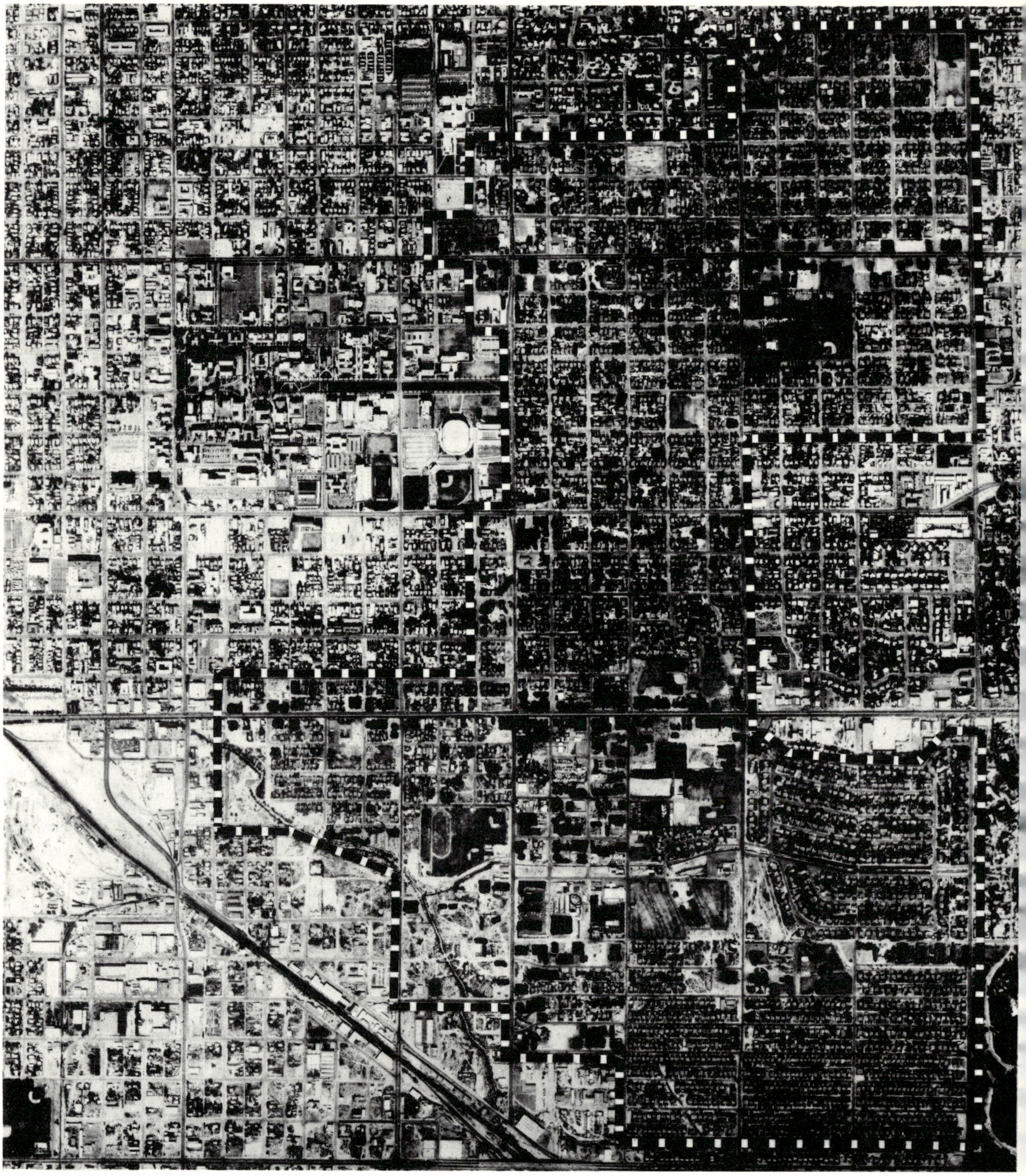

8.3 *Tucson, Arizona: Existing land use*

These urban design and transportation objectives should be paramount in any review process of both new development and renewal proposals. Through further analysis, one may be able to develop a rating system with these objectives properly weighted for a quantitative evaluation of large-scale private and public development proposals.

CASE STUDY: CENSUS TRACTS 6 AND 7

Tucson's characteristics

The existing characteristics of Tucson appear in the land use diagram, **8.3**, which has been superimposed over a 1974 aerial photograph. We can see significant physical differences between Census Tracts 6 and 7 (see **8.4**). Tract 6, closer to the University of Arizona, is basically a single-family housing community. Homes are 5 du per acre, and multifamily units are scattered randomly throughout. Commercial developments follow major east-west mile grids in nodes of greater concentration at intersections with the major north-south

8.4 *Tract 6 and 7 study area, Tucson*

streets. There are two public elementary schools serving these neighbourhoods. The street pattern's rectangular grid has frequent intersections and rear alley access to most lots. Himmel Park is a large public park with tennis courts, a swimming pool, and a public library.

Tract 7 retains a small remnant of the Tract 6 pattern, but it has two significant variations. There is a large quadrant of single-family residential areas, fully built up in long suburban subdivision blocks. There is also a significant area of business or industrial development. This tract is also served by two elementary schools and by a strip of commercial buildings along Broadway. A dry river arroyo runs diagonally across the bottom of the tract, and other drainage ways are also defined throughout the neighbourhood. The Tucson High School athletic fields are in this area, and some vacant lots are designed for government housing. There are no significant parcels of vacant architecturally useful property in either tract, but there are vacant lots throughout the residential neighbourhoods.

Tucson's Department of Planning provided land use information and special census data. A graduate student in planning at the University of Arizona tabulated for the study team the commercial, industrial, and institutional building areas as listed in Tucson's tax assessment records.[2] The team also

Table 8.2 Summary of existing annual energy consumption (1975) Tucson (test area population: 9758)

Land use	Units	Energy consumption per year	
		MMB*/unit	**Total MMB**
Residential			
Single-family (SF)	3 520[a]	120.46	424 019
2-family			
Low-rise MF	900[a]	76.12	68 508
High-rise MF			
Subtotal	4 420[a]		492 527
Commercial			
Office			
Retail			
Mall/large			
Attached/small			
Strip			
Hotel			
Food service			
Recreation			
Theatres			
Subtotal	848 053	.182	154 346
Institutional			
Schools			2 465
Hospitals			
Public service			
Dormitory type			
Subtotal	21 630[b]	.114	2 466
Industry			
Light industrial			69 384
Warehouse			
Subtotal	183 338[b]	.349	69 384
Total (Less transportation)			718 723
Transportation: See table **8.4**			246 000
Grand total			964 723

*MF: multifamily MMB: million BTU *a*: du *b*: ft^2

Table 8.3 Energy consumption factors per dwelling unit per year, Tucson (1975)

	MF Low-rise	**MF High-rise**
1 Heating, oil/gas MMB	98	60
2 Heating, elec. MKWH*	23	14
3 Heating, elec. MWB	78	47.6
4 Other, elec. MKWH	7.33	6.8
5 Other, elec. MMB	25	23
6 4.34% of (3) × 100% eff.	3.38	2.06
7 95.66% of (1) × .65% eff.	144.22	88.30
8 Comb. factor (6) + (7) + (5)	172.60	113.36
degree day correction	(−) 52.14	(−) 37.24
use factor MMB/du	120.46	76.12

Degree day correction, Tucson

BNL degree days 5500
Tucson degree days 1776
diff.: degree days 3724
SF residential 14 MMB/1000° d × 3.724 = (−) 52.14
Low rise MF 20 MMB/1000° × 3.724 = (−) 37.24
to be subtracted from above
All other factors:
5500° d − 5.3 MMB/100ft^2
1775° d − 1.7 MMB/100ft^2
difference: 3.6 MMB/100ft^2
(to be subtracted from all basic factors)

*MKWH: thousand kilowatt hours

referred to aerial photography made in 1974 and provided by the Planning Department. Summary values for existing population and land use are tabulated along with the yearly energy consumption in table **8.2**.

Energy requirements

Energy consumption factors for residential uses and degree-day correction factors are shown in table **8.3**. Planners use these factors to determine the 1975 or present energy consumption of each of the land use categories in table **8.3**. The total existing transportation energy consumption for 1975 is developed in table **8.4**. These tables then indicate total present energy consumption by sector as a base by which alternative future scenarios can be compared.

Table 8.4 Existing transportation energy Tucson (4420 du)

	Trip generation rate (person trip/du)	**Daily trips (number)**	**Modal split (per cent)**	**Trip length (miles)**	**Passenger miles (000 000)**	**BTU per pass. miles**	**Annual BTUs (000 000 000)**
Work	2.20	9 700		5.1			
			auto 97		17	6 600	112
			bus 3		1	1 150	1
			rail			4 430	
			walk				
Shop	1.75	7 700		2.8			
			auto 97		8	6 600	53
			bus 3		—	1 550	
			rail				
			walk				
Other	1.88	8 300		4.9			
			auto 97		14	5 750	80
			bus 3		1	1 550	—
			rail				
			walk				
Total	5.83	25 700		4.4	41		246
(Actual)		(31 000)				BTU/passenger miles = 6000	

Table 8.5 Pima Association of Governments' existing and projected population, dwelling units and land use

	1973[a]	1995[b]	2000[c]
Census Tract 6			
Single-family residential	309	294	389
Multifamily residential	71	135	106
Business	29	44	40
Industrial	1	2	2
Vacant	27	22	2
Population	5739	6987	6057
Dwelling units	2528	2777	2019
Census Tract 7			
Single-family residential	277	246	286
Multifamily residential	29	97	44
Business	73	63	93
Industrial	14	31	47
Vacant	129	61	12
Population	5723	7182	3198
Dwelling units	2008	2394	1066

Sources:
[a]TAZ estimates
[b]PAG transportation planning projections
[c]PAG alternate mode study projections

Table 8.6 Census Tracts 6 and 7 study area proposal

	Existing 1975	Energy-sensitive Planned 1995	Laissez-faire Unplanned 2000
Census Tract 6			
Population	4896[a]	5986	6057[d]
Dwelling units	2414[a]	2777[c]	2019[d]
Commercial	26.86 acres	30.9	30.9
	259 489ft² [b]	349 489	349 389
Industrial		Negligible	
Institutional	22.46 acres	24.46	24.46
Planned and unplanned scenario:	—363 additional dwelling units (assume 420 new multifamily units to allow for some replacement of existing single-family) —4.01 acres or 90 000ft² additional commercial —2 acres of 50 000ft² additional community facilities		
Census Tract 7			
Population	4862[a]	6023	3198[d]
Dwelling units	2007[a]	2394[c]	1066[d]
Commercial	66.53 acres	79.36	66.53
	588 564ft² [b]	868 564	588 564
Industrial	32.68 acres	62.09	62.09
	182 012ft² [b]	432 012	432 012
Institutional	60.63 acres	62.63	60.63
Planned scenario:	—387 additional dwelling units —assumed 500 new multifamily units to allow for some replacement of existing single-family by both multifamily and industrial —12.83 acres or 280 000 additional ft² of commercial uses —29.41 acres or 250 000 additional ft² of industrial uses —2 acres or 50 000 additional ft² of community facilities		
Unplanned scenario:	—941 decrease in dwelling units —commercial and institutional uses unchanged —industrial uses increase same as above		

Sources:
[a]1975 Special Census
[b]Robert Blair, Tucson Land Use Data Compiler
[c]PAG transportation planning year 1995 projections
[d]PAG alternate modes study year 2000 projections

Growth projection scenarios

Through a study of planned energy-sensitive growth and unplanned *laissez-faire* growth, one can provide a basis for developing land use and for projecting population growth for two scenarios. The energy consequences of these two approaches are then calculated based on a feasible urban design proposal for the former and then evaluated against each other and the existing level of consumption.

The city of Tucson and its metropolitan area are growing. According to the draft of the comprehensive plan prepared for community review in 1975, 'it is obvious that the Tucson area has not been growing in harmony with national demographic trends, but at a much more rapid rate'.[3] Statistics from the 1970 Census show Pima County with a population of 351 667. This has grown to 448 926 according to the 1975 Special Census, a five year increase of almost 28 per cent which is very close to the projected 1975 figure in the comprehensive plan report.[4]

In the population growth section of the plan, a regional population forecast for the year 2000 has been set at 800 000. A study of the costs of alternative growth patterns to accommodate the population increase to 800 000 or 1 million was prepared for the comprehensive planning process by Booz, Allen, and Hamilton, Inc, in July 1974. Their conclusions support a policy of contained growth for preservation of the land, for economies of capital expenditure, and for the improvement of residential life for Tucson residents.[5] The present population trend in the study area, Census Tracts 6 and 7, is one of decline in contrast to the overall Pima County growth which is presently primarily on the outskirts of the urban area. Census Tract 6 indicates a population of 5402 citizens in 1970 and 4896 citizens in 1975, a 9.4 per cent total decrease. Census Tract 7 indicates a population of 5598 citizens in 1970 and 4862 citizens in 1975, a 13.1 per cent total decrease.[6]

A reversal of the downtown population decline is a necessary component of a contained growth policy for the urban region. Development of new housing units in inner-city areas will have to offer amenities and advantages competitive with those features currently attracting residents to new low-density single-family homes outside the downtown area. The search for such solutions creates an equal opportunity for energy savings in building systems and design and in land use planning. The goals of contained growth and of energy conservation are highly compatible.

The selection of this portion of Tucson, a presently declining inner-city section of a small and expanding metropolitan area, provides an excellent opportunity to explore solutions for urban revitalization while analyzing the energy saving components. For the purpose of this analysis, we have referred to certain projections by the Pima Association of Governments (PAG) which were developed for use in transportation planning programs. Two different studies were undertaken. The first, prepared in 1962, projected land use, population, and dwelling units for the year 1995. This study showed optimistic population growth for the area, and in land use development it favoured multifamily housing.[7] The second projection, for the year 2000, completed in 1975 for use in the Alternate Modes Study, reflects the actual declining population trend. Despite the total population decline, the projections also note increasing acreage for multifamily and single-family housing units and favour the latter.[8] In table **8.5**, the existing conditions data devised for traffic assignment zones (TAZ), corresponding to the tracts, is presented along with the two sets of PAG projections. Whereas the projections for the year 2000 may represent current trends extended to the end of the century, the goal of both this energy study and contained growth policy requires the trend's reversal. Therefore, the

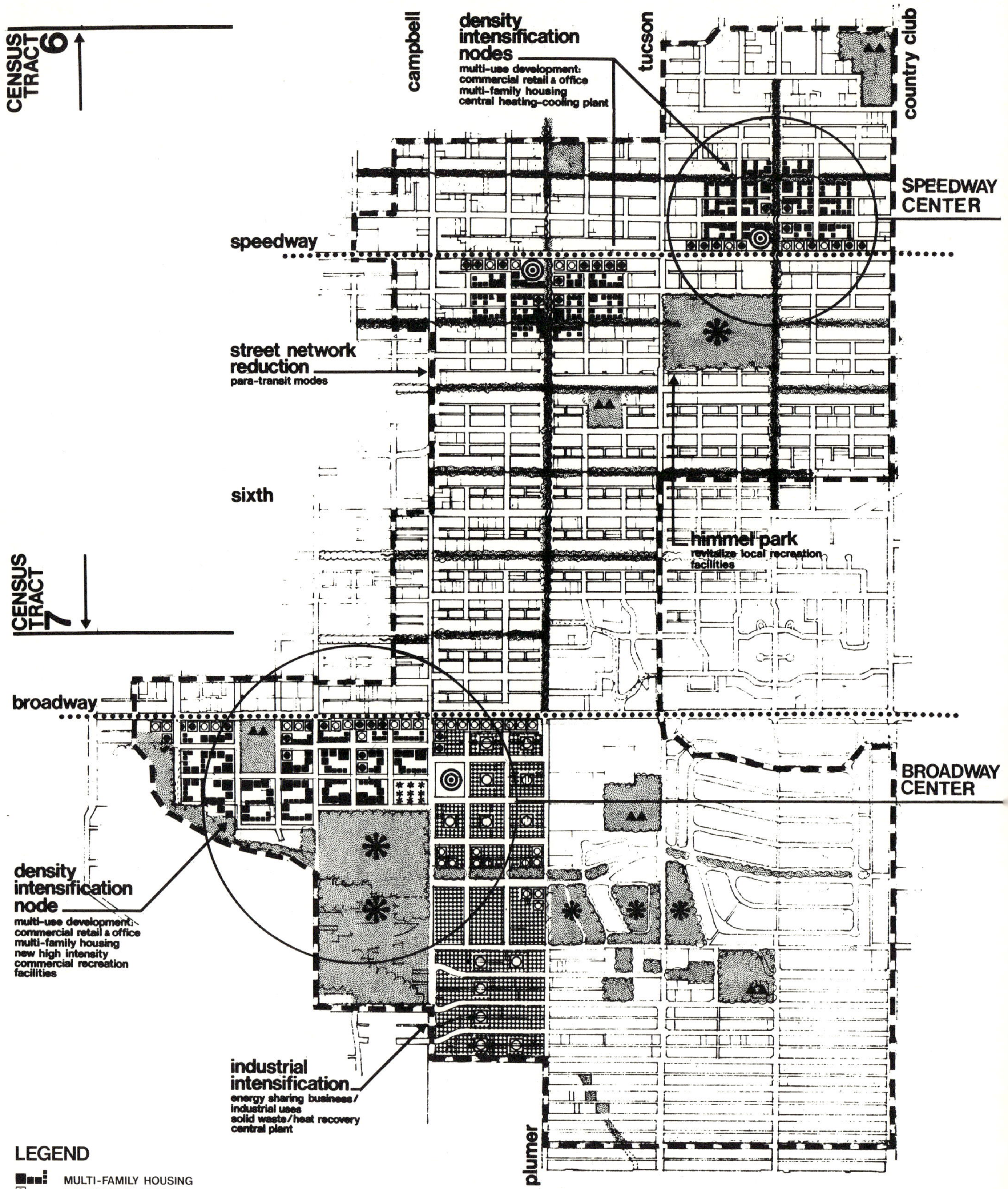

8.5 *1995 planned energy conserving development plan for Tucson*

projected number of du derived in the projections for 1995 has been selected and compared to the existing number of units from the 1975 Special Census. This produces an increase of 363 du in Census Tract 6 and 387 in Tract 7. These statistics are used as the basis for planning the new housing units in each of these areas.

Table **8.6** describes the programme which this study uses, comparing consequences of an energy conserving planned growth scenario with projected continuation of present unplanned *laissez-faire* trends. For the unplanned growth scenario, Census Tract 6 theoretically would follow the same

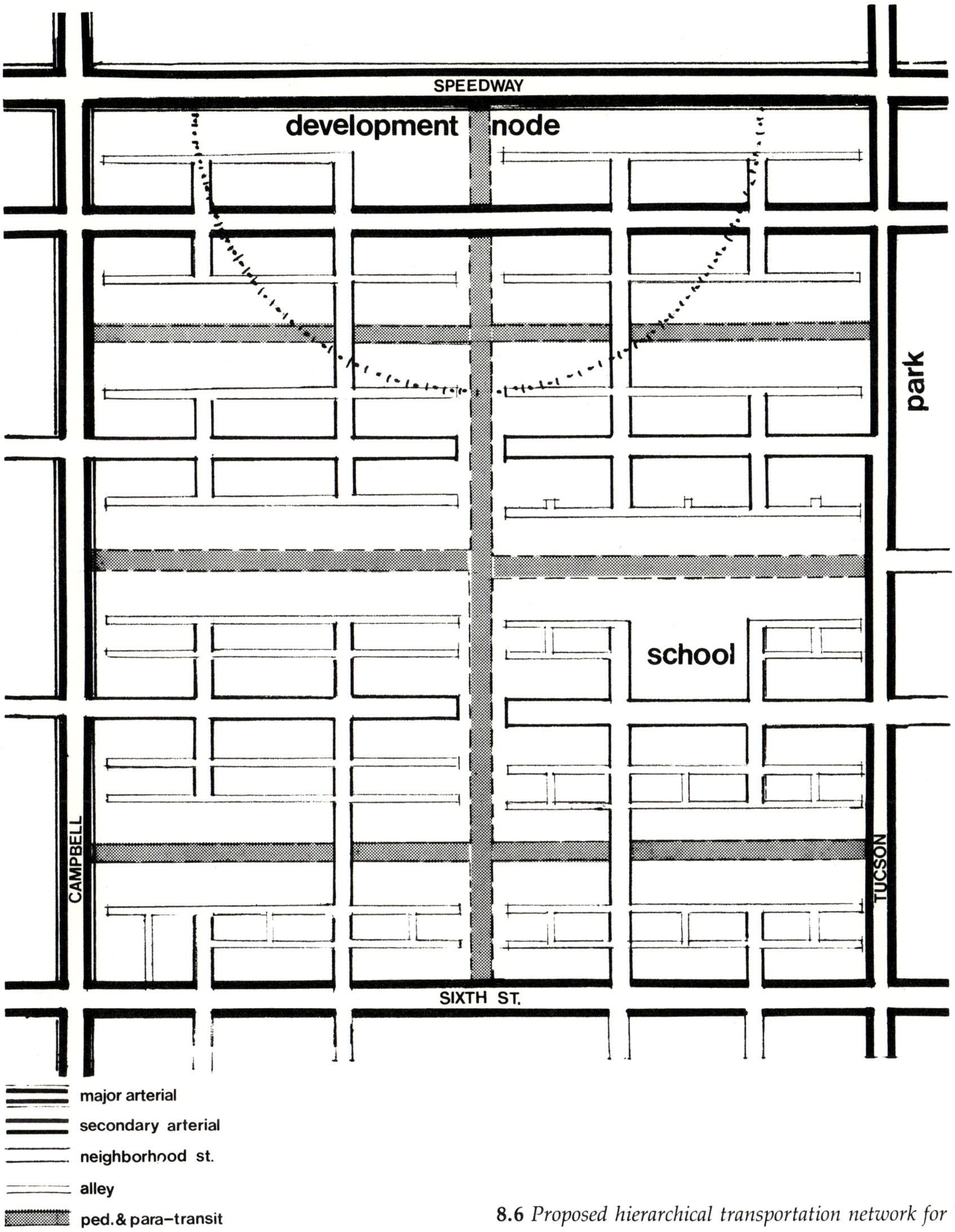

8.6 *Proposed hierarchical transportation network for Tucson*

growth but would develop in unplanned patterns. Census Tract 7 would decline in population according to the PAG year 2000 projection. This scheme would anticipate the same industrial development as in the planned proposal, but it would do so at the expense of lost du and without the amenities for multifamily housing and recreation incorporated in the planned proposal.

Tucson Tract 6 energy sensitive urban design proposals

The Tucson urban design proposals are graphically depicted in **8.5–8.8**, with the first diagram, an area-wide plan view serving as a locational diagram for the Tract 6 Speedway Center seen in **8.7** and Tract 7 Broadway Center seen in **8.8**.

Most new housing in inner-city areas would be in multifamily groupings. It is proposed that these be at densities ranging from 12 du/acre for town houses to 25 du/acre for garden apartment and midrise units. New construction, in order to take advantage of potential energy systems' sharing and efficiency, would be concentrated in density intensification nodes. The program of 420 units in Tract 6 would be divided into two groups of 210 du, each with a common central heating and cooling plant. Design of these housing units would provide unit sizes, open space, and views and facilities competitive with single-family homes. Nodes would be located on either side of the shopping street—in this case, Speedway; new commercial uses would be developed in available parcels within the node's spine as seen in **8.5**. The nodes are situated at the ends of north-south pedestrian bikeways developed within this tract area. A small additional number of single-family units may be built on the remaining scattered vacant properties and on some lots created by street closings.

In an effort to reduce reliance on individual automobiles and to enhance energy saving transportation via walking, biking, and going by public para-transit systems, the street pattern in this tract would be revised to reduce the total amount of paved roadway and of through east-west and north-south thoroughfares and to create a series of limited or no automobile greenways throughout the community. The proposed heirarchical pattern of roadways is based on the mile-grid bounded by Broadway, Speedway, Campbell and the country club. These streets would remain as principal through-traffic routes; by closing cross streets, the number of interruptions by lights and turning cars would be greatly reduced. Each square mile would be also bisected in both directions by secondary through-streets, here Sixth Avenue and Tucson Boulevard.

The square mile thus divided into four quadrant neighbourhoods is served by fewer existing streets but gives automobile access to each house via some streets and all back alleys. The north-south streets of Plumer and Treat become pedestrian bikeways which collect from several east-west similarly closed streets and feed the new multifamily nodes and intensified neighbourhood retail centres. A typical quadrant pattern is shown in **8.6**.

In order to discourage travel to dispersed recreation centres and to encourage living in central city locations, city recreation areas such as Himmel Park would have added facilities and activity programmes. Parking would be replaced with bicycle racks and new recreation facilities and activities, such as tennis courts, swimming pools, nature centres, fairgrounds, theatre and dance pavilions, art and food centres. All commercial areas would remain on principal automobile routes. Additional commercial growth to accommodate planned population increases would be limited to development in these existing areas and integrated into the higher density nodes of new residential units. Land for institutional uses would be limited to schools and community facilities. It is assumed that the school sites are sufficient to handle the population changes, but pedestrian access to these sites would be improved and developed. New community facilities would be integrated into the radially developing nodes. If in the future greater quantities of new multifamily housing and related activities would be needed beyond the projected development, similar nodes would develop on the opposite sides of the major avenues from the presently proposed clusters.

Through the proposed development combination of new construction concentration and street pattern changes, a new neighbourhood identity should develop. Commercial areas would be reinforced with greater numbers of residents in walking range of shopping. Bikeways and walkways would also encourage this. New development nodes would also be located at bus stops for public transit on major thoroughfares to distant work locations or other destinations. Para-transit systems could also be developed to move along the nonautomobile routes connecting neighbourhood retail, educational, recreational, and public functions with the housing units. Figure **8.7** is a conceptual sketch of a typical density intensification node, Speedway Center. Of the 13 energy conserving design measures described earlier, the following have been employed in Census Tract 6: land use densification and revitalization featuring older housing communities and multi-use centres; local auto travel reduction featuring street network reduction and alternate transportation modes; energy system integration into mixed land uses with an energy efficient infrastructure; and regional growth management to achieve trip reduction by featuring local recreation revitalization.

Tucson Tract 7 energy sensitive urban design proposals

Business and commercial land uses are presently found in Tract 7 in several locations between Cherry Avenue and Tucson Boulevard. Some parcels are adjacent to the arroyo at the southwestern edge of the tract and are separated from existing residential areas to the north by high school athletic fields. In the blocks south of Broadway and bounded by Campbell and Plumer, present business and industrial uses are mixed with mobile homes, single-family houses, and vacant properties. This entire band is proposed for commercial business and industrial uses, thereby creating an intensified industrial use in a coordinated, energy efficient and system sharing concentration. Within this band is located a new solid waste incinerator and central energy plant which will serve not only the new industrial facilities, but adjacent new commercial uses developed along Broadway and the new multifamily housing complex to the west. The programme of 500 new multifamily housing units for Tract 7 would be developed in a single major cluster labelled the Broadway Center surrounding the Miles Elementary School.

This area was selected for several reasons. It is presently isolated geographically by the arroyo to the south and by Broadway to the north. The proposed industrial intensification would eliminate some remaining single-family homes to the east. The entire southeastern portion of the tract would remain a homogenous single-family neighbourhood. The several blocks surrounding the Miles School hence would become an isolated pocket of remaining single-family housing units with some scattered multifamily units and vacant lots ideally suited to density intensification. Single-family units would be replaced with new multifamily units. New housing would also tie into the incinerator-central plant located in the adjacent industrial/business area. The school would serve as the nucleus of the renewed neighbourhood, and there would be easy access for children. The location adjacent to Broadway would allow a continuation of existing commerce and would provide a suitable area for new retail construction.

Presently, there is no public recreation facility in this neighbourhood. With redevelopment of the Broadway Center in the Miles neighbourhood, a linear park development is proposed along the arroyo bed as an inner-city recreation park. This area would be linked with the Tucson High School playing fields and would be connected to a future recreation development in the vacant lands held for government housing to the east of the industrial zone. This would also serve as a recreation focus for the present single-family homes in that sector (see **8.5**).

An overall diagram of the Tract 7 proposal appears in **8.5**. A conceptual sketch of the Broadway Center, the tract's major density intensification node with the adjacent business industrial zone can be seen in **8.8**. Of the 13 energy conserving design measures already described, the following have been employed in Census Tract 7: home-to-work travel reduction through multifamily housing near employment and services and housing for all incomes near employment; land use densification and revitalization featuring multi-use town centres and industrial intensification; energy system integration into mixed land uses featuring energy efficient infrastructures and shared residential/industrial energy uses; and regional growth management to achieve trip reduction by new locations for recreation.

TUCSON GROWTH PATTERNS AND ENERGY PROPOSALS

Growth increases at the periphery of Tucson, while older areas nearer the central business district flounder and even decline. Despite the appeal to the finances of 'inner-suburb' living, the periphery will continue to grow faster. New population centres along Speedway and Broadway can only be proposed if a total approach to their replanning takes place; by total, is meant a comprehensive approach to the design of housing, recreation, shopping centres, and the interaction among them. Most assuredly, the parks, shopping, and resi-

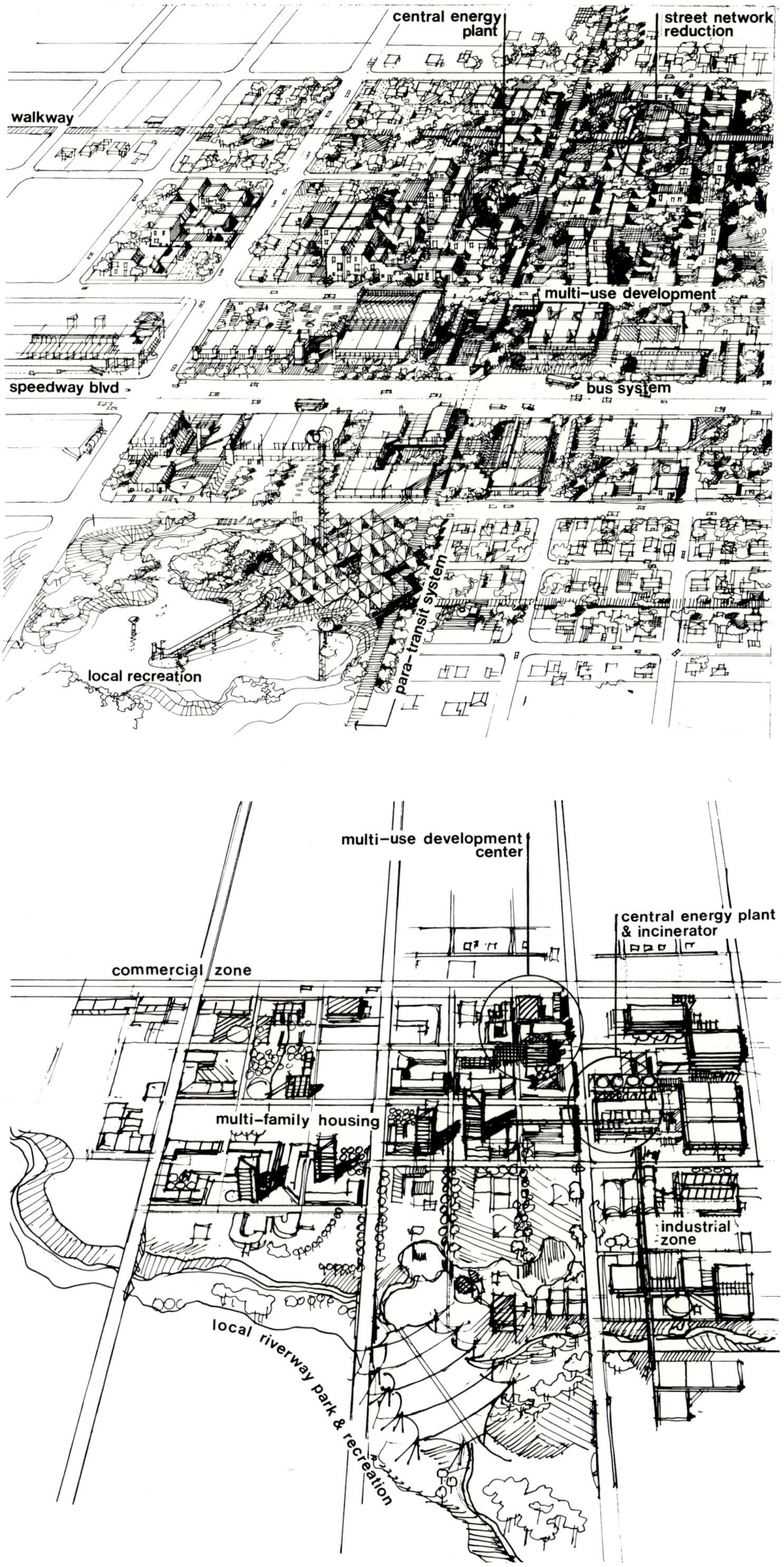

8.8 *Broadway Center's proposed physical concept (Tract 7)*

dential areas should be planned as one. Schools also should be incorporated and worked into the new dense activity centres. Each component must be made to feed off the total so that there will be energy savings in all areas. It has been characteristic of areas like Speedway and Broadway in Tucson to disperse all daily normal activities.

Tucson's peripheral areas dramatically grow because inner-city areas have few of their appealing recreational and prestigious qualities. Areas like Speedway and Broadway can only attract the needed growth if, in addition to the built-in locational advantages and economic advantages (including energy conservation benefits), they also have prestige and desirable amenities.

When we look at the land development pattern of Tucson as a whole, we see that it is essentially a copy or adaptation of the type of land development in areas in the East and midwestern plains: Nassau County, New York, or St Louis, Chicago, Toledo, Cleveland. There we find green lawns, rich deciduous shade trees, relatively plentiful water, and a more or less cool, moist climate. Though not one of these natural elements is present in Tucson, the land development style is all too familiar. Nowhere in the historic development of cities is clustering, dense development, and economy in the use of space more evident than in population centres in desert areas like North Africa where city dwellers huddle together to try to save every drop of moisture and every step of movement in the intense dry heat. In Tucson, however, the single-family bungalow with its parched yard, suburban roads, and parking lots baking in the sun reach endlessly to the horizon.

Table 8.7 Summary of projected annual energy consumption according to the 1995 unplanned growth scenario for Tucson (test area population: 9255)

Land use	Units	Energy consumption per year	
		MMB*/unit	Total MMB
Residential			
Single-family	2 522[a]	120.46	303 800
2-family			
Low-rise MF	1 320[a]	76.12	100 478
High-rise MF			
Subtotal	3 842[a]		404 278
Commercial			
Office			
Retail			
Mall/large			
Attached/small			
Strip			
Hotel			
Food service			
Recreation			
Theatres			
Subtotal	938 053[b]	.182	170 725
Institutional			
Schools			
Hospitals			
Public service			
Dormitory type			
Subtotal	71 630[b]	.114	8 165
Industry			
Light industrial			
Warehouse			
Subtotal	432 000[b]	.349	123 800
Total (Less transportation)			706 968
Transportation: (3842 du)*			150 000
Grand total			856 968

*See table **8.9**; a: du b: ft^2

Table 8.8 Summary of projected annual energy according to 1995 planned energy conserving development scenario for Tucson (test area population: 12 000)

Land use	Units	Energy consumption per year	
		MMB*/unit	Total MMB
Residential			
Single-family	3 350[a]	120.46	403 541
2-family			
Low-rise MF	1 820[a]	76.12	138 538
High-rise MF			
Subtotal	5 170[a]		542 079
Commercial			
Office			
Retail			
Mall/large			
Attached/small			
Strip			
Hotel			
Food service			
Recreation			
Theatres			
Subtotal	1 218600[b]	.162	197 413
Institutional			
Schools			
Hospitals			
Public service			
Dormitory type			
Subtotal	121 630[b]	.114	13 866
Industry			
Light industrial			
Warehouse			
Subtotal	432 000[b]	.349	123 900
Total (Less transportation)			877 258
Transportation: (5170 du)*			166 000
Grand total			1 043 258

*See table **8.10**; a: du b: ft^2

Future energy requirements for 1995

Energy requirements for the projected land uses for the 1995 unplanned and energy conserving planned scenarios for Census Tracts 6 and 7 in Tucson are shown in tables **8.7** and **8.8**. Factors were adjusted to reflect changes in density and composition in the land use configurations, such as the shift from strip to centralized commercial retail facilities.

Related transportation energy requirement estimates are calculated in tables **8.9** and **8.10** for the 1995 unplanned and planned scenarios. Transportation consumption for the two scenarios was influenced by the following considerations: unplanned—the simple extrapolation of present trends (but including improved gasoline mileage for vehicles); planned—the doubling of the industrial floor space increasing industrial employment opportunities to 6000 jobs. (It is assumed that 1C per cent of these jobs (commercial plus industrial) are available to local residents, changing automobile transportation from 97 per cent to 80 per cent; increasing bus transportation from 3 per cent to 10 per cent; increasing walking from 2 per cent tc 10 per cent; and closing off half of the streets in the area to vehicles to provide better access to shopping opportunities.) It is assumed that there is no change in the trip length, for trips are already short in the inner-city area. The resultant mode

Table 8.9 Projected transportation energy according to 1995 unplanned growth scenario for Tucson (3842 du)

	Trip generation rate (person trip/du)	Daily trips (number)	Modal split (per cent)	Trip length (miles)	Passenger miles (000 000)	BTU per pass. miles	Annual BTUs (000 000 000)
Work	2.20	8 400		5.1			
			auto 97		15.1	4 600	69
			bus 3		7.3	800	
			rail				
			walk				
Shop	1.75	6 700		2.8			
			auto 97		6.7	4 600	31
			bus 3		.1	1 100	
			rail				
			walk				
Other	1.88	7 200		4.9			
			auto 97		12.5	4 000	50
			bus 3		.3	1 100	
			rail				
			walk				
Total	5.83	22 300			35		150

BTU/passenger miles = 4 300

Table 8.10 Projected transportation energy according to 1995 planned growth scenario for Tucson (5170 du)

	Trip generation rate (person trip/du)	Daily trips (number)	Modal split (per cent)	Trip length (miles)	Passenger miles (000 000)	BTU per pass. miles	Annual BTUs (000 000 000)
Work	2.20	11 400		5.1			
			auto 80		12.0	4 600	78
			bus 10		2.1	800	2
			rail				
			walk 10		2.1		
Shop	1.75	9 200		2.8			
			auto 60		5.7	4 600	26
			bus 10		.9	1 100	1
			rail				
			walk 30		2.8		
Other	1.88	9 900		4.9			
			auto 80		14.3	4 000	57
			bus 10		1.8	1 100	2
			rail		1.8		
			walk 10				
Total					48.4		166

BTU/passenger miles = 3 500

shifts for shopping are as follows: automobile transportation from 97–60 per cent; bus transportation, 3–10 per cent; and walking, 10–30 per cent. Since the other trips are considered to be longer and less amenable to shifts toward walking, the resultant shifts are as follows: automobile transportation from 97–80 per cent; bus transportation 3–10 per cent; and walking 0–10 per cent.

Technology related energy conservation

The configuration of land uses in the planned energy conserving development proposals for Tucson were arranged to make use of the technology possible for influencing energy consumption. These potential savings are additive to those resulting from only the optimized land use design itself. Two central heating and cooling plants are proposed for Tract 6 in conjunction with the two new density intensification nodes (see **8.5**). Solid waste heat recovery does not appear to be practical because of problems of location. Solar energy for domestic hot water is being tested.

In Census Tract 7, one central plant together with a heat recovery incineration system is proposed for the industrial zone. The possible balance of energy requirements with supplies is shown in tables **8.8** and **8.9**. When considering both Tracts 6 and 7, the summer cooling requirement (which does not include existing commercial uses) is larger than the winter requirement. If existing commercial uses could be heated (but not cooled) from the central plant, better use of the primary equipment would result (see heating totals in parentheses in tables **8.11** and **8.12**).

Solid waste generated on site produces one-quarter of the total peak requirements; there is no reason why the incinerator could not be sized to serve the solid waste generated by an area broader than Tracts 6 and 7 if the economics and the institutional arrangements make it feasible. Energy savings with the use of central plants are generally in the range of from 5–10 per cent. Since some of the development in 1995 will not be connected to the central plant, there should be a saving of 5 per cent. The comparison is made to the total (without transportation) in table **8.8**.

Energy savings from solid waste heat recovery are estimated at 25 per cent for Tract 7. When we consider Tracts 6 and 7 as a whole, a 12.5 per cent reduction would be possible if the plant were appropriately sized to handle only solid waste generated on site. This reduction totalling 17.5 per cent compares very well with a Modular Integrated Utilities Systems (MIUS) plant evaluation in 'Final Environmental Impact Statement Application of MIUS Technology' by the United States Department of Housing and Urban Development in October 1975. Here a 19.1 per cent reduction was found between conventionally heated du and those heated by a MIUS plant.

Additional savings are possible with the use of solar energy. In a central installation for a group of du, a state-of-the-art collector together with adequate storage could provide 70 per cent of the annual domestic hot water required in Tucson. Actually, hot water represents approximately 25 per cent of the residential energy needed annually. Using the residential subtotal of 542 079 MMB/yr from table **8.8**, we can see that domestic hot water would represent approximately 135 520 MMB/yr. A saving of 70 per cent would represent a reduction of 94 864 MMB/yr.[9] Compared to the initial total, this is a reduction in energy required of 27.68 per cent based on the application of appropriate technology (see table **8.13**). These are highly significant savings, especially since conservative savings are assumed for the central plant. Similar opportunities for savings exist throughout the nation, for the Tucson conditions and development patterns are typical of United States cities of similar size and age.

IMPLEMENTATION

In order to facilitate energy conservation, the administrative mechanism of an Energy Conservative Zone (ECZ) is proposed. The boundary of such a zone would be determined at the local level and could be enforced through its official map, comprehensive plan, zoning ordinances, subdivision regulations, or even building codes. The objectives would be to designate an area with acceptable growth potentials within which development would be encouraged but where the added costs or constraints imposed by energy conservation could be offset by proper incentives; worthy of investigation would be zoning density bonuses, tax abatement or preferential assessment, transferable development and energy rights, utility rate structures, and a variety of private and public financing techniques. Energy rights could be established by setting energy budgets of a certain number of $BTUs/ft^2/yr$ for a given land use in a given climate. ECZs could also be used for testing, in a properly designed experiment, the effectiveness of the mechanism for implementing energy conserving design measures on a nationwide scale.

NOTES

1 All figures and tables are from a study *(Reading the Energy Meter on Development)* undertaken for the Federal Energy Administration. This work was supported by the US Department of Energy, Office of Conservation and Environment, under contract (grant) No. CO–04–50351–00, published in November, 1976. Figures and tables are used with the permission of the Department of Energy.

2 The graduate student was Robert Blair.

3 City of Tucson, Town of South Tucson, Pima Association of Governments, and its Transportation Planning Program, 'The Comprehensive Plan', unpublished draft, 1975.

4 Ibid.

Table 8.11 Summary of demands for Census Tract 6, Tucson, according to the 1995 planned energy conserving development scenario

Land use	Solid waste lb/day	Electric KW	Heating MBH	DHW* MBH	Cooling TR*
Residential					
New		630	10 500	1 680	700
Total	24 000				
Commercial					
New		360	1 800	135	300
Total	10 500		(7 000)		
Institutional					
New	1 000	125	1 250	100	167
Totals	35 000	2 155	13 550	1 895	1 167
			(18 750)		

Solid wastes (sw):
35 500lb sw/day = 17.75 tons per day

$$35\,500 \times 3\ \text{MB*/lb sw} = \frac{106\,500\ \text{MB/day}}{16\ \text{hrs}} = 6656\ \text{MBH*}$$

Cooling:
1167 TR × 16lb steam/TR/hr = 18 670lb steam/hr
or 18 670 MBH

*DHW: domestic hot water; TR: tons refrigeration; MB: thousand BTUs; MBH: thousand BTUs/hr

Table 8.12 Summary of demands for Census Tract 7, Tucson, according to the 1995 planned energy conserving development scenario

Land use	Solid waste lb/day	Electric KW	Heating MBH	DHW MBH	Cooling TR
Residential					
New		750	12 500	2 000	835
Total	24 000				
Commercial					
New		1 120	5 600	420	935
Total	26 000		(17 376)		
Institutional					
New	1 000	125	1 250	100	167
Industrial					
Total	32 000	2 000	10 625	2 800	1 440
	83 000	3 995	29 975	5 320	3 377
Use	83 000	4 000	30 000	5 320	3 400
			(47 345)		

Solid wastes (sw):
83 000lb/day = 41.5 tons per day
83 000 × 3 MB/lb SW = 249 000 mb/day
249 000/16 hrs = 15 562lb steam/hr = 15 562 MBH

Cooling:
3400 TR × 16lb steam/hr/TR = 54 400lb steam/hr

Table 8.13 Energy reduction through technology in Census Tracts 6 and 7 according to the 1995 planned energy conserving development scenario

Line		MMB/yr
1	Total from table **8.8** without transportation energy consumption	877 258
2	Central plant (−)5%	(−)43 862
3		833 396
4	Solid waste heat (−)12½%	(−)104 174
5		729 222
6	Residential subtotal from table 8	542 079
7	Domestic hot water 25%	135 519
8	Solar savings 70%	94 864
9	Line (5) − line (8)	634 358
10	Saving re: line (1)	27.68%

[5] Booz, Allen and Hamilton, Inc, *Cost/Revenue Analysis: For Alternative Plans for Growth*, a study for the Comprehensive Planning Process of the City of Tucson, Town of South Tucson, Pima Association of Governments, and its Transportation Planning Program, July 1974.

[6] US Department of Commerce, Bureau of the Census, *1970 and 1975 Census Data by Tract* (Washington, 1976).

[7] Tucson Area Transportation Agency, *Transporatation Systems Planning Forecast* (Tuscon, Arizona, 1962).

[8] Pima Association of Governments, *Alternate Mode Study* (Tucson, Arizona, 1975–6).

[9] One MMB equals 1 million BTUs. One BTU (British Thermal Unit) is the energy needed to raise one pound of water one degree Fahrenheit at 60°F.

SUGGESTED READING

American Society of Heating, Refrigerating, and Air-Conditioning Engineers. *Energy Conservation in New Building Design*. New York: ASHRAE, 1975.

Bosselman, Fred, and David Callies. *The Quiet Revolution in Land Use Control*. Washington, DC: The Council on Environmental Quality, 1972.

Brookhaven National Laboratory and the State University of New York. *Land Use and Energy Utilization*. Interim report for Office of Conservation and Environment, Federal Energy Administration. Stony Brook, NY, October 1975.

National Bureau of Standards, Division of Energy, Building Technology, and Standards. *Application of Modular Integrated Utilities Systems Technology*. Draft Environmental Statement. February 1975.

Palmedo, Phillip, Robert Nathans, and Owen Carrol. *The Planner's Energy Workbook*. Stony Brook, NY: Brookhaven National Labs for the Federal Energy Administration, October 1976.

Real Estate Research Corp. *The Costs of Sprawl*. Report prepared for Council on Environmental Quality, Department of Housing and Urban Development and Office of Planning and Management, Environmental Protection Agency. Chicago, April 1974.

Regional Plan Association and Resources for the Future, Inc. *Regional Energy Consumption*. Report sponsored by the Ford Foundation. New York, January 1974.

Roberts, James S. *Energy Land Use and Growth Policy: Implications for Metropolitan Washington*. Report prepared for Metropolitan Washington Council of Governments. Chicago, July 1975.

Technology and Economics, Inc. *An Overview and Critical Evaluation of the Relationship Between Land Use and Energy Conservation*. Volume I—*Main Report*. Cambridge, Mass, November 1975.

9 Managing a finite groundwater supply in Arizona's Santa Cruz Basin

KENNITH E. FOSTER

Abundant sunshine and scarce water are two factors common to all of Arizona. During an average year, Arizona receives about 80 million acre-feet (a-ft) of water from rain and snow; 2.3 million a-ft of this precipitation are captured in surface water reservoirs or as aquifer recharge. It is estimated that people in Arizona use about 4.8 million a-ft of water per year. Thus, the approximate 2 million a-ft received through precipitation each year must be augmented by pumping about 2.5 million a-ft of water from groundwater reservoirs.

SOURCES OF WATER IN ARIZONA

Arizona can be divided into three principal water provinces: 1 the Plateau Uplands including the northern part of the state; 2 the Central Highlands, a mountainous area extending diagonally across the state; and 3 the Basin and Range Lowlands, or desert, which include the heavily populated southern portion of Arizona (see **9.1**).

The Basin and Range Lowlands contain more than 80 per cent of the state's population and consist of isolated mountain blocks jutting from alluvial sediments that form the broad desert basins. The sediments in the alluvial desert basins constitute large storage areas for groundwater. The occurrence of these large water reservoirs has made it possible for man to live prosperously and comfortably in the desert. The arid basins range in altitude from about 100ft above sea level at Yuma to about 4000ft in the southeastern part of the state. The climate in the lowlands is hot and arid; the rainfall is light, with precipitation averaging less than 10in annually. No large surface water reservoirs are in this province because the runoff is minimal. Average runoff from the driest 15 000 square miles in the lowlands is less than 0.1in. The headwaters of the San Pedro, Santa Cruz, and other large tributaries yield only about 0.5in per year. Streams draining several of the higher mountain ranges yield moderate amounts of runoff along the base of the mountains; but when the water reaches the desert basin, it evaporates quickly and leaves only a small part to recharge groundwater reservoirs.

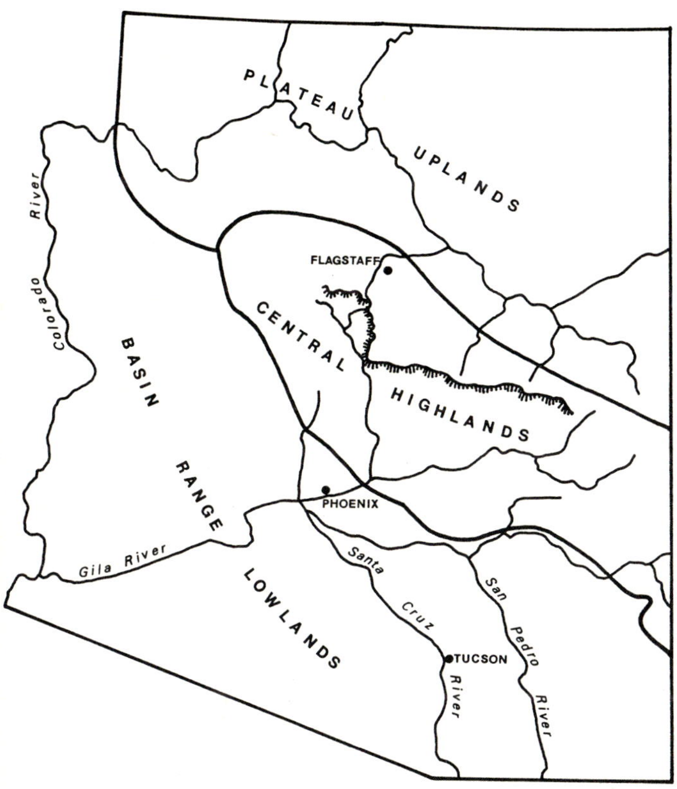

9.1 *Water provinces of Arizona*

SOURCE *Arizona: Its People and Resources* (TUCSON: THE UNIVERSITY OF ARIZONA, 1960), p. 385 USED WITH THE PERMISSION OF THE PUBLISHER

Stream flow in the lowlands occurs mostly as flash floods following thunderstorms. Storms originating in the Gulf of Mexico extend into the southeastern corner of the state and bring precipitation to the mountainous areas. As flood waters move downstream from the mountainous areas, the flood volume decreases rapidly. Factors contributing to this depletion are infiltration, evaporation, channel storage, channel retention, and stream bank runoff.

As in all arid or semiarid regions of the world, the dominant features of the rainfall in most of Arizona are scantiness and extreme variability from one year to the next. On the desert floor in the vicinity of Tucson, at a mean altitude of about 2500ft, the average precipitation is only about 10in a year. Forty per cent of the precipitation occurs in July and August (see **9.2**). A large portion of the remaining 60 per cent falls as heavy showers which are scattered almost randomly throughout the rest of the year.

Although the same rainfall regime is present at higher altitudes, amounts are characteristically greater. For example, in the Catalina Mountains, the Mount Lemmon rain gauge, at slightly over 7500ft, registers an average of 30in of precipitation each year—almost exactly three times that falling on the basin floor. However, monthly amounts (see **9.2**) are extremely variable from one year to the next, perhaps even more so than at lower altitudes.

Most of the rainfall in southern Arizona can be attributed to one of four sources, depending mainly on the season of the year. A large percentage of the summer thundershowers is associated with very warm, moist, and unstable air which sweeps around the southern margins of the Atlantic Ocean high pressure cells and advances into Arizona from the Gulf of Mexico. This air, in passing over the highly heated land masses, is made even more unstable. When it is forced to ascend over the numerous mountain ranges of southern

Arizona, rain showers result. These showers have a very marked diurnal variation, which is most intense over the mountains during the midafternoon when surface heating and the general convergence of air associated with the upslope mountain winds are maximal. In the valleys, the heaviest summer rains usually do not occur until the later afternoon or early evening, when the desert floor is considerably warmer than the surrounding cloud covered mountains.

Not all the warm season rainfall is the result of simple convective activity. A small, but important part is associated with tropical disturbances which form off the west coast of Mexico at about 15°N latitude. These storms usually dissipate as they move northward into middle latitudes, but they are normally still intense and extensive enough when they reach the 30th parallel to produce heavy rainfall in southern Arizona. This type of rainfall differs from the normal convective type in several respects. It is more widespread, has less intensity but longer duration, and is only rarely associated with thunder and lightning. Some of the heaviest rainfalls on record, particularly in September, are associated with these tropical disturbances.

Winter or cool season rains are generally less intense but more widespread than those of summer. They also vary less with ground elevation, but sometimes they are heavier on the desert floor than in the mountains. Part of this precipitation is associated with the middle latitude storm belt which occasionally moves far enough towards the equator in winter for its southern margins to affect Arizona. It is only when these cyclonic storms move in directly from the Pacific Ocean across the northern and central parts of the country that measurable amounts of rain occur. When a storm's path is more nearly north to south, east of the 105th meridian, about all southern Arizona can expect are wind and subnormal temperatures.

Probably the heaviest rainfall of winter is associated with the so-called 'cold lows' of the subtropical Pacific Ocean. These intense disturbances form in the vicinity of the Hawaiian Islands and move very slowly eastward to the coast of southern California. In this region or slightly inland, they often remain stationary for several days. But once caught in the strong upper-level westerlies, they move rapidly northeastward across the United States. These storms normally pass directly over Arizona, frequently advancing very slowly and retaining most of their moisture supply. When they move in from the Pacific, they can produce several days of moderate to heavy precipitation.

THE SETTING

The Santa Cruz Basin, as shown in **9.3**, has several groundwater districts. This basin is bounded by the crests of the Tortolita and Santa Catalina mountains on the north; the Rincon, Santa Rita and Patagonia mountains on the east; and Atascosa, Tumacacori, Sierrita, and Tucson mountains on the west. The total drainage area as described contains about 2 240 square miles.

The Santa Cruz River has its source in the San Raphael Valley where it drains the eastern slopes of the Patagonia Mountains, the south slopes of the Canelo Hills, and part of the western slope of the higher Huachuca Mountains on the east. It flows south across the international boundary into Mexico in a 35 mile loop before reentering Arizona about 6 miles east of Nogales, flows northwest and finally, about 12 miles southwest of Phoenix, joins the Gila River at a point 225 miles from its source in the San Raphael Valley.

Structurally, the Santa Cruz Valley is a typical example of the basin and range province of the southwestern United States. Northward trending mountain ranges border the broad, flat, alluvium filled valley (see **9.4**).

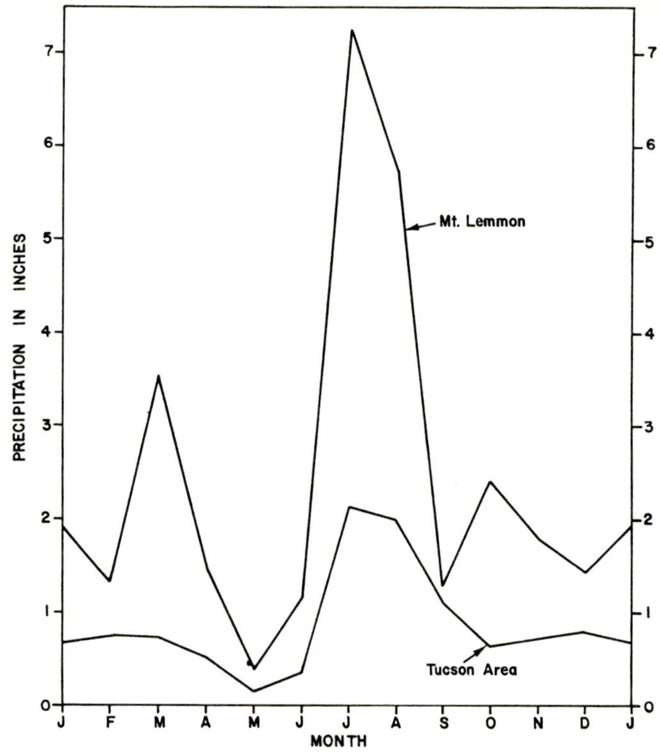

9.2 *Mean monthly precipitation in the Tucson area and at Mount Lemmon in the Catalina Mountains*

FROM THE RILLITO CREEK HYDROLOGIC COMMITTEE AND THE US GEOLOGICAL SURVEY, *Capturing Additional Water in the Tucson Area*, OPEN-FILE REPORT (TUCSON, 1959), p. 59

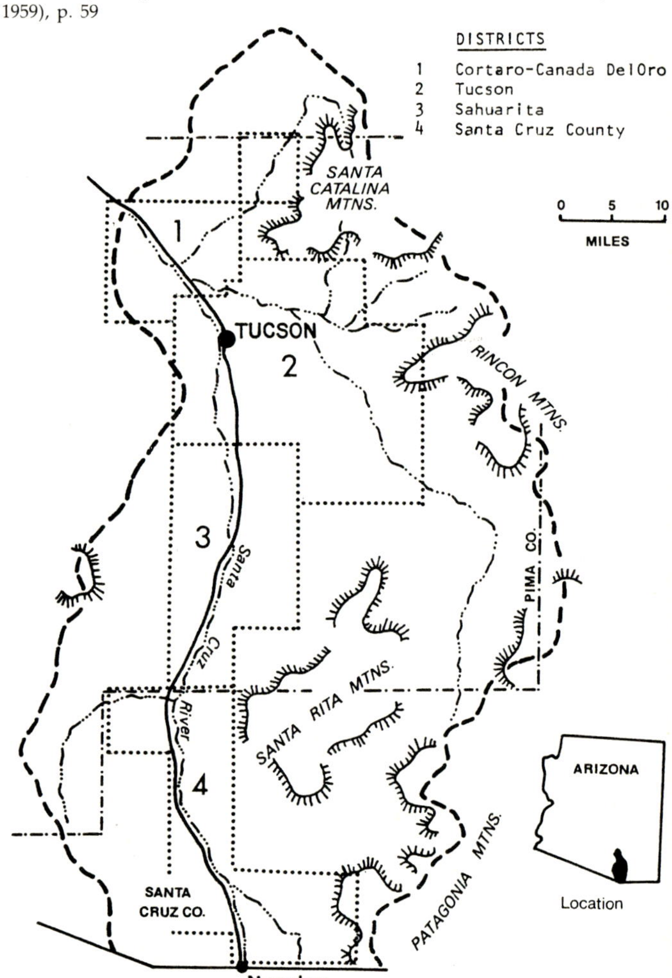

9.3 *Santa Cruz Basin*

SOURCE: W. G. MATLOCK, AND D. R. DAVIS, *Groundwater in the Santa Cruz Valley*. AGRICULTURAL EXPERIMENT STATION, TECHNICAL BULLETIN 194 (TUCSON, 1972)

Hydrologic characteristics

Runoff

Surficial runoff in stream channels is the most important source of recharge to the groundwater reservoirs of the area. The comparatively silt-free runoff from snow melting in the early spring provides maximum recharge through seepage into the Rillito Creek stream bed. The flows are more uniform than summer floods and may be continuous over a period of several months at a season of the year when recharge is most effective because of minimal losses to evaporation or transpiration. Still the short, intense summer storms of small areal extent result in short-lived, silt-laden flows which are relatively ineffective as a source of recharge to the groundwater reservoirs.

Natural recharge

Matlock and Davis calculated annual groundwater recharge for the Tucson-Sahuarita districts over a ten year period, 1960–9. Their study indicated an average annual recharge rate for the period of about 55 000 a-ft.[1] Earlier studies (Matlock, Schwalen, and Shaw) suggested a possible range of 56 000 to 79 000 a-ft for the average annual recharge.[2] The Arizona Water Commission estimated natural recharge in the Upper Santa Cruz Basin to be 65 000 a-ft per year (AFY).[3]

Current water use

Agriculture

Matlock and Davis estimated the average agricultural consumptive use of water for approximately 15 000 a in the Santa Cruz Basin, excluding the Cortaro District, to be 46 000 AFY. Demand for water is leading to retirement of agricultural land, however, through purchase of these lands for water rights by mining companies. Since 1970, over 8000 a in the Santa Cruz Valley have been retired from production to use the water for mining purposes.[4] Less than 1800 a of irrigated crop land remain in the basin near the city; and south of Tucson along the Santa Cruz River, there are only approximately 6800 a of crop land, principally in pecan trees. Water use in the basin today for agriculture is estimated to be 30 000 AFY, assuming a consumptive water use of 3.5 a-ft per a.

Mining

Within the Santa Cruz Basin are five major open-pit copper mines plus a sixth new mine under development. These mines are primarily in the foothills of the Sierrita Mountain Range. They include the new San Xavier Mine, operated by the American Smelting and Refining Company; the Mission Mine under the same management; the Pima Mine, operated by the Cyprus Pima Mining Company; the Twin Buttes Mine, operated by the ANAMAX Company; and the Esperanza and Sierrita Mines, operated by the Duval and the Duval-Sierrita Corporations. The daily tonnage mined in this area is estimated to be 400 000 tons. Approximately 200 000 tons of sulphide-copper ore are processed through the mills.[5] The remainder is either overburden (150 000 tons assumed) or stockpile for oxide-copper ore leaching (50 000 tons assumed).

Leaching is particuarly applicable to low-grade oxide ores. Three types of leaching are commonly used: 1 leaching in place, 2 heap or dump leaching, and 3 vat leaching. Oxide-copper minerals are dissolved by a diluted ferric sulphate solution. Copper contained in the pregnant solution is precipitated by passing the solution over shredded scrap iron. During leaching, evaporative losses and seepage must be replaced by adding new water to the solvent. If ferric sulphate (formed by the action of water on certain sulphides) is present in the solvent, some of the solvent in the system may have to be removed to prevent an undesirable ferric ion buildup.

Water is a necessity in mining. The various kinds of water needs in the mining area are shown in table **9.1**. In flotation, sulphide copper is recovered in a froth from a pulp containing suitable reagents and water in proper quantities. Optimum density of the pulp ranges from 15 to 25 per cent weight solids. A 15 per cent pulp contains 5.67 tons (1360 gal) of water per ton of ore, and a 25 per cent pulp has 3 tons (720 gal) of water. Table **9.2** illustrates the relative water requirements for the flotation method and the leaching method of recovering copper. The additional water requirements for smelting are not treated here since this copper operation is not performed in the Santa Cruz Basin.

Current water use by the mining district is estimated to be 45 million gal per day.[6] An estimated 210 gal of water per ton of ore is consumed in the mining and flotation concentrating processes. Assuming a copper content of 0.5 per cent, 20 gal of water per pound of copper are consumed. Table **9.2** shows that about 9 gal of water per pound of copper were consumed in 1960. The increase in water consumption in 1977 is attributed to greater water use in the mine for dust control and the necessity to mine more tonnage of lower grade ore. Ore grades now mined average about 0.5 per cent copper as compared to 1 per cent in 1960. On daily and annual bases, average water consumption would amount to 139 a-ft per day or 50 735 AFY, assuming 200 000 tons per day of ore are mined. The above consumption figures assume no recharge to the aquifer through the tailing dumps.

9.4 *Cross-section of typical Arizona valley*

SOURCE: H. C. SCHWALEN AND R. J. SHAW, *Water in the Santa Cruz Valley*, AGRICULTURAL EXPERIMENT STATION BULLETIN 288 (TUCSON: UNIVERSITY OF ARIZONA, 1957)

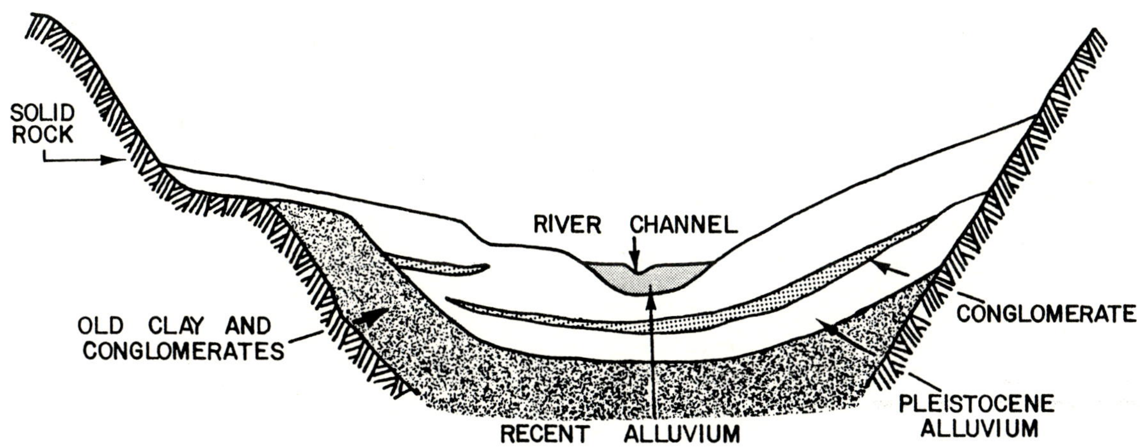

Table 9.1 Water use in sulphide and oxide copper ore treatment

Mining	Bank sprays
	Wet down haulage roads
	Leach and precipitate copper from waste dumps
Crushing	Cool crushing machine
	Collect and slurry dust
	Spray ore to minimize dusting
Concentrating	Crush ore for grinding
	Particle size classification
	Manufacture milk of lime
	Flotation process
	Transport concentrates
	Transport tailings
Leaching	Dilute ferric sulphate solution

Source: Gilkey, M. M., and R. T. Beckman. *Water Requirements and Uses in Arizona Mineral Industries.* US Bureau of Mines Information Circular No. 8162. Washington, DC: Government Printing Office, 1973.

Table 9.2 Average amounts of water required and consumed in producing copper in Arizona, 1960[a]

	Total intake gallons/pound copper	%	Consumption gallons/pound copper	%
Mining	1.1[b]	1.9	1.0[b]	11.1
Flotation concentrating	56.0[c]	96.2	7.5[c]	83.3
Total	57.1	98.10	8.5	94.40
Leaching	29.7	96.4	15.1	96.8
Total	29.7	96.4	15.1	96.8

[a]Excludes water required for power plants, domestic and miscellaneous purposes
[b]Based on average grade of ore mined
[c]Based on average grade of feed at concentrators

Source: Gilkey, M. M. and R. T. Beckman. *Water Requirements and Uses in Arizona Mineral Industries.* US Bureau of Mines Information Circular No. 8162. Washington, DC: Government Printing Office, 1973.

Other uses

The Water Resources Division of Tucson's Water and Sewers Department estimates current urban/recreation groundwater withdrawal to be 80 000 AFY. Currently, the Tucson metropolitan area receives approximately 60 000 a-ft of its water from the Santa Cruz Basin and 20 000 a-ft from Avra Valley. Of this amount 43 000 a-ft are consumptively used or depleted, and 37 000 a-ft are returned to the sewage treatment plant for treatment.[7] The treated water is placed in the Santa Cruz River; eventually it recharges the aquifer. However, most of this occurs outside the city's pumping influence and does not constitute significant abatement of the local water level. The 80 000 AFY are thus treated as depletion.

The total *per capita* water use within the region is approximately 150 gal per day. This includes residential, commercial, industrial and park usage. Current population is placed at 450 000 for the Tucson metropolitan area. Projected growth figures through the year 2000 place Tucson's population between 675 000 and 880 000 persons. These population increases will produce an estimated depletion by the year 2000 of 101 000 a-ft to 132 000 a-ft respectively.

Santa Cruz water balance

The Santa Cruz Basin water balance is a function of inflow (natural recharge and underflow) and outflow (pumping).
Inflow: Inflow into the basin consists of two components: natural recharge and underflow into the basin from the Santa Cruz River. These components are estimated to be:

1 Natural recharge		65 000 AFY
2 Underflow		2 000 AFY
	Total	67 000 AFY

Outflow: Current depletion resulting from pumping by the agricultural, mining, and municipal sectors in the Santa Cruz Basin is estimated to be:

1 Agriculture		30 000 AFY
2 Mining		51 000 AFY
3 Municipal		60 000 AFY
	Total	141 000 AFY

The general problem in the Santa Cruz area is that current water use is exceeding natural recharge of the aquifer by 74 000 AFY. As a result, the Tucson and Sahuarita-Continental districts have been experiencing significant declines in the water table. For the period 1947—65, water levels in the Tucson District have declined over 100ft. The maximum decline was 172ft in one location. Over the 22 year period (1947—69), wells in the Sahuarita-Continental District have declined 1—3ft per year. For the six-year period (1969—75), average annual declines of 6—12ft per year are common. Potential recharge to the aquifer of Tucson's waste water will not significantly reduce water level declines because most city wells are not located within the recharge zone.

The city of Tucson water use demands could reach a figure of 100 000 AFY by 1985 and 130 000 AFY by 2000. Water requirements for the mining industry are estimated to stabilize at 75 000 AFY even though expansion is likely to occur. Intensive conservation efforts, water reuse, and the sealing of tailings ponds will account for the water savings. Farm requirements in the basin are projected to decline to about 20 000 AFY by 1985 and to stabilize at that level. Total water-use requirements in the basin will reach 195 000 AFY by 1985, and 225 000 AFY by 2000.

MANAGEMENT ALTERNATIVES

Various alternatives for water use patterns in the Santa Cruz Basin to reduce water level declines have been discussed by several researchers and governmental agencies over the past 20 years. Among alternatives most frequently discussed are: 1 importation of water from the Colorado River; 2 municipal exchange of sewage effluent with mines or farms for fresh water; 3 interbasin transfers of water; 4 retirement of farmlands for water rights; and 5 implementation of a combination of alternatives.

Alternative one: importation of water from the Colorado River

The history of large-scale diversions from the Colorado River dates back to the early 1900s when farmland in the Imperial Valley of California was irrigated with river water. Frequent flooding of farmland resulted in today's regulation of the lower Colorado River through a system of dams and diversion canals. Water allocation was accomplished through the Colorado River Compact and later the Boulder Canyon Project Act which invited the states of the lower basin to come to an agreement on the final diversion of water. Nevada is allocated 300 000 a-ft annually; Arizona, 2.8 million a-ft; and California, 4.4 million a-ft.

To move the water from the Colorado River to the central part of the state will require an aqueduct system. This system is today carefully planned; but as early as 1947, the first Central Arizona Project (CAP) proposal was submitted to Congress for such a project. The final version of the CAP was enacted in 1968. CAP consists of an aqueduct to carry water from Lake Havasu on the Colorado River to the Phoenix and Tucson areas (see **9.5**). The initial water volume would be about 1.6 million AFY, declining to 380 000 to 600 000 AFY by 2035. Current projections indicate that 100 000 AFY of CAP water will be available to the Tucson area by 1985. Plans call

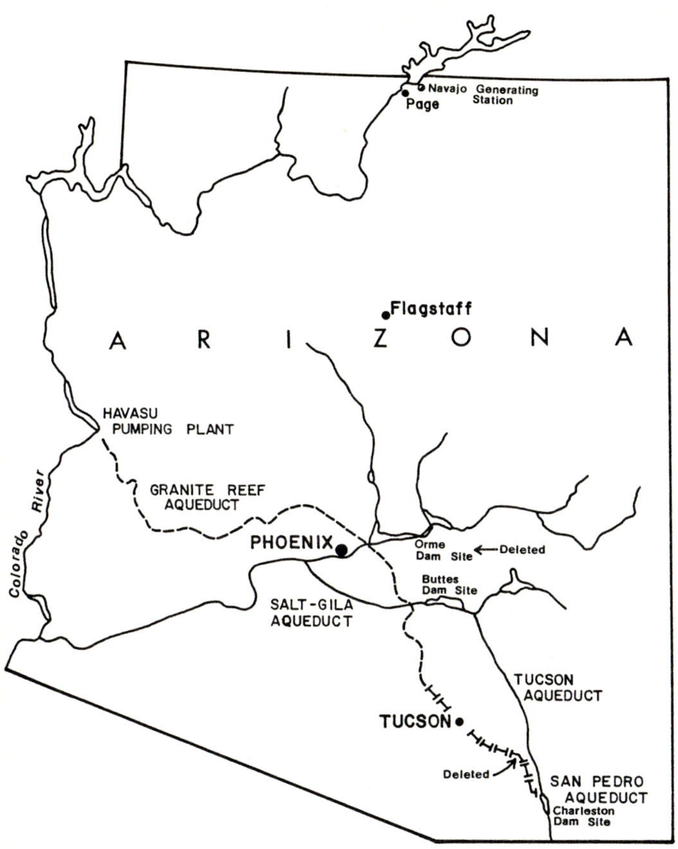

9.5 *Central Arizona Project's major facilities*

SOURCE: CENTRAL ARIZONA PROJECT, *Staff Report to the Metropolitan Utilities Management Agency Board and the Mayor and Council of the City of Tucson* (TUCSON, 1974)

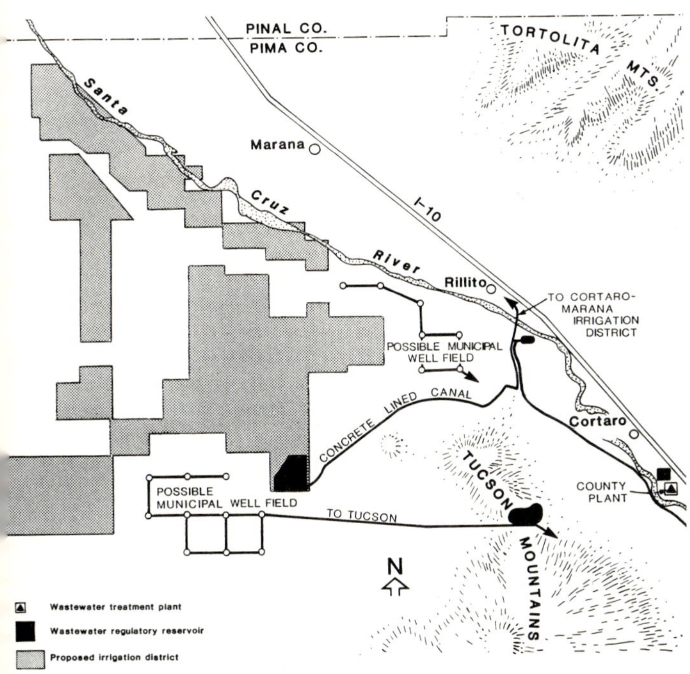

9.6 *Waste water for irrigation of farmland, Avra Valley–Marana region near Tucson*

SOURCE: C. B. CLUFF AND K. J. DECOOK, 'METROPOLITAN OPERATED DISTRICT FOR SEWAGE EFFLUENT-IRRIGATION WATER EXCHANGE'. IN *Proceedings of the 1974 Meetings of the Arizona Section, Arizona Water Resources Association, and Hydrology Section, Arizona Academy of Science*, 4 (1974)

for the city of Tucson to receive 50 000 a-ft, leaving 50 000 a-ft to be divided between the mines and farms.

By 1985, water use in the Santa Cruz Basin could total 195 000 AFY; the groundwater overdraft will exceed natural recharge by 130 000 AFY. The 100 000 a-ft of CAP water would almost alleviate the overdraft if all CAP water were kept within the Basin. This may not be the case, however, because the agricultural water (assuming 25 000 a-ft) will be diverted to agricultural users out of the basin, thus leaving a deficit of 55 000 a-ft annually.

By 2000, annual water use in the Santa Cruz Basin might total 225 000 AFY; and the groundwater overdraft will exceed natural recharge by 160 000 AFY. Projected water delivery by the CAP in 2000 is unknown; however, less water is anticipated. If 80 000 AFY of CAP water is assumed, the basin will remain in an overdraft posture by 80 000 AFY.

Alternative two: municipal exchange of sewage effluent with mines or farms for fresh water

As has already been mentioned, 37 000 a-ft of Tucson's annual groundwater pumpage currently recharges the aquifer after passing through primary and secondary sewage treatment. This annual recharge is projected to reach 65 000 AFY by 2000. Historic reuse of this water, however, has been by agricultural interests in the lower Santa Cruz Basin, north of Tucson and out of the area in which Tucson pumps its water.

A possible alternative would be to pipe the 65 000 AFY to the mines to meet milling process needs of 75 000 AFY. Thus the mines would pump only 10 000 AFY from the aquifer. Tucson, in turn, could withdraw that amount not pumped by the mines in meeting its needs for 130 000 AFY, and there would result a net reduction of groundwater pumping of 65 000 AFY. The estimated water use requirement by 2000 is 75 000 a-ft. Mining use of 65 000 a-ft of Tucson's waste water would also reduce use to 160 000 a-ft annually, but would represent a continued groundwater overdraft of 95 000 AFY. Fisher concluded that such a substitution was technically feasible but that sewage effluent caused a measurable decrease in recovery of both copper (2.4 per cent) and molybdenum (16.2 per cent). Thus the mines may be hesitant to implement such an alternative.[8]

Cluff and DeCook proposed the establishment of a metropolitan-operated district to provide waste water for irrigation of farmland in the Avra-Marana area (see **9.6**). Exchanging waste water for groundwater to be used in the city water system would be a viable alternative to Tucson's present practice of purchasing farmland and retiring it to acquire rights to the groundwater. By 2000, the city could pump up to 65 000 AFY of additional groundwater annually in Avra Valley if the farms used the effluent and reduced their pumping by a similar amount.[9]

This alternative is not without its disadvantages because secondary effluent cannot be used for the considerable acreage of lettuce irrigated. The nitrate concentrate also would have to be diluted with fresh water before applying it to some agricultural crops. Such a solution would require up to 50 per cent fresh water, thus decreasing the potential groundwater savings. The 1980 Arizona Groundwater Management Act would also require amending to allow such action.

Alternative three: interbasin transfer of water

Since the early 1970s, the City of Tucson and the mines have purchased and retired over 18 000 a of irrigated farmland. About 10 000 a of this are in Avra Valley, west of Tucson. Once the land was purchased, the City of Tucson set up a well field and now transfers about 20 000 AFY (25 per cent of Tucson's annual demand). Current plans call for Tucson to pump up to 40 000 AFY from Avra Valley by 1985. This amount would reduce the Santa Cruz Basin's annual water use requirements to 155 000 AFY by 1985. Today, farms within Avra Valley are presently using 120 000 AFY to meet water use requirements; and as a result, water levels are also declining in this area.

Alternative four: retirement of additional farmlands for water rights

One policy of the City of Tucson and of the mines has been to purchase and retire farmland to obtain its water rights. This policy was in effect for a number of years. About 10 000 a of farmland in Avra Valley have been retired by Tucson. Mining interests also have been purchasing farmlands for water rights, especially south of Tucson in the Santa Cruz Basin. About 8000 a of farmland have been retired by mine purchases.

The Arizona Supreme Court is allowing the city to pump an average of 2.4 a-ft of groundwater per a of farmland retired. Assuming this to be an average groundwater pumping amount for the mines as well, almost 43 000 AFY is being diverted from agricultural use by the city and mines. However, in 1973, 41 000 a were still under cultivation in the Santa Cruz and Avra valleys. Assuming a retirement of all but 6000 a of this acreage over a ten year period, an additional 84 000 AFY could be diverted for city and mining use by 1985. And if farm water requirements stabilize at 20 000 AFY as projected, the Santa Cruz Basin's water use requirements through 2000 could still not be met through natural recharge (65 000 AFY) and the 124 000 AFY diverted from farm use and interbasin transfer (see table **9.3**).

IN PERSPECTIVE

The distribution of water within the Santa Cruz Basin has never had the benefit of a long-range water plan. Ironically, the city of Tucson was sited near the Santa Cruz River; in the late 1800s, it was a perennial stream and a means of serving the local water supply. Not until the 1930s did the river dry up because of the advent of modern deep well turbines which are capable of pumping water in excess of the recharge rate. Water consumption by the copper mines did not begin in earnest until the mid-1950s, about the time the high-grade copper had been mined and open-pit operations began to mine the ever-increasing volumes of ore required to extract the low-grade copper.

Table 9.3 Synthesis of alternatives: possible inputs and transfers to the Santa Cruz Basin

Source	Alternative	1985 amount (AFY)	2000 amount (AFY)
Natural recharge	—	65 000	65 000
CAP import	1	75 000	40 000
City effluent exchange	2	50 000	65 000
Interbasin transfer	3	40 000	40 000
Farmland retirement	4	84 000	84 000
	Total	314 000	294 000

Estimated demands are listed below:

Sector		1985 demand (AFY)	2000 demand (AFY)
City of Tucson		100 000	130 000
Mines		75 000	75 000
Farms		20 000	20 000
	Total	195 000	225 000

Furthermore, the combined uses of water are clearly exceeding the natural recharge rate; and unless some combination of the four alternatives is implemented, water levels will continue to decline. The CAP, discussed in alternative one, is now under construction; so the possibility of Colorado River water being diverted to central Arizona becomes an ever-increasing probability. By 1985 alternatives two and four will not be totally implemented. Thus the farmland retired in Avra Valley will supply only the 40 000 AFY described in alternative three. The 1985 demand of 195 000 AFY cannot be met with CAP, interbasin transfer waters, and natural recharge. These sources total 180 000 a-ft, 15 000 a-ft short of a balance.

As demand for groundwater increases, some trade-offs must occur by 2000. Purchase of farmland for the water rights may lead to the widespread farm retirement discussed in alternative four. This alternative would afford the opportunity to divert more of the CAP allotment into the basin. Should this occur, alternatives one, three, and four plus natural recharge could provide over 200 000 a-ft by 2000.

By 2000, waste water use by the mines (alternative two) may become a reality as new techniques are found to minimize copper loss. This alternative would also provide for an effective use of the effluent within an area where Tucson and the mines have extensive well field development.

CONCLUSION

Overdraft of a limited water resource has created debate among developing countries. The debate centres on the advisability of restricting short-term water use to reserve a finite source for future generations. The Santa Cruz Basin is obviously a case of short-term water use; and now as water levels decline, alternatives must be considered to maintain the local economy.

After a distinction between potential and available water supplies is made and practical limitations are established on how much water can be withdrawn, it is not apparent that in arid areas policies should be written to totally restrict short-term use of valuable water resources. Almost all policy making places a greater emphasis on near-term solutions rather than on long-term solutions. Thus, it seems more appropriate to stress long-term conservation of water for the benefit of future generations while at the same time improving the standard of living of people in the present age. Water policies used to guide developments in many Third World countries include two statements based upon perceived resource availability:

1 Immediate urban, industrial, and agricultural demands at inland locations will be supplied by groundwater.
2 Immediate urban and industrial demands at coastal locations will receive supplies from accelerated desalination developments.

Not all arid regions will sustain continued development of water resources. While a more conservative water policy may be in order, the resource should be used in the most efficient manner. We might use a physical indicator of efficiency based upon a comparison between the volume of water actually used and the amount which would have been used if the most advanced technology were to have been applied. Using this criterion, the method of flood irrigation for agricultural development would have a much lower physical efficiency than sprinkler or trickle irrigation techniques. In a similar manner, a desert city without a sanitary sewer system would have no possibility of effluent reuse and would approach zero efficiency.

Water policy in the Santa Cruz Basin, as well as in developing countries, is probably much more affected by political considerations than by economic feasibility. When a government has written a law or established a policy in broad generalized language, such as creation of the CAP, the next step is to translate this policy into action by referring it to a smaller unit of government. Here is where the difficulty may begin, for the entrustment of the enactment of the policy may be delegated to either a particular agency or may even be broken up into several components and turned over to a series of agencies for implementation.

No simplistic or incomplete approach can be used for the formulation of sound water policy. Instead, the economic and political aspects of the problem, as well as the physical aspects, must be seriously considered. Water policy based upon an efficiency approach is possible in the physical and economic realms, but much additional definition and development will be needed to apply the same concept to the political realm. Probably the most important variable contributing to the success of any water policy is the characterization of social well-being; yet this is precisely the variable that appears to be the least understood.

The tendency to do something quickly in response to perceived problems, without a serious consideration of the whole problem must be thoughtfully resisted. However, the need for making short-term policy decisions in desert environments is self-evident, and those decisions were made in the past initially for self-preservation and later for economic returns.

NOTES

W. G. Matlock and D. R. Davis, *Groundwater in the Santa Cruz Valley*, Agricultural Experiment Station Technical Bulletin 194 (Tucson: University of Arizona, 1972).

[2] W. G. Matlock, H. C. Schwalen, and R. J. Shaw, *Progress Report on Study of Water in the Santa Cruz Valley*, Agricultural Extension Station Technical Report 233 (Tucson: University of Arizona, 1965), pp. 26–27.

[3] Arizona Water Commission, *Phase I—Arizona Water Plan, Inventory of Resources and Uses* (Phoenix, 1975).

[4] J. Armstrong, Pima County Agent's Office, Tucson, Arizona, to Foster, 1977.

[5] J. S. Nobeker, 'Water Use in Arizona's Copper Industry', paper presented at Arizona Water Pollution Control Association Conference (Tucson: 1973), pp. 15–16.

[6] *Tucson Citizen*, 6 August 1975.

[7] F. Brooks, Water Resources Division, Water and Sewers Department, Tucson, Arizona, to Foster, 1977.

[8] W. W. Fisher, *Utilization of Clear Water Sewage Effiuent in Mineral Processing*, Water Resources Center Project Completion Report A–046–ARIZ (Tucson: University of Arizona, 1975), p. 14.

[9] C. B. Cluff and K. J. DeCook, 'Metropolitan Operated District for Sewage Effluent—Irrigation Water Exchange', in *Proceedings*, 1974 Meetings of the Arizona Water Resources Association and Hydrology Section, Arizona Academy of Science 4 (1974), pp. 94–98.

SUGGESTED READING

Cochran, G. F., et al. *Quality of Water Considerations in Arid Area Water Management*. Center for Water Resources Research Publication No. 16. Springfield, Va: National Technical Information Service, 1973.

Floyd, T. 'Survey of Western States—Underground Water Management Provisions'. *The Cross Section*, 17, no. 7 (1971), 1, 3.

Peirce, H. W. 'Arizona's Water Supply—Some Reflections'. *Field Notes*, 6, no. 2 (1976), 1–4, 16.

Walker, W. R., et al. *Institutional Requirements for Optimal Water Quality Management in Arid Urban Areas*. Springfield, Va: National Technical Information Service, 1973.

Walker, W. R., et al. 'Water Supply, Treatment, and Distribution, and Reuse Optimization in Arid Urban Areas'. Paper presented at the Ninth American Water Resources Conference, Seattle, Washington, 21–23 October 1973.

PART THREE

DESIGN OF HUMAN SHELTER

10 Southwestern Pueblo architecture: some implications for modern housing design*

JONATHAN ERIC REYMAN

The extraordinary cold which enveloped much of the United States during the winter of 1977 and 1978, the resultant fuel shortage due to increased consumption, and our inability to move supply barges through ice-locked inland waterways are prompting engineers to reassess the adequacy of our housing in terms of future energy supplies. This reassessment and suggestions for remedying the various deficiencies were given wide coverage by the news media. Scarcely a day passed without someone publicly providing a helpful hint on how to combat the cold. Many of these suggestions were impractical, others, ingenious. For instance, a family in Armington, Illinois, built a new, cave-like house in the earth and reduced their December (1976) and January (1977) heating bill to less than $12.[1]

CURRENT REASSESSMENTS OF HOUSING DESIGN

Of course, although it is possible to build similar structures in other hilly regions, not all of us can live in cave houses. Nor is the current reassessment of our housing needs, strengths, and weaknesses peculiar to 1977–8 and solely a result of that winter's cold. Indeed, such reassessment has been taking place on a large scale since the Arab oil embargo of 1973, and the entire problem has been developing from the immediate post-World War II period.[2] Thus, our current dilemma can be viewed within the larger context of housing design and energy conservation problems during the last three decades. The present cold and increasing evidence of drought merely exacerbated existing conditions and emphasized the need for workable, efficient, long-term solutions. We must reexamine our ideas regarding our environments and restructure our winter ecosystems.[3] Appropriately, recent issues of *Science* were devoted to various aspects of this general problem.[4]

In terms of building design for both residential and commercial use, two factors—heating systems and insulation—seem to be receiving the greatest attention. The shortage of natural gas, real or artificially created, and the high costs of this fuel and petroleum products for heating are necessitating the consideration of coal, solar, and nuclear power as alternative energy sources for the future.[5] Other sources, such as the wind and geothermal heat, have also been considered.

Yet not all of these energy sources are equally recommended. Given our present technology, coal is our most abundant and easily accessible source; but the use of coal leads to massive pollution and environmental degradation. Unless we develop the means for retrieving coal with minimal environmental damage and devise an economical system for burning it cleanly and efficiently, we face major problems in increasing our coal use.

The expansion of nuclear energy also faces stiff opposition from environmentalists, scientists, and technocrats. The problems with nuclear energy are threefold. First, the public is unconvinced that the operation of nuclear reactors, especially breeder reactors, can be safely controlled. Thus, people are lobbying, with considerable success, against the construction of new reactors, and President Carter declared a moratorium on the construction of fast breeder reactors. Secondly, the problem of disposal of radioactive wastes has not been adequately solved, delaying still further the construction of new nuclear power plants. Thirdly, the huge costs of building such plants require increases in energy rates to consumers. State power commissions, under pressure from citizen lobbies, are often reluctant to grant the full requests of the utility companies. Whether the various concerns and technological problems can be overcome in the near future is unclear, but I doubt that nuclear energy will become a major power source within the next ten to fifteen years.

Solar energy, on the other hand, holds greater promise. If we could harness it (and technological advances already made predict our future success), we could use this energy on a large scale, especially for residential housing. In fact, the needed basic technology has existed for over 1000 years in many native American cultures like that of the Pueblos in the American Southwest.

Similarly, adequate insulating techniques and materials have been available in certain areas for about the same time period; if employed, these can effect significant savings in energy consumption. Moreover, in certain regions, like the semiarid American Southwest, these techniques and materials are integral parts of the basic wall structure, rather than add-on items such as fibreglass batting, blown-in cellulose fibre, and various polyethylene and polyurethane foams. As such, these integral components not only contribute to the basic stability of the structure, but they also conserve more energy and thus are more efficient.

THE PUEBLO MODEL

The Pueblo peoples, now numbering about 45 000 and living in about 30 scattered villages in northeastern Arizona and in western, central, and northern New Mexico, still have much to teach us about architectural design and energy conservation.[6] Several points should be noted.

First, most prehistoric towns and villages are now ruins primarily because of changing climatic conditions, environmental degradation from over-exploitation of local resources, disease (especially that introduced by Europeans after 1540), war, and disasters such as fire. Structural defects in the architecture or errors in town planning *per se* are not so much

*Support for the field research for this paper was provided by National Science Foundation grants GS–2829 and GS–40410 and by Illinois State University and Southern Illinois University. The author gratefully acknowledges the assistance of Rose Schilt, who helped with the research, and Teri Bartholomew, who aided in the preparation of the manuscript.

to blame. Indeed, despite raiding of the pueblos for wood and stone by 18th and 19th century Navajos and whites, the structural integrity of numerous southwestern sites, such as the large towns of New Mexico's Chaco Canyon, remains almost complete. (This, we must remember, is 900 years after the towns were abandoned.) As many who have worked in the Southwest have quickly learned (especially at Chaco Canyon, Mesa Verde, and the Kayenta region of Arizona), with slight repairs many prehistoric sites still would make excellent living quarters. For example, miners working in the Kayenta region of northern Arizona during the late 1800s often repaired and occupied prehistoric Anasazi buildings and found them to be far superior quarters to the usual miners' shacks.

Secondly, several southwestern pueblos, like Acoma, Oraibi, and Taos, have been occupied since prehistoric times, the first two for perhaps as long as 1100 years. This continuous habitation is not due merely to Pueblo conservatism, as some have argued, but is best explained as a result of efficient and successful adaptation to the environment by these native American peoples.[7]

Thirdly, that these people have been able to occupy the sites for so long despite major environmental changes and enormous pressures from outside groups, suggests their adaptive strategies are worth studying. By examining their architecture, we can learn about the Pueblo culture. We may also study designs typical of cultures of other native American, Near Eastern, and other arid and semiarid regions; but such consideration is beyond the scope of my examination.

10.1 *Cliff Palace, Mesa Verde National Park, Colorado, as seen from the mesa top to the northwest. The afternoon sunlight of winter bathes the site, heating its walls and those of the surrounding cave and storing heat there which will, after sunset, radiate and warm the inhabitants*

Fourthly and finally, this chapter is basically a synthesis of existing materials rather than a report on new research. Nevertheless, by drawing together data from several areas and by examining certain achievements of earlier cultures, I hope to provide some direction for future energy conservation in terms of housing design, particularly in arid and semiarid regions. With regard to this last point, Knowles states:

> . . . for the life and technology of its time, the Pueblo Bonitan settlement must be recognized as one of the major achievements of pre-Columbian America. Beyond that, and more generally, the pueblos offer an intriguing architectural heritage and a promising legacy that has generally been ignored. The use of form to reduce the consumption of organic fuels and yet achieve a high degree of equilibrium can be studied with profit by any society, however technologically advanced.[8]

Knowles' study represents one of the few attempts to examine critically Pueblo architecture in terms of efficient energy conservation. His analysis, using scale models of three prehistoric sites—Long House at Mesa Verde, Colorado; Pueblo Bonito at Chaco Canyon, New Mexico; and the prehistoric and modern pueblo of Acoma in New Mexico—and a solar simulation technique in a heliodon, demonstrates empirically what many southwestern anthropologists have recognized from their subjective observations: many prehistoric Anasazi pueblos as well as the modern pueblos are highly efficient in energy conservation.[9] Those of us who have worked in the Southwest have long been aware that intact rooms at prehistoric sites are relatively cool during the hottest summer days, warm during the cool nights, and surprisingly warm in winter, even when no artificial heat is used in the room.

Both the prehistoric structures and the modern pueblos are evolutionary products resulting, in part, from the application of ideas and data derived from careful, long-term astronomical

observations. These observations were used to establish solar and lunar calendars in order to plan and implement the subsistence strategy at each pueblo for farming and hunting-gathering activities. The same observations probably were also used to establish celestial referents for directional orientation of the pueblos. This orientation, in many cases, maximized insolation and thereby facilitated energy conservation for solar heat in winter and cooler temperature in summer. In short, many prehistoric and modern pueblos, especially the larger sites, were planned as architectural units and constructed on the basis of astronomical data concerning apparent solar movement.[10]

Long House and Mesa Verde

As Knowles demonstrates, the cave location and southern exposure of Long House, combined with the arrangement of buildings, provides '. . . a high degree of protection from the summer sun . . .' and '. . . more efficient energy collection in winter when such efficiency is of great advantage'.[11] The same is true, although to a lesser degree, for many of the mesa top sites which also face in a general southerly direction.[12]

From as early as the Basketmaker III period (AD 600), Mesa Verde people lived in caves.[13] After AD 750, the number of mesa top sites increased markedly, probably to facilitate access to farming areas which had become increasingly important as subsistence bases. Nevertheless, cave occupation continued on a reduced scale. By AD 1100, however, changing environmental conditions, particularly increasing cultural deforestation and the accompanying erosion, had caused farmland productivity to decline and cave occupation subsequently to increase. This return to the cliff sites expanded, somewhat, the land available for farming; raids by seminomadic peoples or even by other Pueblo groups (among whom there may have been sharpened competition for land), also may have encouraged people to move back into the caves. It was during this period that Long House, Cliff Palace, (see **10.1**) and many of the other cliff dwellings underwent major expansion. It must be noted, however, that the occupation of the mesa tops persisted, and new sites like Sun Temple were being built right up to the general abandonment of Mesa Verde around AD 1300.

This increased occupation of cliff sites and the subsequent expansion of earlier, smaller sites helped energy conservation. Knowles shows that in the case of Long House, the choice of a cave location improved the efficiency of energy collection in winter.[14] The same is true at other sites like Cliff Palace. This is an important point to consider because wood, the primary energy source, was in short supply by the 13th century. The mesa top forests had been badly depleted as farmland was cleared and site construction increased.

Richard Sense demonstrates that there is a 'thermal belt' at Mesa Verde where there is a gradation of cold to warm air from the base to the top of the cliffs.[15] Cliff dwellings were usually located from one-third to one-half way down from the top of the cliff, placing them in one of the two warmest strata in this belt. Such site locations saved energy by reducing the need for wood for heating.

The use of double walls and thick adobe and stone construction, sometimes 1 m in thickness, provided excellent insulation in an area with winter temperatures often below −18°C and summer temperatures frequently exceeding 36°C.[16] From the late spring through early fall, the thick walls kept the pueblo rooms relatively cool. My field research has shown that summer temperatures may be 10°–15°C less inside the pueblo than outside.

The incorporation of astronomical alignments, especially solar ones, into architectural features at several of the cliff pueblos, like Cliff Palace and Balcony House (as well as at mesa top sites), allowed the priests to make accurate predictions of seasonal changes. Therefore, the people had adequate time to make needed repairs to walls and roofs to assure safety and to reduce heat loss which otherwise would occur through the cracks and areas where mortar had fallen out.[17]

The careful placement of ventilator openings for the *kivas* (subterranean ceremonial chambers) took advantage of the air currents within the caves or rock shelters. Thus, while the surrounding earth and rock and the interior masonry provided insulation, the location of the ventilator opening ensured an adequate supply of fresh air. In winter the opening could be closed to reduce heat loss, while in summer the air flow could help to cool the room.[18] Similarly, small openings at the floor levels of rooms in the Mesa Verde pueblos and elsewhere aided in circulating air and cooling interior living areas. Summer research in 1976 among prehistoric sites at Hovenweep National Monument in southwestern Colorado and southeastern Utah indicated that many of the numerous openings or 'ports', especially those at and just above floor level, still function efficiently and therefore introduce and circulate fresh air within the rooms.

The general lack of furniture meant that habitation rooms could be kept small. These rooms could be heated in part by the body heat of the occupants, again reducing fuel needs. The importance of these architectural traits and their adaptive efficiency increased as time passed and the environmental conditions at Mesa Verde became more critical and marginal. The use of sunlight in the upper and rear rooms of cliff dwellings was facilitated by a coating of white paint or plaster around windows and doorways. This coating reflected light into the room and decreased the need for artificial light until after sunset.

Knowles notes that Long House in its setting 'is 56 percent more efficient as an energy collector in the winter than it is in the summer'.[19] Although no comparable tests have been conducted for other large Mesa Verde cliff dwellings, it is not unreasonable to assume that, given their similar construction, proportionately comparable results should be achieved when variations in cave locations are taken into account; that is, while the southerly exposure of the Long House cave is unusual and accounts for some of the energy collection efficiency, the architecture of Long House is not significantly different from that at other Mesa Verde cliff houses. Thus, other sites should show roughly the same winter-to-summer ratios in energy collection efficiency, even though, because of less suitable cave locations, they may be somewhat less efficient than Long House. In short, Long House's location, directional orientation in the cave, and the thick or double wall construction were useful and efficient for energy conservation.

Pueblo Bonito and Chaco Canyon

Pueblo Bonito, (**10.2**) in Chaco Culture National Historical Park in northwestern New Mexico, was even more efficient than Long House in terms of energy use and conservation.[20] Made of sandstone masonry and originally covered with adobe plaster, this D-shaped pueblo contained more than 830 rooms and reached a height of five stories in its rear (north) section. The pueblo was oriented toward the south, increasing the amount of sunlight available for use at its latitude, N36° 03′ 40″. The orientation of the pueblo also made possible the incorporation of winter solstice alignments into architectural features for the inhabitants' use.[21] Other astronomical alignments may have been present.[22] Pueblo Bonito (**10.3**) was well built and has some of the most beautiful masonry found in the Anasazi area.[23] The other large towns in Chaco Canyon are

10.2 *Pueblo Bonito, Chaco Culture National Historical Park, New Mexico, as seen from the top of the cliff to the north of the site*

similarly constructed; this development within the canyon and the surrounding area represents, to many archaeologists, the apex of Anasazi architecture. In places, the solid and rubblecore masonry walls of Pueblo Bonito are more than 1 m in width, and the original adobe plaster was several cm thick. This plaster was usually applied to both sides of the walls; thus, it provided additional insulation of up to ⅓ m in thickness and eliminated the problem of rock dust from the unplastered sandstone. Here again, the orientation of the pueblo in conjunction with the thick walls resulted in an efficient energy collection and conservation system.

From the pueblo's inception (circa AD 828)[24], the builders gradually improved the overall efficiency of their structure by approximately 5–7 per cent. By the end of the occupation (AD 1150), they were taking maximum advantage of the available sunlight in terms of both daily and seasonal variations. This is demonstrated by the absolutely flat energy curve for the site from sunrise to sunset at the winter solstice.[25] Although comparable studies have not been undertaken for the other large Chaco towns, I think that most of these sites should approach the efficiency of Pueblo Bonito.[26]

In short, the architects of Pueblo Bonito, Peñasco Blanco, and other Chaco towns had carefully formulated plans for their communities prior to construction and continued to improve upon these designs during subsequent rebuilding periods. They apparently understood the adaptational problems facing them in terms of the interrelationships between the earth and sun, and they solved these problems in their architecture and approached a 'steady state' system.[27] Unfortunately, like the inhabitants a century later at Mesa Verde, the Chacoans were unable to view and understand clearly the larger environmental context; therefore, they severely damaged their environment, largely through the over-exploitation of the mesa-top forest resources.

Acoma and other historic pueblos

Acoma, in west-central New Mexico, has been continuously occupied since about AD 1000–1100. Thus, like Oraibi, a Third Mesa Hopi village, Acoma was contemporaneous with the Mesa Verde and Chaco Canyon occupations. It is not surprising, therefore, that Acoma shares certain architectural traits with these other prehistoric sites. Nor were the lessons of the past lost on the builders of the other historic pueblos; for, after all, these people were the descendants of the prehistoric Anasazi and Mogollon. Most of the pueblos occupied today have existed at least since the Spanish, led first by Marcos de Niza and later by Coronado, made their first *entradas* into the area in 1539.

Situated atop a high (125+ m), steep-sided mesa, Acoma today is sparsely populated; although village capacity exceeds 1500 and most families maintain a house at Acoma for major ceremonial events, there are probably less than 200 permanent residents. Most people have moved to the outlying villages like Acomita and McCarty's where they have greater access to wage labour opportunities, water for irrigation, and commercial shopping areas. A paved road leading up the mesa now makes Acoma more accessible to tourists, and several families derive a relatively good income from selling the generally fine Acoma pottery to visitors.

The basic ground plan of the village consists of three more or less parallel elongated house blocks extending in a general east-west direction; the house blocks are segmented at uneven intervals by north-south walkways. This general layout is common to a number of other pueblos, eg, San Juan, Jemez, Santo Domingo, and several of the Hopi villages.[28] The buildings at Acoma range from one to three stories in height, and the house blocks extend across the mesa for about 310 m. Made from stone or adobe bricks and covered with adobe plaster, these units are generally stepped down in terrace fashion from the north to the south, with most doors and windows opening in a southerly direction. The walls vary in width from ⅓–1 m, with a mean width toward the lower end of the range. In at least one location, an original window made from a rounded slab of mica is still in place. As at Mesa Verde and Chaco Canyon, the thick walls provide excellent insulation. While doing field work at Acoma in 1968, I found the rooms to be cool and comfortable despite an outside temperature in excess of 38°C.

As I noted earlier, Knowles tested Acoma by placing a scale model of the village in the heliodon.[29] Given the site location and available technology, he found that the architectural layout and site configuration of Acoma represent the best adaptation to the environment under the circumstances; energy transmission is 50 per cent more efficient in winter than in summer.[30] Although Acoma is not as efficient as either Pueblo Bonito or Long House, its layout is more flexible in terms of any future expansion of the site.[31] In short, Knowles states: 'The study reveals what the Indians must have understood for the past thousand years: the pueblo is an efficient energy system that tends to equalize internal energy profiles over the extremes of season and day'.[32]

Although similar studies have not been conducted for other modern pueblos, an examination of existing ground plans suggests that such analyses would produce results comparable to Knowles'.[33] As noted above, a number of other villages have an overall design and a directional orientation similar to those for Acoma. Despite the deteriorating condition of many of these towns and the accompanying fragmentation of their original architectural design (both due in large measure to the effects of the enormous acculturative pressures to which the Pueblos have been subjected), the basic structural integrity of these towns is still discernible. Santo Domingo, for example, has a southerly orientation, and so do Jemez, San Juan, and several other villages. At the approximate 35°–36° north latitude at which all of the pueblos are located, a southerly orientation maximizes insolation. Even at sites such as Walpi, a First Mesa Hopi village where the narrowness of the mesa requires a generally north-northeastern/south-southwestern orientation, the buildings are still oriented to permit maximum exposure to the sun. Furthermore, all of these pueblos originally were of heavy stone and adobe construction, thus using insulation as an efficient response to insolation and thereby conserving energy.[34]

An interesting test comparison would be between Taos and Pueblo Bonito since the two sites have ground plans which are markedly similar. Such a comparison should yield useful data concerning the evolution of this particular Pueblo architectural design as an adaptational device over a period of several hundred years.[35]

Thus, the work of Knowles has shown that, through the deliberate directional orientation of the site and 'by the overall shape of the buildings and . . . the selection and use of materials', the Pueblos designed efficient architectural adaptations to their environments.[36] To be sure, the prehistoric Anasazi and their modern descendants did not design perfect buildings. They often failed to bond the walls at points of intersection. Then, when the building settled, the walls quickly pulled apart and had to be patched. In some cases, whole walls collapsed soon after they were erected. Nevertheless, Pueblo construction was usually quite good, given the available technology; and their designs were excellent. Our task, then, is to adopt their ideas, to use them, and to improve upon them for our benefit.

Knowles discusses some applications of Pueblo design principles with regard both to large buildings and also to single-family homes.[37] His discussion, however, is basically theoretical, whereas I want to concentrate on a few specific suggestions.

10.3 *Masonry construction at the southeastern end of Pueblo Bonito where the fine masonry and thick, tapering walls are still clearly visible and well preserved 900 years after the initial construction*

APPLICATIONS OF THE PUEBLO MODEL

We are faced with a threefold problem: reducing energy use, using energy more efficiently (thus further reducing use), and developing new sources of energy. Pueblo architectural design, if properly applied, can facilitate solutions to the first two facets of this problem and, less directly, can aid toward solving the third by making the use of solar energy easier and more practical.

I find it curious that, especially in the Southwest, closer attention has not been paid to the practical, as opposed to the aesthetic aspects of Pueblo architecture; this seems stranger still, given our infatuation with Pueblo arts. Perhaps this is an example of ethnocentrism, our unwillingness to recognize the possibility that less technologically developed people might have something to teach us. It is paradoxical that a city like Santa Fe, New Mexico, has a building code which requires that architecture within the central precinct be constructed in either Pueblo or Spanish Colonial style, but ignores a consideration of the basic building materials. Whatever the reasons for our attitude, it is disadvantageous to continue to ignore the important potential applications of Pueblo architecture in semiarid areas.

Site location

In attempting to apply the Pueblo architectural model to modern construction, one must consider a number of factors. Whether for a single-family house or a larger building, the directional orientation of the structure is of primary importance. Indeed this fact was certainly recognized by both the prehistoric Anasazi and their modern descendants, probably as a result of their astronomical observations.[38] Knowles emphasizes that wherever possible, site orientation should be southerly to maximize insolation.[39] Where space is at a premium, we ought to recontour the land wherever possible to maximize the number of homes with a southerly exposure, the better to utilize both active and passive solar energy systems. Some artificial terracing of flat areas will be helpful, both for increasing the number of favourable construction sites and for lessening erosion. A site location should also be chosen which permits excavation so that the floor level of the house will be from ⅓–½ m below ground. In this way, the surrounding earth not only provides additional insulation, but also adds protection from high winds and blowing dust.

Ground plan

Two basic ground plans, each with several variations, are found among the prehistoric Anasazi and modern pueblos.[40] The first type is based on elongated room blocks, as at Acoma, while the second type features room blocks arranged around a central, open plaza, as at Cochiti and San Ildefonso. Several pueblos, most notably Isleta situated just south of Albuquerque, combine both types of ground plans. Prehistorically, D-shaped, elliptical, oval and almost circular variations of the second type are found within the Anasazi area (the best known examples being Pueblo Bonito (D-shaped) and Peñasco Blanco (elliptical) at Chaco Canyon and Tuyonyi (roughly circular) at Bandelier National Monument). Although only Pueblo Bonito has been studied, I expect that the various other central-plaza layout variations would be comparable in terms of energy efficiency; if so, they provide alternative forms which can add variety in appearance to modern community housing areas. In other words, having demonstrated that form (in this case, the ground plan) is a major factor in determining the degree of efficiency in the use of solar energy by these pueblos, we should consider adopting these forms for our own site locations.[41] For example, the elongated room block layout is obviously better suited to shallow lots, especially those on hillsides, whereas the central-plaza ground plan, in its several variations, is better suited to larger, flat building sites.

Wall construction

Whether over a slab base or a foundation, walls should be constructed from the basic materials native to the area, ie, stone and earth. Several techniques of construction are available: rubble-core masonry faced with mud plaster, coursed stone masonry (also with adobe plaster on both sides), adobe (either in puddled or brick form), and *jacal* (still widely used in Mexico). The walls should be thick, at least ½–¾ m. A full 1 m thickness provides even better insulation, as well as additional strength; this thickness is almost mandatory if the structure will be more than a single storey (the addition of fibreglass insulation toward the inside of the wall between the final courses of adobe brick or stone will further improve the efficiency of the structure).

The principle of a vapour barrier or thermal layer, (see roofs) has long been employed by peoples in northern latitudes and at high altitudes to increase air circulation around the body, thus facilitating heating and cooling. The same principle can be applied in modern wall construction, resulting in an improvement over the basic Puebloan building techniques. Construction of double rather than single exterior walls (a design occasionally found in Pueblo sites, although probably for purposes other than insulation alone—as previously explained), with air trapped between the two, should produce marked increases in energy use efficiency.

I surveyed buildings in the Santa Fe area in 1972 and found that either the adobe brick construction or stone masonry covered with adobe plaster would add about one-third more to the standard basic building cost. In other words, if one were to use Pueblo materials, a $30 000 house would cost about $40 000, a $45 000 house would cost approximately $60 000, and so forth. More recent data (1979–80) indicate that the initial building costs would now be one-half to two-thirds more than for a conventional home; but energy costs have more than doubled since 1972. Therefore, this cost difference would be offset in a relatively short time because the Pueblo-style house needs less powerful heating and air-conditioning systems and uses them more efficiently; in fact, one should be able to eliminate air conditioning entirely in most locations, especially if double wall construction is utilized.

Like many pueblos, modern houses should be step construction with the north wall higher than the south. In summer, the sun is in a northerly position *vis-à-vis* the earth; thus, a higher north wall shades the house and improves cooling. Conversely, in winter, with the sun in a southerly position, the high north wall can be used to reflect heat and light down onto the rest of the structure, thus lessening energy use, especially for heating, Moreover, maintenance costs should be lower for adobe and stone than for other types of wall construction. One problem with both adobe and adobe plaster has been that walls tended to fragment or chip as time passed. This problem could be greatly reduced by the addition of one of the polymer resins currently available. The resin can be mixed with the adobe, or it can be sprayed on the walls after they have been erected. The use of this type of resin would lower maintenance costs still further.[42] Door and window frames could be embedded in the thick walls, thereby almost eliminating the energy loss which normally occurs at their edges. Here again without the use of additional insulation, Pueblo architecture could sharply reduce a major cause of energy loss in conventional homes. In addition, the cost of such insulating materials as weatherstripping would be eliminated. Finally, the thick walls would provide an excellent seating for the installation of solar collectors, thus reducing still further the dependency on fossil fuels.

Roof construction

Traditional pueblo roofs were fashioned from large beams, called *vigas*, over which were layered smaller poles, twigs, bark batting (eg, juniper) and a thick capping of mud plaster. In multiple-storey units, this plaster served as the floor of the upper rooms.[43] The completed roofs were flat and functioned as areas for a variety of activities such as craft production and other work, food drying and preparation, and sometimes cooking (inferred from the presence of hearths found in roofs at prehistoric sites). Some of these same activities still continue at modern pueblos, especially among the Hopi.

Although roofs constructed in the Pueblo style are aesthetically pleasing, such construction may not be practical, either in terms of costs or present timber resources. Southwestern forests cannot supply the wood necessary for large-scale construction of this roof type without serious erosion. Whether sufficient timber resources are available elsewhere and whether they could be shipped into the Southwest at a reasonable cost, are two questions which need further study. Whatever the solution to the problem of roof materials may be, however, we ought to follow the multiple layer technique used in the pueblos; this type of construction works in a way similar to thermal clothing and is an excellent insulator.

Nevertheless, one major change needs to be made: rather than build flat roofs, to maximize insolation we should construct roofs which are slanted to face the sun. Orientation of the roof toward the sun in conjunction with multi-layered construction provides an excellent context for the placement of solar collectors. Solar bubbles or reflectors, unlike conventional roofs, provide additional light to the interior of the structure. They can also be more easily installed and sealed against leakage. Solar reflectors have been used by the Inuit in their igloos, as is visually well-documented in the classic ethnographic film, *Nanook of the North*; likewise, roof openings to admit light are common among the southwestern pueblos. Thus, the combination of Pueblo design and modern technology can provide greater energy efficiency in roof construction, as well as throughout the entire house.

CONCLUSION

Prehistoric Pueblo architecture and modern technology can be combined to provide better, more energy-efficient housing for the Southwest and other semiarid regions. Although the initial costs of Pueblo-style architecture using traditional materials are somewhat higher than for other types of construction now in wide use (especially prefabricated structures like mobile homes and modular houses), these higher costs can be amortized in a relatively short time by savings in energy consumption. Since the cost of natural gas (the predominant fuel in the Southwest) is rising, the higher initial cost of Pueblo-style architecture would probably be offset by energy savings in about ten years.

The fact that the Pueblos have maintained themselves *in situ* for as long as 1000 to 1100 years shows that their architecture is adapted to the semiarid environment of the American Southwest. This longevity also indicates the success of their subsistence strategies. They are remarkably stable groups.

Therefore, we need to examine Pueblo architecture more closely and to incorporate its best features into our future housing construction. As Knowles notes, we have ignored for too long what the Pueblos have to teach us about the use of form in the conservation of energy.[44] Our unwillingness to learn from the earlier inhabitants of the region has been an example of myopic ethnocentrism, as well as uneconomical energy use.

NOTES

1 See *Daily Pantagraph* (Bloomington, Illinois). 12 December 1976, 3 January 1977, 11 July 1977.
As of July 1980, a check of local advertisements and business directories indicated that there were at least four building contractors in central Illinois specializing in the construction of subterranean dwellings. The main building materials used for the shells of these dwellings are reinforced concrete and stone.

2 Jack E. Snell, Paul R. Achenbach, and Stephen R. Peterson, 'Energy Conservation in New Housing Design', *Science*, 192 (1976), 1305–1311.

3 T. R. Lakshmanan, Lata Chatterjee, and P. Roy, 'Housing Requirements and National Resources', *Science*, 192 (1976), 943–949; Richard L. Meier, 'A Stable Urban Ecosystem', *Science*, 192 (1976), 962–968; Lee Schipper and Allan J. Lichtenberg, 'Efficient Energy Use and Well-being: The Swedish Example', *Science*, 194 (1976), 1001–1012.

4 American Association for the Advancement of Science, 'Energy', *Science*, 184 (1974), 245–389; 'Food', *Science*, 188 (1975), 501–653; 'Materials', *Science*, 191 (1976), 631–776.

5 John A. Duffie and William A. Beckman, 'Solar Heating and Cooling', *Science*, 191 (1976), 143–149.

6 Edward P. Dozier, *The Pueblo Indians of North America* (New York: Holt, Rinehart and Winston, 1970), p. 122; Anne M. Smith, *New Mexico Indians: Economic, Educational, and Social Problems*, Museum of New Mexico Research Records, No. 1 (Santa Fe: Museum of New Mexico Press, 1969), pp. 3–7.

7 Edward H. Spicer, 'Spanish-Indian Acculturation in the Southwest', *American Anthropologist*, 56 (1954), 663–678; Florence Hawley Ellis, 'Comment on Spicer's, "Spanish-Indian Acculturation in the Southwest"', *American Anthropologist*, 56 (1954), 678–680; Stanley A. Stubbs, *Bird's-eye View of the Pueblos* (Norman: University of Oklahoma Press, 1950), pp. xi–xv, 3–20; Jonathan E. Reyman, 'Southwestern Pueblo Conservatism: A New Look at an Old "Myth"', paper delivered at Thirty-Fifth Meeting of the Society for American Archaeology, Mexico City, 30 April 1970, pp. 1–4; Jonathan E. Reyman, *Mexican Influence on Southwestern Ceremonialism* (Ann Arbor, Michigan: University Microfilms, 1971), pp. 112–337.

8 Ralph L. Knowles, *Energy and Form: an Ecological Approach to Urban Growth* (Cambridge: Massachusetts Institute of Technology Press, 1974), p. 46.

9 Since Knowles' study is readily available, I shall only summarize his data and conclusions. For a better understanding of Knowles' work and how his arguments and mine are interconnected, the reader should examine *Energy and Form* in detail.

10 Reyman, *Mexican Influence*, pp. 112–138, 183–337; Jonathan E. Reyman, 'Astronomy, Architecture, and Adaptation at Pueblo Bonito', *Science*, 193 (1976), 957–962; Jonathan E. Reyman, 'The Emics and Etics of Kiva Wall Niche Location', *Journal of the Steward Anthropological Society*, 7 (1976), 107–129; Jesse Walter Fewkes, *A Prehistoric Mesa Verde Pueblo and its People*, Smithsonian Institution Annual Report for 1916 (Washington, DC: United States Government Printing Office, 1917), p. 469.

11 Knowles, pp. 24–25.

12 Reyman, *Mexican Influence*, pp. 188–241.

13 Alden C. Hayes and James A. Lancaster, *Badger House Community, Mesa Verde National Park*, Archaeological Research Series, 7–E (Washington, DC: National Park Service, 1975), pp. 1–5, 182–186.

14 Knowles, pp. 20–27.

15 Richard G. Sense, 'Microclimate and Archaeology', paper delivered at the Thirty-Sixth Annual Meeting of the Society for American Archaeology, Norman, Oklahoma, 8 May 1971.

16 On double walls, see Sense.

17 Jonathan E. Reyman, fieldnotes for National Science Foundation Research at Mesa Verde, Hovenweep, and Chaco Canyon during 1975 and 1976.

18 Reyman, *Mexican Influence*, pp. 190, 232.

19 Knowles, p. 26.

20 Ibid, pp. 43–45.

21 Reyman, 'Astronomy, Architecture, and Adaptation', pp. 957–962.

22 Reyman, *Mexican Influence*, pp. 270–272.

23 Neil M. Judd, 'The Architecture of Pueblo Bonito', *Smithsonian Miscellaneous Collections*, 147, no. 1 (1964), pp. 14–16, 27–29.

24 Bryant Bannister, *Tree-ring Dating of the Archaeological Sites in the Chaco Canyon Region, New Mexico*, Southwestern Monuments Association, Technical Series, Vol. 6, pt. 2 (Globe, Arizona: Southwestern Monuments Association, 1965), pp. 179–187.

25 Knowles, pp. 39–43.

26 In my recent fieldwork, I discovered that Peñasco Blanco, a mesa top site on the south side of Chaco Canyon about 4.8 km west-northwest of Pueblo Bonito, is situated in the only location in the general area from which all four solstice points are directly observable. Such a location is advantageous for: 1 making the astronomical observations used to establish and maintain the

calendars and their conjunctive socio-ceremonial and subsistence adaptation systems; and 2 making the maximum use of available sunlight. Perhaps the latter hypothesis can be adequately tested if the site of Peñasco Blanco is excavated.

[27] Knowles, p. 39.

[28] Stubbs, pp. 24–119.

[29] Knowles, pp. 27–33.

[30] Ibid. p. 31.

[31] Ibid, p. 46.

[32] Ibid, p. 31.

[33] Stubbs, pp. 24–119.

[34] In recent years, there has been a trend among the Pueblos to modernize, eg, to build new homes using prefabricated materials such as cement blocks and to build away from the central pueblo. Thus, they have reduced structural integration and have substituted mobile homes and modular homes for traditional houses. Such innovations are seen, especially by certain federal bureaucrats, as desirable solutions to various housing problems. Therefore, federal agencies push strongly for this new housing. Yet, whatever problems these innovations appear to solve, they all lessen the efficiency of the original pueblo architectural form and contribute to a marked increase in energy consumption. For instance, a metal-sided mobile home must be air conditioned in summer and almost continuously heated in winter in order to be livable in the Southwest. In the original pueblo, no summer cooling system was necessary, and far less heat was needed for warmth during the winter months.

The blame for these changes in Pueblo housing, however, does not lie solely with federal officials. Many Pueblo people adopt such housing as a means to demonstrate their awareness of the importance of appearing up-to-date in terms of the mainstream of United States culture. They also view the adoption of nontraditional housing as a means of lessening the power of the traditional Pueblo leadership which they view as oppressive and self-serving. Nevertheless, the adoption of these modern house types remains ecologically less satisfactory than retention of traditional Pueblo architecture, regardless of the perceived benefits.

[35] Judd, pp. 26–27 and Stubbs, pp. 18–19.

[36] Knowles, p. 135.

[37] Ibid, pp. 111–193.

[38] Reyman, *Mexican Influence*, pp. 112–337; Reyman, 'Astronomy, Architecture and Adaptation', pp. 957–962.

[39] For further discussion, see Knowles, pp. 189–191.

[40] Stubbs, pp. 10–119.

[41] Knowles, pp. 1–80.

[42] This is one example where the use of modern materials and technology can improve upon the basic pueblo structure.

[43] Judd, pp. 22–31.

[44] Knowles, p. 46. It might also be noted here that on a season to season basis, the difference between interior and exterior temperatures is less for pueblos than for conventional housing which we tend to overheat in winter and overcool in summer. Although the issue needs considerably more research before any conclusions can be made, it is possible that Pueblo-style housing is healthier than conventional homes because the body is not forced to adjust to as great ranges of temperature differences, either when coming inside or going outside. In short, the pueblos may provide a more stable environment in terms of temperature.

SUGGESTED READING

Judd, Neil M. 'The Architecture of Pueblo Bonito'. *Smithsonian Miscellaneous Collections*, 147, no. 1. (1964).

Knowles, Ralph L. *Energy and Form: an Ecological Approach to Urban Growth*. Cambridge: Massachusetts Institute of Technology Press, 1974.

Mindeleff, Victor. 'A Study of Pueblo Architecture: Tusayan and Cibola'. Bureau of American Ethnology, 8th Annual Report, 1886–1887. Washington, DC: Smithsonian Institution, 1891. pp. 3–228.

11 Building design in hot, dry climates

MILES DANBY

The hot, dry desert is a harsh wilderness of sand, unbearably hot during the day but chilly if not cold at night. Traditionally, the only people who have been found there are the tough nomads like the Bedouin Arabs who travel the wilderness in search of the slight vegetation resulting from the light infrequent rain so that their herds of camels and donkeys can survive; for only if their animals survive, can they make a living in this hostile environment. This way of life remains to this day in many areas of the hot, dry world. In Saudi Arabia, for example, many nomads still resist the temptations of materialism and have not yet traded in their camel for a Toyota or exchanged their tent for a fixed concrete box home.

The Muslim vision of paradise is a garden with water, fountains, trees, and luxuriant vegetation—the very antithesis of the desert where water is more valuable than gold. For many Muslims, life is a symbolic journey through the desert in search of an oasis or the paradise obtainable only after death. Although Islam is not the predominant religion in every hot, dry area, its teachings serve to emphasize the reality of life in such places. Life is only possible where water is present, and a study of arid climates and building design must therefore be concerned with the importance of water. The difficulty of finding and conserving enough water to maintain human, animal, and plant life means that its presence cannot be taken for granted. It is furthermore needed for washing the body, clothes, and utensils; for the Muslim it is needed for performing ritual ablution before every daily prayer (although the use of clean sand is permitted under the harsh conditions of the desert).

WATER AND THE DESERT

If sufficient water does not come from the sky in the form of rain, then it must be sought elsewhere as the nomads do. Permanent settlement can only occur where water can be obtained on a permanent basis. Water is found in springs at oases, on the banks of great rivers like the Nile and the Euphrates which bring water from areas upstream where there is higher rainfall, and in areas into which water is brought through irrigation and the construction of dams on such rivers. The presence of water can account for the presence of settlements, but it cannot account for the shape or form of the structures. Water is also present in the atmosphere as vapour, and the amount of it (usually expressed as relative humidity) together with air temperature affects our assessment of human comfort. Generally speaking, the human body can withstand relatively high air temperatures provided the relative humidity remains low; but if high air temperatures are combined with high air relative humidity, the climate becomes intolerable and heat stroke can result.

In hot, dry climates as the very name implies, the relative humidity is normally low. However, in certain districts in areas which are often regarded as arid on the basis of temperature and rainfall data, the relative humidity may in fact be quite high. In other words, relative humidity is not necessarily directly related to the amount of rainfall. For example, Port Sudan, as **11.1** shows, has a climate which is normally referred to as hot, dry maritime. It has low rainfall and high temperatures, but for seven months of the year it has a relative humidity between 60 and 80 per cent. This is because during this time the prevailing wind blows across the Red Sea absorbing water vapour from its surface. This combination of high air temperature and humidity can be most uncomfortable: the skin becomes moist with sweat, and the surrounding air is only able to absorb a little moisture since it is already almost saturated. The sweat can only evaporate and cool the skin if more air is circulated. It is thus as essential to maintain the possibility of through-ventilation in all buildings in hot, dry maritime climates as it is in hot, humid ones; and this need obviously affects building form and design.

However, for the purpose of this chapter, hot, dry climates are defined as areas of low humidity and high solar radiation where cooling of the skin by evaporation *can* occur and where dehydration and dust are the two main threats to human comfort. Two other climatic factors should also be considered. First, because of the lack of cloud cover in these areas, the ground becomes intolerably hot during the day, quickly loses this heat by radiation to the atmosphere after sunset, and may become cool or even cold at night under a clear sky. The diurnal range of air temperature can thus be as high as 17°C or more. These daily extremes of temperature can be destructive to buildings and building materials as a result of the daily expansion and contraction of the exposed surfaces and especially to any joints between different materials because of different expansion rates. Second, there is a short rainy season in some arid areas, and quite high humidity can be experienced during this period. Through-ventilation is, therefore, necessary at this time. At Khartoum, for example, as **11.2** shows, the relative humidity can be over 50 per cent during the months of July and August and in early September.

BUILDINGS AND SETTLEMENTS

The traditional vernacular form of buildings and settlements embodies the folk experience of many centuries. The adaptation to the climatic and cultural requirements of a particular community has taken place subconsciously over such a long period of time that a building procedure is eventually taken for granted as the appropriate solution to the community's immediate needs. Close study of the vernacular building form in hot, dry climates therefore reveals much wisdom which has only recently been ignored under pressure from the internationally accepted high technology and the politically and economically attractive images of modernism. Moreover, this

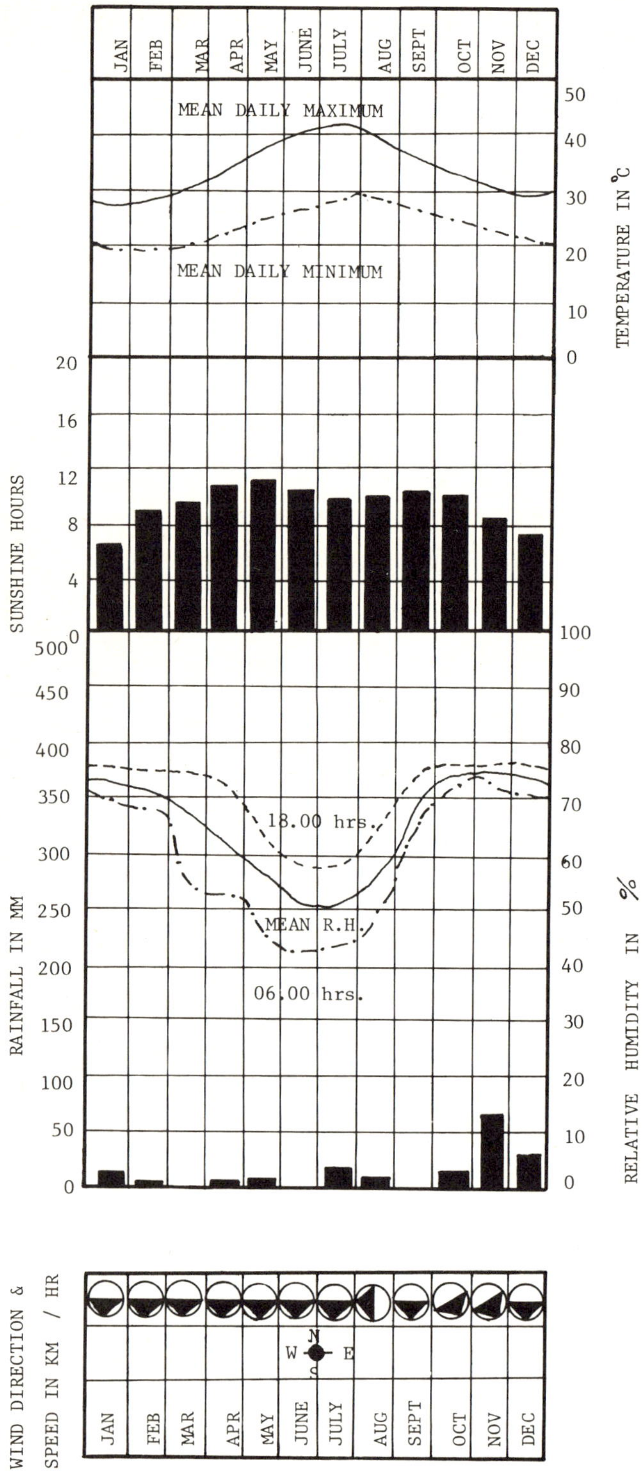

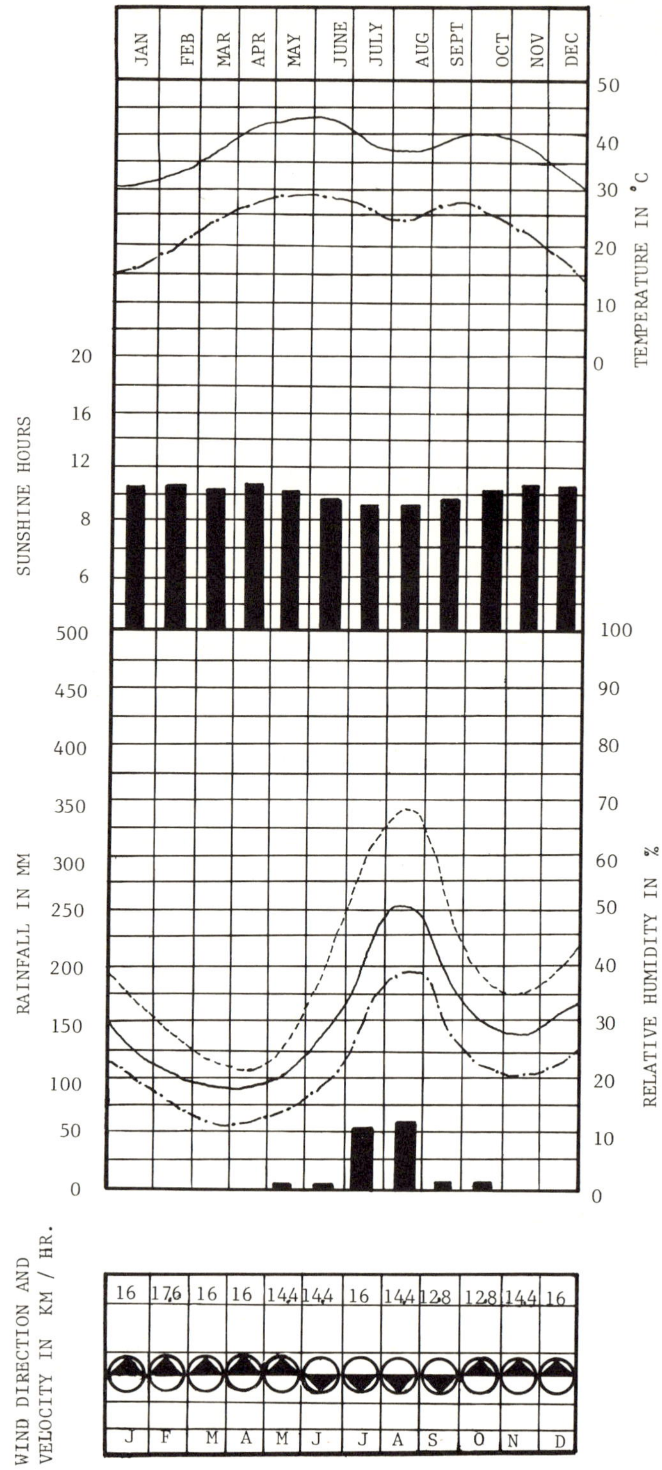

NOTES **(11.2)**

1-The temperature rises from minimum in Dec. to maximum in May and June.

2-The maximum degree of air temperature is near 48°C and the minimun tends to 15°C.

3-The sunshine hours are over 70% of the daily hours all through the year and may reach 95% in four months.

4-The relative humidity is usually below 50% except during the months of rainfall: July, August and September.

5-The total amount of rainfall is 167mm/year. Half of this amount is in July, while six months of the year are dry.

6-Prevailing wind direction is North-South and vice versa.

11.1 *Climatic data for Port Sudan, Sudan (6 m above sea level; 19° 37′ north latitude, 37° 13′ east longitude)*

COURTESY THE UNIVERSITY OF NEWCASTLE UPON TYNE

11.2 *Climatic data for Khartoum, Sudan (380 m above sea level 15° 36′ north latitude 32° 33′ east longitude)*

COURTESY THE UNIVERSITY OF NEWCASTLE UPON TYNE

pressure has become stonger now that the exploitation of vast oil resources has presented many countries in these regions with new and unfamiliar economic power. Unfortunately, the achievement of huge national incomes has not coincided with a concomitant gain in intellectual confidence. This would have produced a realization that the culture of Islam and the traditional wisdom in building design that is its complement evolved forms superior to those recently introduced into the area.

Traditional

Urban

The traditional Arab town house represents a level of sophistication in urban development that has rarely been matched elsewhere. (The 17th century town house in Holland is perhaps a comparable European example.) This house surrounded an open space (the garden and a fountain forming a cool, pleasant microenvironment) and presented a blank face and entrance door to the narrow public streets. The domination of the camel, horse, donkey, and pedestrian have resulted in a closed network composed of these streets, occasionally punctuated by larger spaces used for social and religious purposes. Narrow streets together with the high-density buildings of two to four storeys created a situation where, for most of the day, the buildings shaded each other as well as the streets (as **11.3** shows). Furthermore, because the buildings look inwards and the majority of rooms face the courtyard, they are closely packed and share common boundary walls (often on three sides), thus reducing the amount of wall surface exposed to solar radiation.

The traditional urban layout, although suitable for the climate, is incompatible with the free circulation of motor traffic. It can cope with horse traffic, donkeys, bicycles, and even motor bicycles; but the streets are far too narrow and the intersections too restricted for motor vehicles. What is needed is a new pattern of urban development allowing for the controlled circulation of motor traffic while maintaining the density of the traditional layout wherever possible.

Rural

The inward-looking character of traditional urban housing is also present in rural development, although the density of rural housing is lower. Either the courtyard layout is used, or a wall is constructed along the perimeter of each individual site before the construction of any room within is begun. In this latter case, rooms are added within the wall when needed, gradually reducing the area of the contained open space. Where irrigation water is not available, waste water is used to grow one or more trees and some flowers and shrubs within the enclosed space to provide shade and thereby improve the microclimate wherever possible.

A typical example of a traditional rural house can be found in the irrigated lands of the Gezira in the Sudan.[1] Since the early 1920s, irrigation has steadily expanded, making possible the well-organized production of cotton (the main national export crop). While the canals and fields were laid out by engineers and surveyors in rigid geometrical patterns, the village houses in contrast developed informally—haphazardly—with public access to them by a network of lanes and alleys ending at the entrance door to the *hosh* or courtyard.

Once the boundaries of these rural house plots were determined, a mud wall was built to enclose each of them. Room units were then built as and when they were required according to the needs of the family. The plans of individual room units were usually consistent. As in **11.4**, there was a main room of about 4.5 × 4.5 m with an attached store (4.5 × 1 m)

11.3 *Narrow street in Khan El Khalili district, Cairo*

ALL PHOTOGRAPHS ARE BY THE AUTHOR

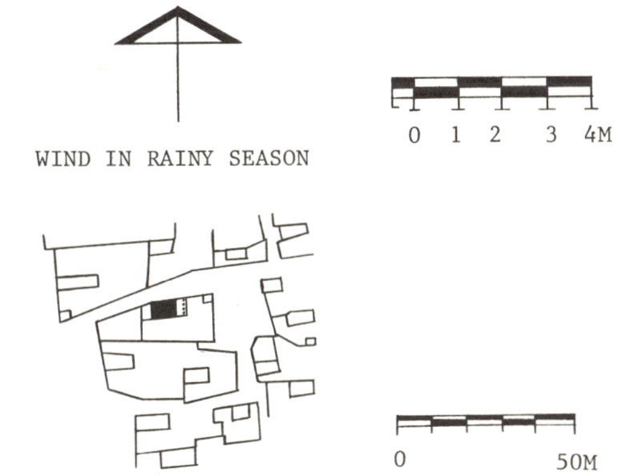

11.4 *Plan of typical rural house in the Gezira, Sudan*

COURTESY THE UNIVERSITY OF KHARTOUM; PLAN DRAWN BY STUDENTS IN THE DEPARTMENT OF ARCHITECTURE

on its west side and a sloping lean-to roof, often on its east, to form a veranda. There were four windows and one door. The door was usually in the east wall and shaded by the veranda roof, but windows were on the north and south walls to allow through-ventilation during the rainy season.

Transition from traditional to modern

An annual population increase of between 2 and 3 per cent is common to most countries of the hot, dry region; but it is creating a vast housing need that is not being met, except perhaps in the smallest and richest oil states. Then only their own citizens benefit, not the guest workers from poorer countries upon whom they now depend for all labour in the less desirable occupations (including the building industries). The real need is not for housing but for economical housing on a large scale. The gap between the cost of a house large enough for family living and reasonable rent *vis-à-vis* the income of the poorer section of society is large, even in a country such as Iraq that enjoys a high national income due to oil revenues. If the gross national product there is divided by the number of families, the resulting figure equals one-quarter of the cost of an average house produced by the private sector in Baghdad.[2] A low-income family is therefore unable to pay for a house large enough for its needs unless there is either a large government subsidy or organized 'self-help' construction. The size of the problem is staggering: in 1980 housing in Iraq was estimated as 1.25 million dwellings of which approximately 400 000 were of unacceptable quality. In spite of considerable investment by the government and the private sector, there was nevertheless a shortfall of .5 million dwellings according to the most optimistic estimates. As a result, barely half the families were adequately housed in 1980. Because it is a massive problem, many argue that it can only be solved by mass production. Acceptance of this argument has resulted in the emphasis on production problems.

The prefabrication production methods that have produced unacceptable solutions both in environmental and human terms in many Western countries are now being sold to many of the oil states to solve their massive problems. These reinforced concrete panel and box systems are being suggested as solutions, with the implicit assumptions that providing housing is predominantly a production problem and that high technology can produce the necessary units in a volume which will put the operation on an economic basis. It is taken for granted that the high technology must be purchased from the technically advanced countries. Without the widespread use of high technology in the so-called advanced countries, there would be less demand for oil and therefore less wealth in the oil-producing countries. Because the new found wealth is based on the need of the technically advanced countries for more and more petroleum to feed their cars, lorries, buses, factories, and power stations, high technology is seen as the necessary philosophical basis of all modern problem solving—the way to improve and extend the operations of the building industry—regardless of the possible unsuitability of the final product from the user's point of view.

The simple fact that 6in of reinforced concrete is the worst possible building skin to put between a human being and the external environment of the hot, dry regions is usually brushed aside (if it is considered at all) with the assertion that this can always be compensated for by lining the shell with insulation, although no experiments are made to test the validity of this assumption. The economic assumption that building on a large scale is, by definition, bound to be cheaper per unit than smaller scale operations using traditional methods has also been found to be untrue when total costs are taken into consideration. In addition, it is very difficult for the designer using mass-produced buildings to avoid creating a dull, repetitive, and inhuman environment. Moreover, many of these buildings have been found to have very high maintenance costs because of the difficulty of achieving proper air- and watertight junctions between the assembled elements. These difficulties would be accentuated in arid zones where there is a large diurnal range of air temperature and dust-bearing winds.

Modern

Building design

The inner suburbs of every urban centre in the arid world now contain square mile after square mile of boxes made from reinforced concrete and solid burnt brick, built solely for structural stability and outward appearance and designed by builders who never thought of the environmental consequences to the occupants. In the more affluent suburbs, the problem is ostensibly solved by the use of electric air conditioners in almost every window. These noisy, rusting, ugly boxes are greedy in their consumption of electricity and are already becoming too expensive even for the middle class. In some cities, the main electrical supply is not reliable, and the occupants of even these air-conditioned buildings must suffer during periods of electrical failure.

The economics of building in hot, dry climates is at the moment a complex and relatively unexplored field. The integrated assessment of the cost of a variety of design solutions including estimated capital, maintenance, and running costs for each is never made. The affluent and their (very often foreign) designers operate, rather, under two assumptions. First, modern and, therefore, acceptable buildings are made from reinforced concrete and are in some international style, with occasional exceptions such as mosques which can feature pastiche elements of Islamic architecture and airports with domes and Saracenic arches (usually in reinforced concrete). Second, all modern and, therefore, acceptable buildings are air conditioned; and the owners must accept the high cost as a fact of modern life.

It would be wrong to assume that the opposite of these assumptions is true and that a reinforced concrete building with air conditioning is never acceptable. For the residences of the affluent few and some public and commercial buildings, such a solution might well be suitable, provided the design reduces the heat load to a minimum and that life is not impossible within such a structure when the system breaks down or the electrical supply fails. If these failures occur, life can be miserable.

The writer recently spent a miserable few hours between flights in an airport in the Persian Gulf during an evening when, for some unknown reason, the air conditioning was not working. The building was designed to rely totally on the mechanical system; with the system not functioning, the exposed reinforced concrete surfaces were transferring the solar heat load into the interior. Because of the fixed glazing on the interior walls, the heat was retained inside, although the temperature outside was many degrees cooler. The unfortunate waiting passengers were not allowed to sit outside because of reasons of security. The terminal building features external pastiche elements of Islamic architecture, which must have looked attractive to the client when a model of it was presented to him; but the interior looks like the standard, characterless international airport. The designers had obviously made the two assumptions and used only the outward and visible elements of Islamic architecture, not its basic philosophy which is in sympathy with the hot, dry environment.

Urban design

The recently adopted and alien nonarid pattern—a wide street with the house forming a rectangular block set in the middle of its own plot, sharing only a fence or boundary wall with its neighbour—means that all the external walls of the modern house are exposed to solar radiation. Similarly, the garden is exposed to the sun and requires an enclosing wall to protect it from the dust and marauding goats. This type of layout results in a lower density and greater distances to work and commerce. Its adoption coincides with the increasing use of the automobile and bus. We thus have all the characteristics of suburban areas in Europe and America. Unfortunately, the investment in infrastructure typical in Europe and America does not always follow. Wide spaces are allocated to roads, assuming that tarmac (macadam) surfaces for motor traffic, paving for pedestrians, and strips of public space for planting and grass will be 'built'. Even if the tarmac road is built, the rest usually does not materialize. In most metropolitan areas in the arid world, irrigation for public areas is not available; unless special precautions are taken, the goats will destroy any trees or shrubs planted. In the end, such wide spaces lack shade and are hot, dusty, and uninviting.

CRUCIAL CONCEPTS FOR BUILDING DESIGN IN THE ARID ZONE

Enclosure

The exclusion of the outside air is often desirable in hot, dry climates when it is considerably hotter than the interior air temperature and, of course, when it is dust laden.

The screen

If there are openings in the external surfaces, then we must use devices to exclude the direct penetration of solar radiation into the interior. Regardless of what shading device is used, it is easier to shade openings if the angle of the sun's rays is high relative to the horizontal. The easiest time of day is midday when the sun is directly overhead. The most difficult time of day to exclude the sun from an opening on a vertical surface is in the early morning or evening when the sun is at a very low angle *vis-à-vis* the horizontal. Consequently, a building should be oriented in such a way that all windows are placed on either the north or the south walls and that no windows at all are placed facing east or west. It is often difficult, however, in planning to maintain a square housing plan with windows only on two sides (namely the north and the south); therefore, frequently the east and west sides are shortened.

One possible way of shading is to hang vertical lattice screens. Traditionally, these were made from hardwood with great skill and sophistication. These skills are now lost, and the only suitable modern substitute would appear to be the fixed aluminium sun-shading devices of which there are a number of models on the international market. These are expensive, but they both shade the external walls and allow the occupants a reasonable amount of visual access to the outside world. In addition, they need little or no maintenance.

In some parts of the Middle East, an effective system combining the use of outward-opening louvred wooden shutters with inward-opening glazed windows has evolved. Both the shutters and the windows are fixed to the same timber frame. This has become a modern vernacular solution to the problems posed by openings in the external skin of buildings. When the exclusion of solar radiation and the circulation of air is required, the shutters are closed and the windows opened. When it is necessary to exclude the outside air, the windows are closed. After sunset, when the cool outside air is welcome, both windows and shutters are opened unless the shutters are kept closed for reasons of security or privacy. This solution, although environmentally effective, must be combined with thermally appropriate external wall and roof construction to be totally effective.

Although enclosure has environmental advantages in arid areas, there are also social and religious reasons for it. In Muslim communities, privacy is essential for the family, whether it is extended or nuclear (consisting of a married couple and their children). Hospitality to strangers and visitors is proverbial in the Islamic world; but it is not considered proper for visiting males to have contact with the women of the household, and houses are thus planned accordingly to prevent this. A form of sexual segregation is practised in household design with one zone allocated for the women and children and another for the adult males and friends. Even nomads with their movable dwellings practise such a formal segregation of the sexes. It is common in Saudi Arabia, for example, for the tent to be divided by a partition into two sections with separate entrances facing opposite directions (see **11.5**). Another example of this segregation is found in the multi-storey town houses of Old Cairo where segregation was maintained by allowing male visitors access to the ground floor reception halls only. Usually, there were several of these, the main reception hall often rising through two floors. The ladies of the house were confined to the upper floors from which they were able to view the courtyard, and from a gallery they could see into the main hall. From these vantage points they could observe but at the same time remain invisible to the males by the use of pierced screens into which were fixed small, hinged shutters.

Screens were usually also part of an elaborate cantilevered balcony made of hardwood (usually teak) and including a seat raised above the floor level upon which cushions were placed. If the screened window was in an outside wall facing the courtyard or garden, there was often a special box there with a hole in it to support a porous pot of drinking water kept cool by evaporation. This type of balcony window is known in Arab countries as a *mashrabiyya* and is found in a variety of forms from the Far East to South America.

The *mashrabiyya* allows air to circulate and at the same time prevents the ingress of direct solar radiation; it permits a certain amount of natural daylight into the interior and reduces the glare to a minimum. Some of the large examples of

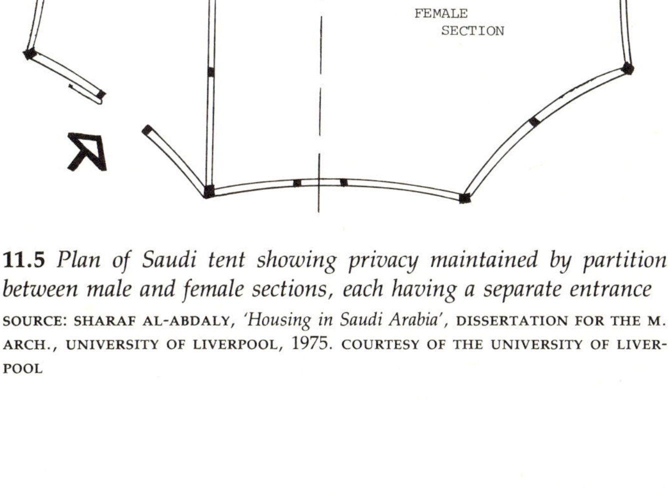

11.5 *Plan of Saudi tent showing privacy maintained by partition between male and female sections, each having a separate entrance*

SOURCE: SHARAF AL-ABDALY, *'Housing in Saudi Arabia'*, DISSERTATION FOR THE M. ARCH., UNIVERSITY OF LIVERPOOL, 1975. COURTESY OF THE UNIVERSITY OF LIVERPOOL

11.6 *Window in Beit El Siheimi, Cairo*

11.7 *Courtyard of Beit El Siheimi, Cairo, viewed from the first floor balcony*

the *mashrabiyya* use coloured glass panels set at a high level. Here the distance between the slats is often greater to allow more light into the room at a height where it is not possible to see into the interior. Figure **11.6** shows that the carpentry of a *mashrabiyya* was beautifully executed and extremely complicated, often involving the use of sophisticated geometrical patterns.

The courtyard

The courtyard, such as the one depicted in **11.7**, is basic to the traditional architecture of hot, dry areas. Its environmental advantages have already been partially described. The ancient Graeco-Roman house of modest or palatial proportions was arranged around a courtyard or a series of interrelated courts. Initially, these houses were single-storey, and later they became two or three storeys high. It was, however, only in the multi-storeyed congested tenements of later urbanized Rome and Ostia that the courtyard disappeared. Throughout its existence, the courtyard was related to water collection and gardening and therefore often contained a fountain. Similarly, throughout the arid regions from Morocco to Delhi, the courtyard was predominant. In fact, the first mosque was the courtyard of Mohammed's house at Mecca.

In the early period of Islam, there were, as **11.8** shows, no enclosing walls or doors between the courtyard and the surrounding interior of the mosque. Rather, as the Mosque of Ibn Tulun at Cairo and the mosque at Cordoba demonstrate, the entrance court is open to the sky, extends on three sides directly into the interior space below the roof of the mosque, and is then punctuated by the many regularly spaced columns and closed by an end wall in whose center is the *mihrab* or niche indicating the direction of holy Mecca. With the rapid expansion of Islam, many came to listen to the discussion of the holy men within the mosque. The attentive students would squat on the carpet around their favourite wise man who would sit on a raised stool or 'chair'. This practice gave rise to the religious school or *madrasa* which also followed the courtyard form, with many small rooms facing into a rectangular garden. This basic design spread into Europe during the time of the Crusades. In time, universities were founded, professors appointed to 'chairs', and colleges built—all on the courtyard or quadrangle principle and just as suitable to northern climates because there was shelter from wind, privacy, and security.

The use of the courtyard form can be an efficient way to reduce the amount of solar heat, provided that walls are

11.8 *Arcade at entry from courtyard to Mosque of Ibn Tulun, Cairo*

shared with other buildings on at least two sides. Many traditional examples that share two walls can be cited such as the Mzab in Algeria (**11.9**), which has been described by Etherton.[3] This is a two-storey house built around a small courtyard, and it shares walls on two sides. The flat roof is used for sleeping and drying food; consequently, it is surrounded by a wall to ensure privacy. Access from the street is on one side into the ground floor and is generally used by the males and their visitors. (The main family rooms are on second floor.) Inspection of the plans reveals that the surfaces exposed to solar radiation are kept to a minimum. Even the roof is shaded for part of the day by the walls mentioned above; since the street is narrow, the exposed vertical surface on the entrance side is shaded for much of the day by the buildings on the other side of the street.

The courtyard design is often, although not necessarily, associated with multi-storey living, acceptable in most countries to only a small minority of the population. Most families prefer, if possible, to have a home directly in contact with the ground and featuring a small garden. This preference does not preclude the use of medium-rise development, but it does mean that multi-storey housing, if used at all, should provide only a small proportion of the total housing. Multi-storey housing is often promoted because of the need to house many people in limited space, and it is often assumed that it is the only possible answer to this need. This is not necessarily so; the traditional two- or three-storey courtyard design indigenous to arid zones can be built to a relatively high density (Mzab is a case in point) because there is generally no need for through-ventilation. Where there is a need for through-ventilation during the rainy season, a wind catch can be built into the structure and screened windows provided on the street side to provide openings on both sides of the part of the house surrounding the courtyard.

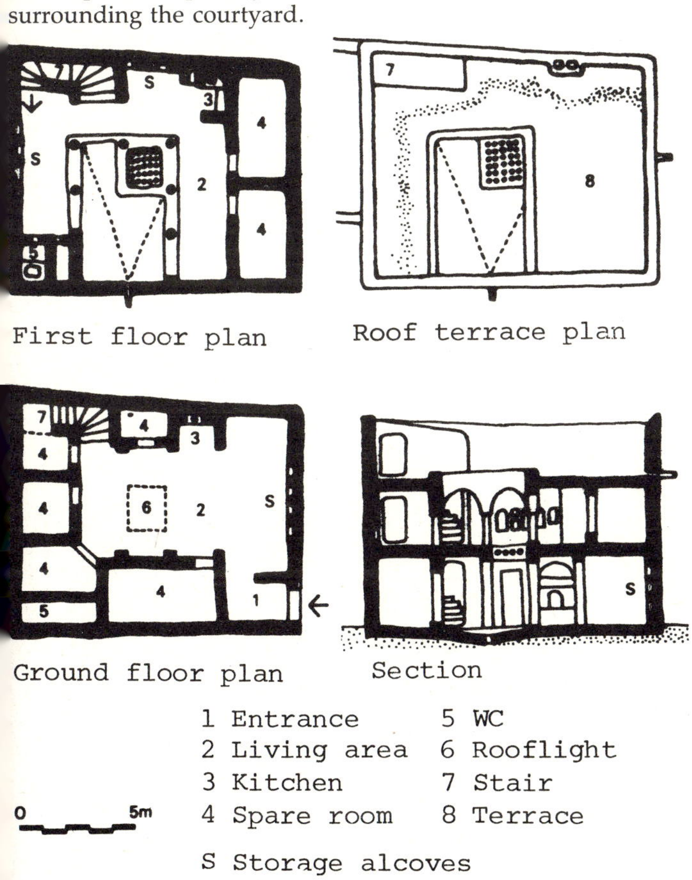

Orientation

So far, we have only touched upon orientation which depends upon the only absolutely predictable climatic fact: movement of the earth around the sun allows us to predict the angle of the sun's rays at any time and at any point on the earth's surface.

Knowing this, whatever shading devices we use, it is simpler and more economical, as we noted above, to use an east-west building orientation wherever possible, although an adjustment of up to 20° off the axis may be possible in some places. For example, solar radiation on vertical and horizontal surfaces in Khartoum (assuming a rectangular, 3—1 ratio building) does not significantly increase if the building is turned up to 20° clockwise.[4] This is because the amount of solar radiation falling on a particular surface is controlled by both its position and the atmospheric conditions at the time. Solar radiation can be reduced by the presence of water droplets and dust particles in the atmosphere. In arid areas, clear skies usually predominate except during the short rainy season when dust is blown into the atmosphere during the storms preceding the rainy season or during the day by thermal currents. The particular situation in Khartoum is probably explained by the atmospheric conditions. The dust content in the air is significantly higher in the afternoon, and this causes the asymmetrical distribution of solar radiation around solar noon. Similarly in many other arid areas, the dust content in the atmosphere is greater in the afternoon due to upward thermal currents during the day.

Exposed surfaces

Once again, traditional construction in arid regions is found to demonstrate an ingrained knowledge of the thermal performance of the walls and roofs when exposed to maximum solar radiation. Thick heavyweight construction is normal, with small wall openings to the environment usually placed at high level unless facing into an interior courtyard and/or garden. A thick heavy wall delays heat passage from the outside surface. The temperature of the inside surface of the exposed walls does not start to rise until the outside air has cooled after sunset. If it is too cool outside for comfort, then the heat stored in the structure may be used to maintain a higher and more comfortable temperature inside by simply keeping the windows closed. If, on the other hand, it becomes too warm inside, then the windows can be opened and the cooler outside air used to reduce the inside temperature to a more comfortable level. Thermal capacity of a mud wall is shown in **11.10**.

Flat roofs are common because there is so little rainfall and because it is possible to sleep outside under the clear starlit sky for many months of the year. Traditional roof construction combines timber logs, branches, matting, and a mixture of mud and straw with animal dung sometimes used as a part of the mud.[5]

However, now it is often considered necessary to use materials more permanent than the traditional ones in order to reduce the frequent maintenance needed to extend the life of many of the traditional buildings. New materials, however, are often adopted unthinkingly without an analysis of their thermal properties. A particular material may be introduced because of its durability, strength, or even modernity; but rarely is thermal performance considered when a modern substance is substituted for a traditional material.

For example, a simple, flat, solid reinforced concrete slab with a waterproof membrane is now often used for roof construction. This new roof has a longer life and, if properly built, needs little maintenance, whereas the traditional roof needs annual maintenance and often leaks if there is a rainy

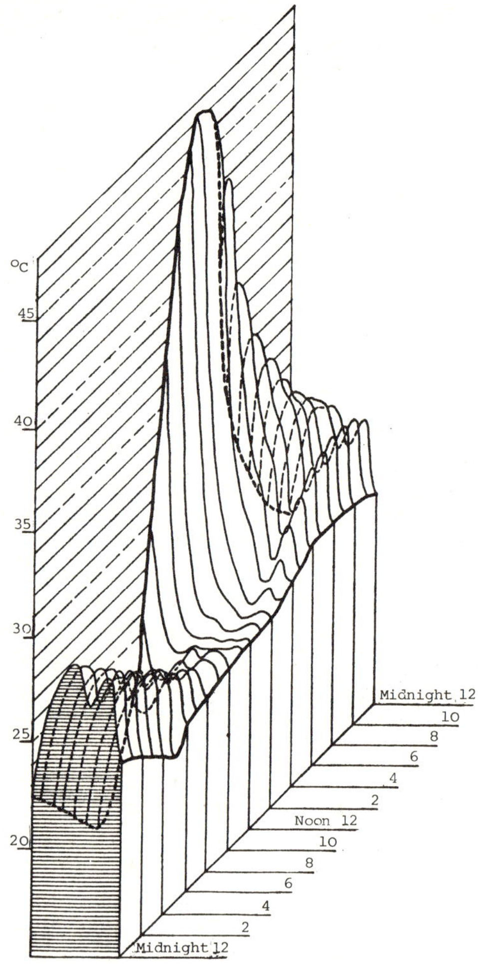

11.10 *Three-dimensional diagram showing the hourly temperature distribution in November of a solid mud wall 42 cm thick facing west in Khartoum*

SOURCE: GELLERT KUBA, 'CLIMATIC EFFECT ON BUILDINGS IN HOT ARID AREAS', DISSERTATION, UNIVERSITY OF KHARTOUM, 1970, p. 169

season. Thermally, however, the traditional roof is much superior, and life beneath the reinforced concrete slab is made tolerable only by air conditioning or construction of a secondary shade roof using either galvanized iron sheets or bamboo.

The roof

The largest amount of solar heat is, as **11.11** shows, received by a horizontal surface. In the case of a cube, the flat horizontal surface (the roof) may receive only 20 per cent less than the total of all the vertical surfaces. The thermal performance of roofs is, therefore, of vital importance; the twin aims of minimum heat transfer and maximum delay in transfer to the interior should have priority in design. Experiments with a model building in Khartoum have demonstrated that these goals can be achieved by using external insulation and shading and by increasing the thermal capacity of the roof.[6] Ten different types of roof were tested in this investigation.

It was found from temperature readings recorded during three days in mid-May 1974 (the time of year when the highest air temperatures are normally recorded) that the coolest internal air temperatures during the day were recorded below a roof with a top surface of roofing felt (two layers painted white). Under this were 50 mm of expanded polystyrene on a reinforced concrete slab 100 mm thick. The maximum shade air temperature was 44.3°C and the minimum was 28.4°, giving a diurnal range of 15.9°. The corresponding maximum internal temperature was 41.0°C (3.3°C cooler), and the corresponding minimum temperature was 37.0°C (8.6° warmer). The interior temperature was thus cooler at the hottest time of the day (about 3 pm) and considerably warmer at dawn, maintaining a diurnal range of only 4°C. The windows were kept closed throughout the day and night for this particular test.

11.11 *Direct yearly mean solar irradiation on vertical and horizontal surfaces normal to an east-west direction at Khartoum*

SOURCE: GELLERT KUBA, 'CLIMATIC EFFECT ON BUILDINGS IN HOT ARID AREAS', DISSERTATION, UNIVERSITY OF KHARTOUM, 1970, FIG. 6 PLACED OPPOSITE p. 10 OF APPENDIX 3

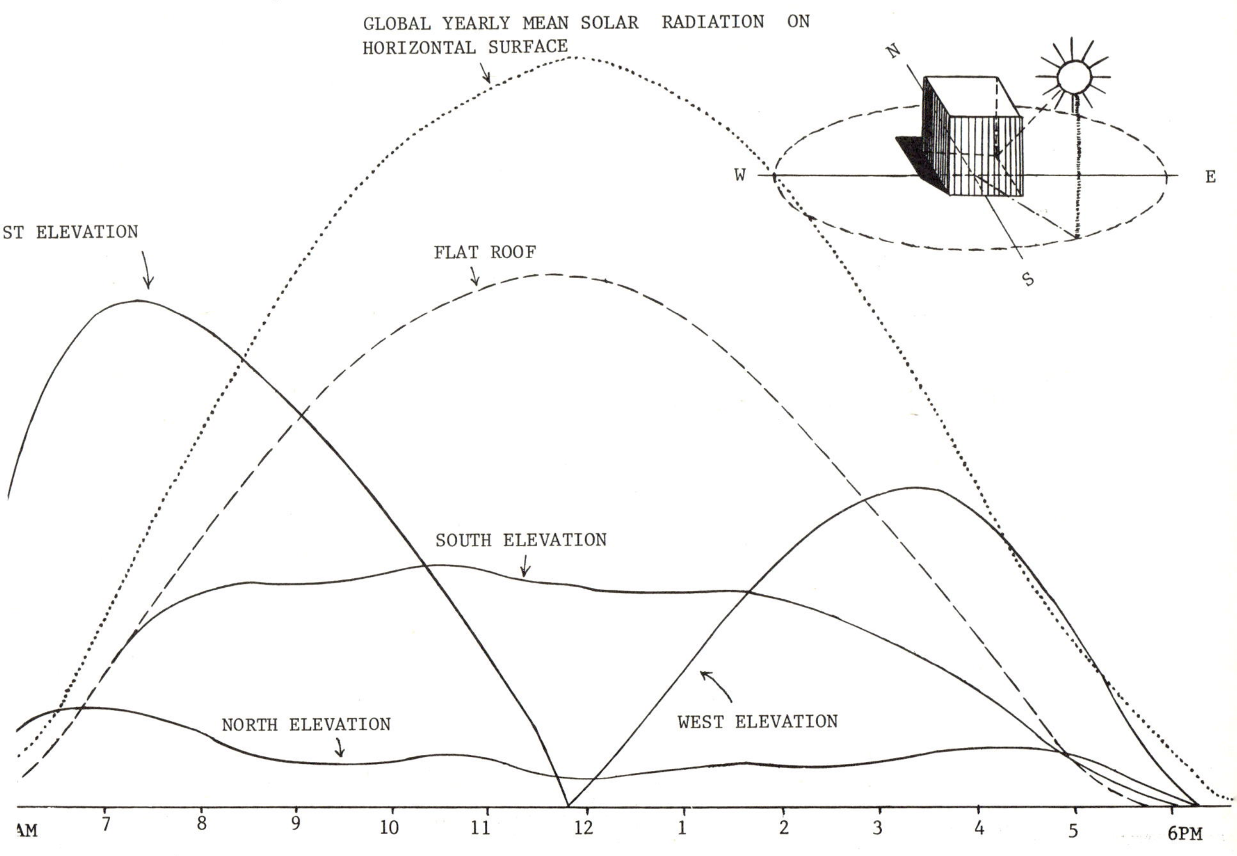

Almost as good results were obtained with a roof of corrugated galvanized iron sheeting over a ventilated air space and a reinforced concrete slab 100 mm thick. This roof permitted a diurnal range of 6°C with a maximum internal temperature of 41.4°C and a minimum of 35.4°C. The worst result was obtained using a reinforced concrete slab 100 mm thick; it permitted a diurnal internal temperature range of 13.2°C and a maximum internal temperature of 47.4°C, 3.1°C higher than the maximum external shade air temperature.

Although the best results were obtained using expanded polystyrene as insulation, it is necessary to issue a warning about the use of this material in such a situation. There is sufficient evidence from the use of polystyrene in situations where it may get hot to conclude that its life-span is limited. In the Khartoum experiment, there were only two layers of felt protecting it. The underside of the felt will of course become very hot during the middle of the day and transfer this heat to the top layer of polystyrene below. It is likely that this will, with the passage of time, cause the polystyrene to deteriorate.

Another test was made using the same Khartoum building to investigate the effectiveness of painting the horizontal surface of a traditional mud roof with lime wash. In many countries in arid regions, walls and roofs are painted annually. It was found that the difference between the internal air temperatures beneath two roofs of the same traditional construction—one left in the natural colour of finished mud and the other limewashed—was 10°C. The dust in the atmosphere, however, soon dulls the brightness of the limewash, and its effectiveness in reducing the amount of solar radiation absorbed is consequently reduced. This is why frequent reapplication is necessary to attain the best results.

Walls

If the exposed solid walls are too thin, there will also be problems. The delay in heat transfer will be too short, and the interior temperature will rise too high too soon. The interior will heat up in the afternoon while the outside air is still far too hot to bring any relief. This is frequently the case with 'modern' walls constructed of burnt brick or poured concrete.

The experiments with different types of roof construction confirm the research into the thermal performance of the traditional mud walls mentioned earlier. Such research leads us to the theoretical conclusion that a thermally ideal wall would have an outer leaf which absorbs small amounts of solar radiation, resists the inward flow of heat, and has a low thermal capacity, and an inner leaf which cools rapidly and has a large thermal capacity. Furthermore, the desirability of both mass and thermal capacity, provided the surface under consideration is not directly exposed to solar radiation, would suggest that, as far as possible, internal partition walls should be heavier than is customary in modern buildings and used as load-bearing elements in the structure. A cross-wall system of construction would thus be particularly suitable in arid climates and also would help to improve acoustic insulation between neighbouring rooms.

Some designers have attempted to ameliorate the problem of interior heat gain and avoid dependence on air conditioning or other expensive mechanical means of cooling by using cavity walls, originally developed in colder climates to improve insulation and resistance to dampness and driving rain. Experiments made in Khartoum with a special test building showed that a cavity wall performed no better thermally than a solid wall of the same total thickness.[7] Both leaves of the experimental walls were made of solid mud, and it was found that the surface temperature of the inside face of the outside leaf rose so high that the heat was radiated across the cavity to the inside leaf. Experiments with sealing the cavity and ventilating it to either outside or inside air did not significantly modify the performance of the cavity wall.

The experimental cavity wall was made, as most modern walls are *not*, of mud. Most are built using reinforced concrete. It is easy to criticize the use of reinforced concrete but difficult to offer alternative materials. In some areas, good building stones are available; but these are very dense and must, therefore, be used in some thickness to obtain an adequate thermal lag, thereby adding to the building's cost. Where stone is not available, burnt bricks are often used for load-bearing walls. Again, unless an adequate thickness is obtained, the thermal lag will be too short, passing heat to the interior of the building before the outside air has begun to cool. Both stone and brick would be suitable, however, if they were shaded or satisfactorily insulated externally; but it is difficult to suggest economic ways of achieving these goals.

In traditional construction, the external insulation is obtained by adding a coat of mud mixed with straw and animal dung. This adheres well to a wall made from native mud or sun-dried brick but would not adhere to burnt bricks or stone. Moreover, this traditional rendering is now becoming very expensive since animal dung is now more in demand as a fertilizer. It would seem that a sprayed cement-based rendering over a straw or fibre base might well provide suitable insulation, but the writer knows of no attempt to create such a material.

The most attractive material, however for wall (and roof) construction would in itself combine insulative value, thermal capacity, and load-bearing strength. There are many surplus fibres available in agricultural areas where sugar and grains are produced by irrigation. These fibres could be mixed with lightweight concrete and reinforced with steel where necessary to make wall panels (and roof slabs), but there would still remain the need to develop a suitable method of waterproofing and to design joints that would stand the strains caused by a wide diurnal range of temperature.

Besides the faulty construction of external walls and roofs, many modern house and apartment designs use re-entrant and projecting forms for aesthetic effect. This, however, increases the amount of external surface exposed to solar radiation and consequently the total heat load of the building. It follows from an analysis of this fault in modern designing, therefore, that an important planning principle is to reduce the exposed surface area of a building to the minimum. The logical extension of this principle would be an underground cave, and from a purely thermal perspective this would be a correct conclusion. But there are obvious problems of convenience, economics, lighting, and ventilation. Assuming then the need to build above ground but without undue exposed surface, the best theoretical solutions are a cylinder, a cube, or at least a building that is square in plan.

Cooling

If suitable principles are followed in the design of the form and fabric of a building, the resulting internal environment may be such that it is not considered necessary to use any mechanical means to improve it further. Even if it is found desirable to use some form of air conditioning or cooling, the heat load would have been reduced to the minimum and significant savings will result in the capital and running costs of the mechanism installed. That is, with intelligent design of the form and fabric, many buildings which would otherwise be uninhabitable at certain times without full air conditioning may now be reasonably comfortable with a cheaper mechanism (eg. fans or air coolers). Other buildings, particularly those with interior spaces accommodating large numbers of closely-packed people (eg, lecture halls and theatres) or machinery, would

11.12 *Department of Architecture building at the University of Khartoum (under construction). Note installation of air coolers and fixed aluminium louvres used as a combined screen and sun-shade*

probably need some form of full air conditioning because a considerable amount of heat is unavoidably generated in these circumstances. Now because of the reduced heat load, such buildings would require less expenditure on the refrigeration plant—the most costly part of the plant to buy and run.

Mechanical methods of cooling the air inside arid-zone buildings are not new, although 'environmental engineering' is as yet relatively unexplored. The use of fans operated by hand (usually a servant's hand) was popular many centuries ago in India and other countries. Honoria Lawrence, wife of the first British ruler of the Punjab, wrote to her son at home in England in 1851 describing the new house into which she and her husband had just moved (later to become Government House, Lahore).[8] In each of the principal rooms there was a fireplace and a flue. The fireplace was necessary for heating during winter evenings, but in the summer it was used to accommodate a device known as 'thermantidote' in which a fan, turned by a man or bullock, forced air through wet matting. The air was drawn down the flue from the outside and cooled by evaporation as it passed through the matting. This was an improvement on a traditional cooling device used in a variety of forms in Arab countries, Iran, and Pakistan. There, a vertical built-in flue opened above the roof in order to catch the prevailing cool wind and direct it into the interior of the house. This is known as a *malkaf* in Arabic and is called a *badgir* in Iran. In some cases the flue went down into the lowest room where a porous pot of water placed at the bottom of the flue cooled the flow of air by evaporation before it entered the structure. In some areas, particularly the Persian Gulf where the wind passed over a body of water and gained high relative humidity, the porous pot could be omitted. Depending on the direction of suitable cooling winds, these wind catches can be found in versions with openings on one to three sides.

In many arid countries, a modern version of the 'thermantidote' is used. It is usually known as an 'air cooler' to distinguish it from an air conditioner. The air cooler, of course, operates on the evaporation principle and uses no compressor for refrigeration. It consists of a variable speed electric fan which blows air from outside the building through a filter of specially treated wood shavings kept moist by a spray operated by an electric pump. The apparatus can deliver the cooled air directly into the building; an adjustable louvred grating gives the resident some control over the direction of the flow or, alternatively, a limited length of ducting may be used. The air cooler is a useful device for use in the home, office, and school, especially during the hottest part of the day. It is much cheaper to buy and run than the standard 1.5–2 hp basic air conditioning unit. In larger spaces, however, the required air cooling units become large and unwieldy; the ducts needed are bulky and the fans noisy. Full air conditioning is probably more suitable for these large spaces. In arid areas where there is a rainy season, the pump operating the spray in the air cooler unit may be switched off, allowing the fan alone to operate.

CONCLUSION: THE NEED FOR RESEARCH

It would be interesting to know the precise thermal performance of a variety of housing designs through an average year; unfortunately, although appropriate portable instruments to record all the necessary data are available at a reasonable cost, very little research of this kind has yet been carried out. That this is so is unbelievable, considering the huge investment now being made in construction in so many of the oil-rich states in arid regions. For a very small percentage of this investment, a comprehensive research programme could be mounted that would be able to establish with certainty the most suitable forms, materials, and mechanical cooling systems for every building type needed.

The research already done seems to confirm the superior environmental consequences of using traditional methods, but more research and analysis of the on-site thermal performance of buildings is needed to give the planner a truly comprehensive picture.[9] Regrettably, much of the current research related to conditions in the arid zones is undertaken in countries with very different climates, and this research relies heavily on electronic analogues and prediction techniques. Wherever they are checked by thorough field observation, the two sets of results do not agree very well. This strongly indicates the need for further investigation of heat transfer and other topics under the actual conditions of hot, dry climates.

Similarly, it would appear that very little analysis has been made of the performance and operating costs of air conditioning and cooling systems in everyday use in hot, dry regions. The frequency of breakdowns and the very high operating costs should have provided the motive for further research and development of improved apparatus. The type of air cooler presently used is a relatively crude piece of engineering and is both ugly and bulky. It is also, as **11.12** shows, necessary to use screening to make these coolers architecturally acceptable. It should be possible to develop a smaller and cheaper device if more ingenuity and attention were given to the problem.

Housing design in general is likely to be the largest and most difficult task facing the building industry in the future. Thus far, there has been a lack of sufficient analytical thinking. It is particularly important that such analysis begins, now that traditional solutions are being abandoned in most countries and, in many cases, modern techniques developed in other parts of the world are being adopted without due consideration of their suitability. Rational analysis of the technical and thermal problems would probably reveal better solutions. Rational thinking alone, however, is not enough; for housing, in common with all forms of architecture, expresses cultural and emotional needs which cannot be totally satisfied by a technically and thermally correct solution. The satisfaction of cultural and emotional needs is the most elusive and difficult task for the architect, and there can be no handbook with clear and reliable instructions to guide him. An obsession with the image of international modernity or traditional techniques and motifs or the scientific analysis of environmental problems will not provide the solution. Awareness of the environment and of the economics and social history of the area in which he is working, and love for the people who will use his buildings together with the qualities of modesty and sensitivity, would seem to be the essential attributes for a designer facing this seemingly impossible task.

NOTES

[1] Miles Danby, 'Buildings and the Environment', *Overseas Building Notes*, no. 165 (December 1976), pp. 1–8.

[2] M. B. Al Adhami, 'A Comprehensive Approach to the Study of Housing Sector Iraq', dissertation, University of Nottingham, 1975.

[3] David Etherton, 'Algerian Oases', in *Shelter in Africa* ed. Paul Oliver (London: Barrie and Jenkins, 1971), pp. 172, 189.

[4] Gellert Kuba, 'Climatic Effect on Buildings in Hot Arid Areas', dissertation, University of Khartoum, 1970.

[5] It is self-evident that the hot, dry regions do not produce timber in any significant amount. Therefore, timber must be imported when needed in quantity for building. While balance of payments problems do not trouble some countries in these regions, most of them must aim to reduce the amount of imported materials to the minimum. Timber is used in building for roofing, windows, doors, shutters, and lattice work; and it is difficult to find domestically produced substitutes for these purposes except in countries that have their own steel plants. Steel can be successfully used for windows and door frames and roof structural members, but it is not suitable for shutters and doors because its thermal properties are inferior to those of timber. It thus seems likely that imported timber will continue to be used for these purposes.

[6] Younis Abdalla Mukhtar, 'An Investigation into the Thermal Performance of Roofs in Hot Dry Climates', dissertation, University of Newcastle upon Tyne, 1976.

[7] Kuba.

[8] Mark Bence-Jones, *Palaces of the Raj* (London: George Allen and Unwin, 1973), p. 167.

[9] Research into the thermal performance of buildings constructed using traditional methods and materials is rare; in the countries where some research has been made, eg the Sudan, the total picture is still incomplete. When further research is carried out, it is likely that scientific evidence would support the traditional ways of design decision-making that could be applied with the use of modern materials and improved traditonal techniques using natural resources in the contemporary situation. The cost of the portable apparatus needed for field observation is modest.

SUGGESTED READING

Danby, Miles. *Grammar of Architectural Design*. London: Oxford University Press, 1963.

Koenigsberger, O. H. *Manual of Tropical Housing and Building*, Part I, Climatic Design. London: Longman, 1974.

Oliver, P. ed. *Shelter and Society*. London: Barrie and Rockliff, 1969.

Oliver, P. ed. *Shelter in Africa*. London: Barrie and Jenkins, 1971.

Saini, Balwant Singh. *Building Environment: An Illustrated Analysis of Problems in Hot-Dry Lands*. Sydney: Angus and Robertson, 1973.

12 Housing and community development in Australia's arid zone

BALWANT SINGH SAINI

The continent of Australia lies between the latitudes 10° south and 40° south and is geographically, as well as climatically, very similar to North Africa. It is located, as most of the world's desert regions (including the Simpson, the Gibson, the Sahara, the Kalahari, and the desert area of Arabia) are, between the latitudes 15° and 25° either north or south of the equator. The average rainfall in Australia is only 17in per year. This compares with 29in for the United States, its nearest comparable country in size, and an average of 26in for all the land areas of the world. This low rainfall makes Australia the driest continent.

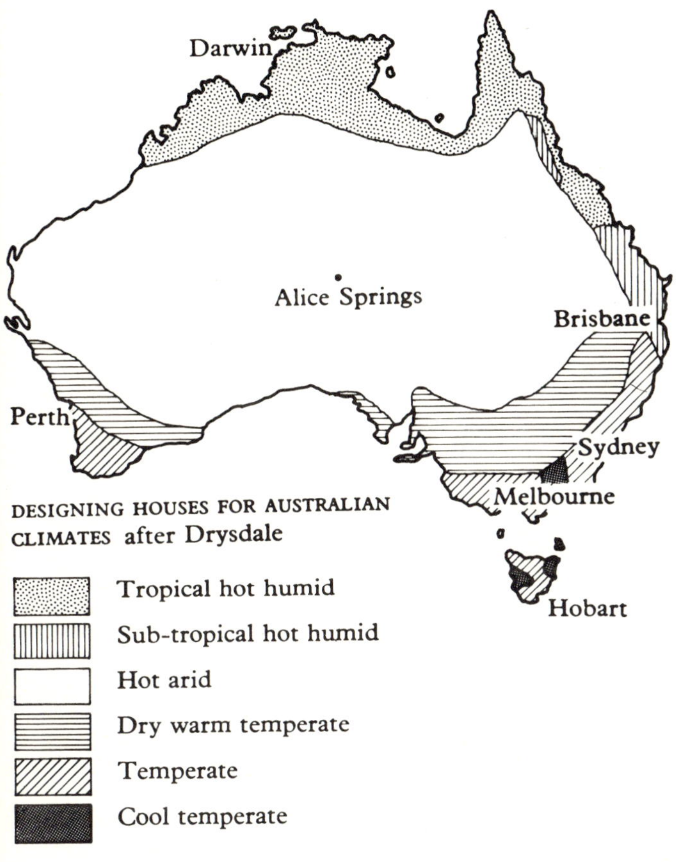

12.1 *Climatic map of Australia, indicating that nearly 70 per cent of the land mass lies in the hot, arid zone*
DRAWN BY THE AUTHOR

INTRODUCTION TO THE ARID AUSTRALIA

The hot, dry region which constitutes approximately 70 per cent of the land mass of Australia is also the least attractive. Much of it is sandy hill country where the rainfall measures less than 10in per year and where there is little or no vegetation. Until the last decade or so, when signs of growth and change began to appear in the area, Australians have shown little interest in settling there. As a result, most of the continent's 15 million people are concentrated in the southern temperate areas as shown in **12.1**. The hot, dry interior is sparsely populated, except for scattered pastoral families, a few single-industry towns (such as mining towns), and small commercial and administrative settlements whose livelihood depends upon the other activities.

The concentration of Australia's population along the coastal fringe can be partially explained by Australia's colonial ties with Britain and partially by difficulties created by distances and the available means of transportation. For many years, at least until the advent of commercial aviation, the major seaports provided the only necessary direct link with the mother country. Having thus established themselves, these seaport cities then continued to strengthen their positions, first as administrative and financial centres and later as foci of state-controlled railway and road systems.

The impact of distance and availability of transportation on Australia's development has been admirably researched by Blainey and others.[1] In the absence any of inland waterways, the only available means of transportation—also the cheapest—was the small coastal ship. According to Blainey, there was, therefore, little alternative but to plant other early settlements on the coast or on the banks of the navigable rivers. To fully exploit inland and regional resources, improved transportation has now made it possible for people to get to the distant countryside but, at the same time, has made it necessary for them to develop self-contained and independent communities there. The accepted policy of developing secondary industries tied to local markets, however, accelerated a process which turned the seaport cities into bigger markets, and the process has continued, ensuring their further growth.

Initially, however, when a settlement was established inland, it had to rely upon the slow and costly bullock dray; apart from cattle which could be moved from place to place on the hoof, the only product which could afford the cost of bullock transportation was wool. In addition, wool did not deteriorate during its journey from the inland to the coast and then on to Europe. As a result, Australia became a great wool growing and exporting country. When gold (a high-priced product) was found, it could also bear the cost of transportation by bullock dray and sailing ships. Other goods had to wait for the development of railways, steamships, motor vehicles, and aviation before they could find export markets. In the case of some goods, they also awaited the widespread use of refrigeration.

The early pioneers who settled in the area thus established a frontier type of economy where risks were high and progress was erratic. This pattern continued until the end of World War II. Today, new discoveries of minerals and the greater flow of capital have altered this picture dramatically. We can see the older townships such as Mt Isa growing rapidly with new discoveries of copper. There are copper and gold in Tennant Creek, nickel in Kalgoorlie, and enormously rich deposits of iron ore in northwest Australia. This last resource is largely concentrated in the Pilbara region where whole new railway systems, deep water ports, and industrial enterprises have, as a result, been established.

All of these developments have generated the establishment of new settlements, many of which have had to be located in very remote and hostile environments. To architects, builders, and planners, these developments have posed a challenge to find new techniques and develop expertise to provide comfortable living conditions for people who are essentially accustomed to cold winters and reliable rainfall.

Although many of the problems of living in the Australian desert are connected to the necessary psychological adaptation by people of European origin, the development of indigenous building forms has been largely dictated simply by the need to minimize physical discomfort in a hot, dry environment, an environment which presents high daytime temperatures and low humidity, surroundings devoid of vegetation, pests (mainly flies), and dust and sand. Although not a health hazard, dust and sand are things that generally lower the morale of the people, particularly the housewife who is forced to clean the home and the clothes frequently; furthermore, domestic equipment and gadgets become clogged with dust, necessitating frequent repairs.

HOUSING DESIGN

Colonial

The traditional homesteads which were built by the pioneers over the last 150 years or so (and which are still being constructed by people engaged in the pastoral industry) probably owe their origin to British colonial architecture of the 19th and 20th centuries. Examples can be seen in countries as far apart as the Indian subcontinent, Malaysia, and the West Indies. A typical colonial bungalow type, depicted in **12.2**, had a series of spacious rooms with very high ceilings, ventilators, and access from all four sides. The important rooms were surrounded by a deep veranda, part of which was enclosed in order to house storage and washing facilities. A pantry adjoining the main dining room was further connected by a covered way to the kitchen built at some distance from the main building. The deep veranda and the few secondary rooms surrounding the main areas provided valuable shade and insulated the walls from external heat. Some of the early colonial structures in India and Pakistan were pleasantly cool in summer because they had a thatched roof topped off by a layer of mud. This practice was, however, later discontinued because of high maintenance costs. The roofs of the later bungalows in the hot, dry parts of these countries were mainly flat with tiles laid on joists and finished with mud and then another layer of tiles.

This colonial heritage, when transplanted onto the Australian soil, was interestingly transformed. The plans, as **12.3** shows, were basically similar; but because of the absence of cheap domestic labour, the kitchen lost its role as an outbuilding and moved into the house proper where it occupied an important position alongside the dining room and pantry. The veranda, which was perhaps one of the original design's features, remained; but very often it was enclosed by fly screens to keep out flies and mosquitoes as **12.4** reveals. These adapted verandas, although they shaded and protected the inner walls, also reduced the amount of natural light entering into the interior. As a solution to living in the hot, dry areas, this traditional homestead, though adequate, was far from ideal.

Later colonial

During the earliest period of inland penetration by Europeans, a number of 'wattle-and-daub' earth and stone structures were constructed. In fact, it is still possible to see the remains of many of these primitive structures in some parts of the Australian desert.[2] With the advent of better transportation, it

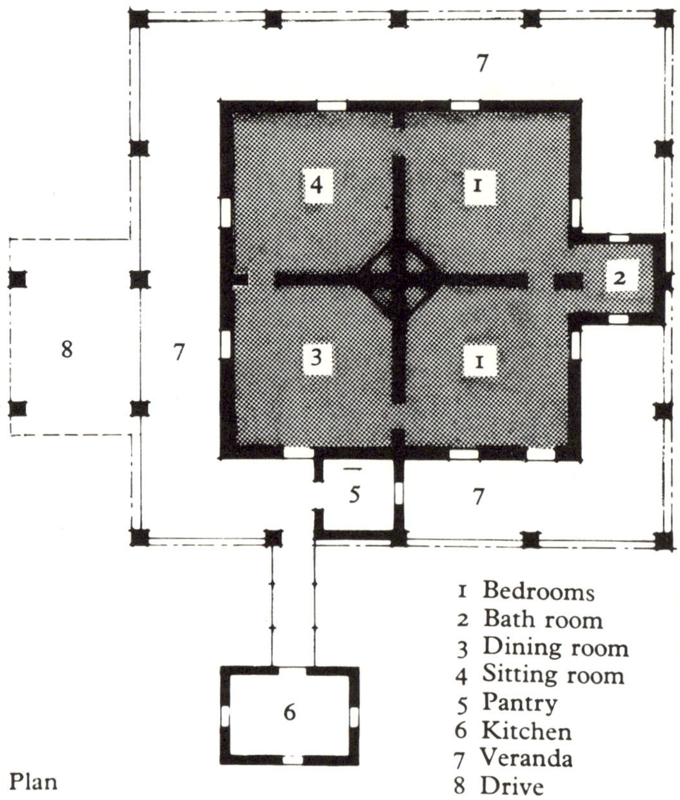

12.2 *Floor plan of Indian colonial bungalow*
DRAWN BY THE AUTHOR

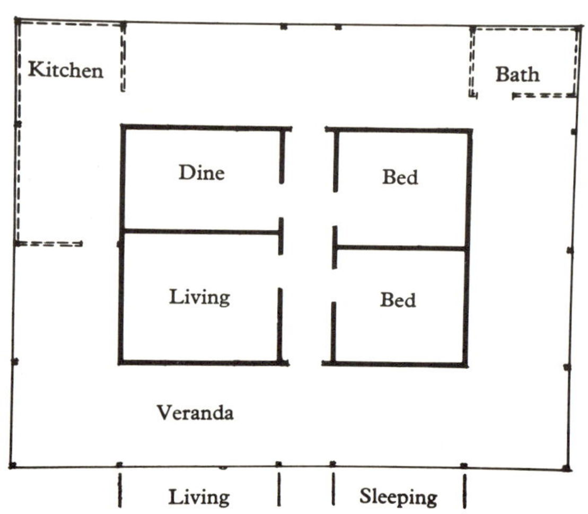

12.3 *Floor plan of Australian colonial bungalow*
DRAWN BY THE AUTHOR

12.4 *Typical Australian colonial house, characterized by its compactness and its deep verandas*

12.5 *Totally air-conditioned prefabricated house, mining township of Mt Goldsworthy, northwestern Australia*

became possible to bring cheaper lightweight building materials to sites in remote areas. Then, the earlier costly and time-consuming practice of masonry construction was slowly abandoned in favour of the building of timber frame structures. Thus, unlike colonial bungalows in India and Pakistan which continued to be built from masonry, these Australian homesteads were constructed using the lightweight materials, which could be easily and cheaply carried over long distances. Because of the lack of sufficient heat storage capacity of such materials as timber boards, asbestos, and galvanized iron sheets, they are definitely disadvantageous in a hot, dry environment. Furthermore, the roofs and often the walls were lined with corrugated iron sheets, and the ceilings were usually uninsulated. The net result of these late colonial techniques is that, during extreme periods of heat, the occupants are often forced to move out of these houses and stay under the trees or go into a 'spinifex cool house', an outbuilding constructed from clumps of spinifex desert grass attached to a chicken wire enclosure which is then continually sprayed with water. Being in the structure is the same as sitting inside a giant evaporative cooler.

Modern

Many of the houses built during the last few decades were devoid of even the few good things which these pioneer houses had. To reduce costs, for instance, the verandas were often omitted, windows were far too large, and there was no alternative (shaded and protected) outdoor living space provided. The layout of many of these modern houses shows a complete departure from the traditional homestead concept. In the process of departing, planners have not only lost the positive qualities of the old homesteads, but have also failed to learn anything from the new approaches taken in many new settlements in other parts of the world.

Contemporary—in combat with the environment

During the last few years, in order to attract the work force to remote areas, the mining industry has paid increasing attention to improving building design and settlement planning in the Australian desert. Often, total air conditioning, as in the case of the house depicted in **12.5**, was offered as the main attraction. However, this approach presented many problems, particularly the additional cost of maintenance and repairs in remote locations. Questions are now being asked whether total air conditioning, which really means turning your back on nature, is the ultimate way to live in hot, dry regions. Social scientists are investigating the psychological effects of living in a closed and confined space for long periods (especially for the housewife and the children) and the consequent tensions which occur within the same family. Air conditioning is furthermore a negative approach which denies the enjoyment of the pleasant aspects of hot, dry climates, particularly the cool evenings and nights when it should be possible to spend a great deal of time out of doors.

One alternative lies in partial air treatments, such as the use of an attic fan to remove the accumulated hot air during the nights.[3] This fan is fairly standard equipment in many American homes, but somehow it has never caught on in Australia. Another alternative would be a blower fan used in conjunction with an evaporative water cooler.[4] This method has been very effectively used in Australian desert townships. It is probably one of the best known methods for cooling which actually changes the quality of the air. Its major advantage is that it is not necessary to make the building completely airtight and closed off from the outside. Parts of Australia's hot, dry region are suitable for this method of cooling. In remote locations an evaporative cooler has a further advantage: its simple equipment (motor, pump and fan) can be easily assembled and maintained by relatively inexperienced workmen.

In Australia, the refrigerant normally used in air conditioners has proved to be fairly expensive. It has been possible to achieve cooling by evaporation at about one-fifth of the actual costs entailed in cooling air by air conditioning. However, in some locations evaporative cooling has presented problems as far as having a supply of clean water is concerned; bore water has a tendency to block the filter pad by depositing salts, and the waters drawn from streams or wells also eventually block the pad due to the growth of algae. River water is, however, improved if it is allowed to settle. It is possible to achieve this by installing 200–300 gal tanks, one of which could be used while the other is filled and allowed to sit. Such tanks would also require sludge cocks to permit periodic cleaning.

Another possible way of counteracting the climate is provided by the mechanisms of heat exchange. One of the simplest methods tested by Commonwealth Scientific and Industrial Research Organization (CSIRO) scientists involves drawing outside air over a cool rock bed and then feeding it through ducts into a building.[5] The cool night air in the desert will chill the stones, and they will maintain sufficiently low temperatures through most of the next day and cool the hot day air on its journey into the building at the times when cooling is most needed. In practice, this method has proved somewhat more expensive than evaporative cooling; but it seems to provide a good starting point for further refinement, particularly in countries where both labour and materials are cheap. It holds considerable promise for the cooling of buildings at a low cost and without the noise associated with mechanical air treatment and evaporative coolers.

Research workers in CSIRO Division of Mechanical Engineering in Melbourne have developed several air cooling systems which combine efficient heat exchangers with evaporative cooling. These could revolutionize cooling methods in hot, dry areas, once their potential is fully realized.[6] So far, these new systems have not been commercially exploited. Laboratory tests, however, suggest that an ordinary cooler will bring a day temperature of 43°C (common in the Australian desert) down to 26°C, while the evaporative cooler with heat exchanger added could achieve as low a temperature as 19°C in similar conditions. The difference of 7° in the temperature of the air supplied to the house can be critical from the point of view of physical comfort, and its importance increases even more when the outside temperature exceeds 43°C. A final refinement combines a very effective heat exchanger with an evaporative cooler in such a way that the process does not significantly add to the moisture content of the air; but all these developments are at an experimental stage and are, therefore, unlikely to affect the market until they have proved that they can compete with existing systems.

Throughout the hot, dry parts of Australia, solar hot water systems are also now extensively used, but solar air conditioning systems are still in an experimental stage. Experiments conducted at the University of Queensland have indicated that the most suitable refrigeration unit is a lithium-bromide-in-water absorption type, designed to run on commercial fuel gas.[7] An experimental solar house was also built some years ago, and the results indicated that solar air conditioning could find a place in the desert, provided the prototype air conditioner is further refined and, of course, made available commercially. The most expensive item is, however, the collector, which at present costs something like US$120/m^2. For cooling a small area of 50 m^2 the cost of the collector alone could be beyond the means of most homeowners.

Contemporary—in sympathy with the environment

Beyond these mechanical means of altering the environment, there are two areas in which an architect and town planner could make useful contributions. There is the building structure itself, which can be designed in sympathy with climatic requirements, and second, community design, which we will discuss later. In designing a structure, the first step is to reduce the flow of heat from the outside; this is mainly achieved by creating heavy shade. If we add to this some means of reducing glare and stopping the entry of dust and sand, we are then involved with all the elements which affect design and, as we shall see later, the orientation of buildings.

Apart from direct sunlight, which can be checked by suitable sun-shading devices such as louvres, verandas, pergolas, and the like, heat enters structures either by conduction or by radiation. Heat from radiation is checked by shading, the use of reflective surface coating, and the planning of nonreflective surroundings and suitable building orientation. Heat flow by conduction is stopped to some degree by insulation and by reduction in the surface areas of exposed walls and roofs. One should thus select materials by assessing their comparative performance as insulators. Apart from insulation, the main consideration in structure design is the heat storage capacity of the building materials used for the walls and roofs. Materials such as mud, brick, or stone have high heat storage capacities: they take a long time to heat up and, once heated, take an equally long time to cool down. In hot, dry regions such materials offer a great advantage because they take a long time to absorb the heat received during the day and pass it on to the inside surfaces. This means that the building interiors stay cool during the hotter parts of the day. At night, when the outside temperature drops, the interior will, however, be too warm for comfort. To overcome this problem, people in traditional desert settlements have moved out to live and sleep in courtyards and verandas and on roofs.

COMMUNITY DESIGN

Landscape

One external factor which affects comfort in desert areas is the use of water and vegetation; they not only cool the air but also stimulate the eye.[8] Water does the trick by evaporation—a process which increases the relative humidity and decreases

12.6 *Massed dwellings of Sanaa, Yemen*
PHOTOGRAPH COURTESY UNESCO/PIERRE SCHNEIDER

the dry bulb temperature of the surrounding air. If we follow the marvellous examples of the planners of the ancient garden city of Isfahan in Iran, we can improve the microclimate as well as the macroclimate by carefully designing stretches of pools, fountains, and jets in and around buildings. Vegetation, whether trees, plants, shrubs, or merely grasses, can also help to reduce the heat load on buildings. Trees and other forms of vegetation have to preserve their leaf temperatures very much like human beings by evaporating moisture. The result is that they, like the water pools, have a great cooling effect. They also act as a sun-break and shield buildings from dust and sand; if located in an open area, vegetation can clean the air of petroleum fumes and other atmospheric impurities. To get the best results, of course, this vegetation must be carefully selected and sited where it can be most effective.[9]

Bare earth, asphalt, concrete, and other types of paving are highly absorptive and therefore become very heated during the day's exposure to the sun.[10] As a result, the air near the ground also becomes heated; when carried into the buildings, such hot air raises the temperature level there considerably. In all such cases, vegetation, particularly stretches of grass, is useful. Vegetation also reduces entry of dust and sand into buildings.

The subject of landscaping is fascinating to study in itself; and it is, of course, difficult to deal with it fully here.[11] One point, however, should be noted: the need to design, plan, and maintain open spaces in hot, dry areas should be weighed against the fact that water is short and vegetation is difficult to grow, resulting in decisions to keep such spaces down to a manageable proportion of the community being designed.

Australian mistakes

The Australian designers of buildings and communities in the hot, dry areas of the nation have followed layouts and created subdivisions more suited to temperate and cooler regions. That such an approach can create very unpleasant environments is obvious to anyone who examines some of the older Australian desert townships where row upon row of small, timber-framed houses with uninsulated iron roofs virtually sit in a dust bowl on plots of land which are far too generous. Roads, even the local ones used only for light traffic, are extremely wide; because they are unsealed and have no grassed areas, they are a source of the dust and sand that plagues buildings in these communities. Pedestrian traffic is seldom separated from vehicular traffic in these poorly designed towns. There is furthermore no attempt to shade either traffic-way from the sun. Such unimaginative layouts make it difficult to orient houses in such a way as to control the sun's effects. These community designs show no signs of having been adapted to meet the stresses of hot, dry environments as is normally done by traditional desert civilizations.

Foreign models

If we consider towns like Sanaa in Yemen (**12.6**), we find that designers tended to pile individual dwellings together in massive forms and thereby reduce their exposed surface to the absolute minimum.[12] Thick walls of the designed houses and courtyards delay thermal effects and also provide shade to living areas. The houses themselves look inward with their backs turned to the hostile external landscape. This inwardly-directed, massed attitude of traditional desert communities is, as we have seen, in remarkable contrast to that of the Australians, who, temperamentally, as well as because of their early European beginnings, prefer large open spaces around their shelters. Such exposure to the surroundings of course creates drastic problems in a hot, dry climate. In response to this problematic Australian preference, an important question for architects there is how to sell the idea of compactly grouped housing.

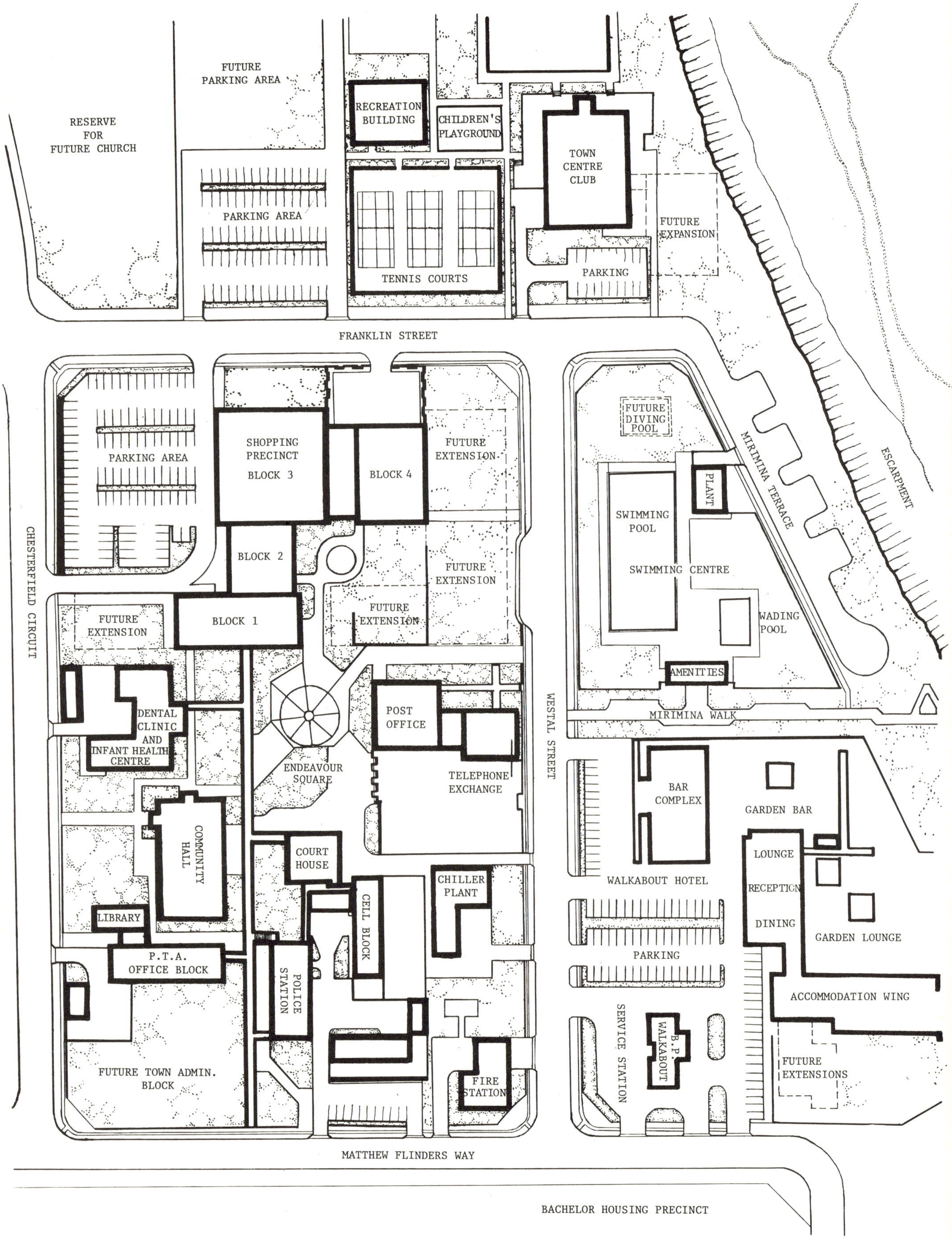

12.7 *Plan of Nhulunbuy town centre*

COURTESY OF THE ARCHITECT, J. AGIUS

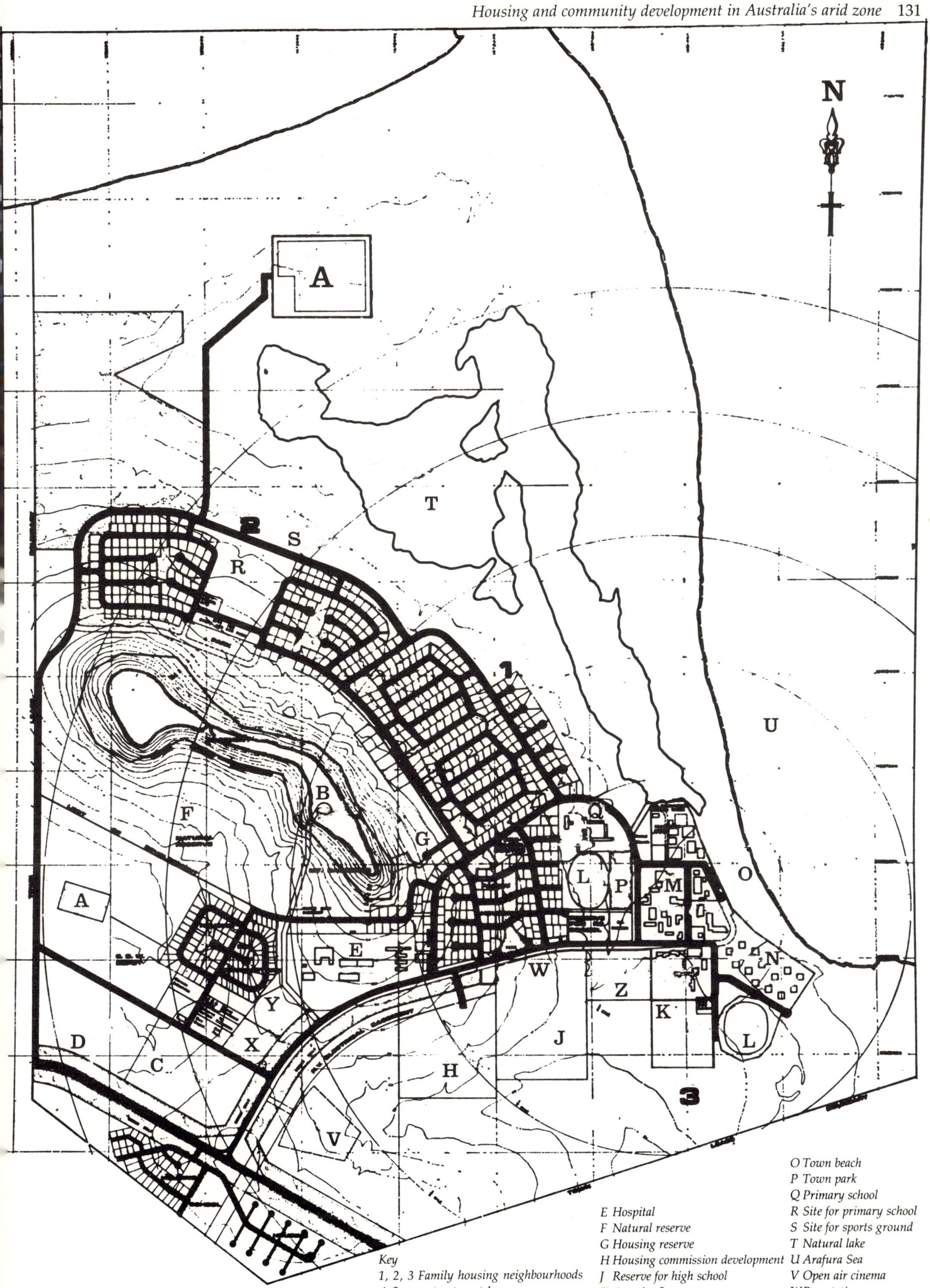

Key
1, 2, 3 Family housing neighbourhoods
A Sewage treatment lagoons
B Town reservoir
C Bachelor accommodation
D Extension to bachelor accommodation
E Hospital
F Natural reserve
G Housing reserve
H Housing commission development
J Reserve for high school
K Area for flats
L Sports ovals
M Town centre
N Family flats
O Town beach
P Town park
Q Primary school
R Site for primary school
S Site for sports ground
T Natural lake
U Arafura Sea
V Open air cinema
W Bus station
X Warehouse complex
Y Nursery
Z Reserve for town centre

2.8 *Development plan of Nhulunbuy*

OURTESY OF THE ARCHITECT, J. AGIUS

12.9 *Staff housing in Nhulunbuy*

Australian advances

In central Australia in many cases, protection from the climate is partially provided by the deep, well-shaded veranda; but even there, the single exposed shelters are still denied the shared benefits which could be available through more compact layouts. Compact planning based on the courtyard concept has proved successful in most hot, dry lands; there is no reason to believe that it would not be so in an Australian setting. An early proposal prepared in the Architecture Department, University of Melbourne, under this author's direction some twelve years ago suggests that courtyard-oriented housing can be designed to meet the needs of typical Australian families in the hot, dry interior of the continent. The principles, which evolved then and are now being applied to some of the new mining townships, will ensure the same amenities as those found in houses located on individual plots of land. If the few open spaces are judiciously planned, such plans will also reduce the total area of the development and therefore reduce the cost of services. A township which combined this planning approach with high standards of technical services at a reasonable cost (by Australian standards) is Shay Gap.[13] This small mining town is located in a remote desert region of Western Australia. The buildings are grouped closely together so as to shade each other and to save on the cost of services such as reticulated refrigerants; sewage and sullage connections; high, medium, and low tensioned electrical cables; television cables; telephone lines; water pipes; irrigation systems; and fire alarm circuits. All are included in a trench enclosed in a plastic wrapping. As the architect of the town, Lawrence Howroyd, puts it, this large 'line' is a sort of umbilical cord which terminates at a concrete block outside each building. The houses are air conditioned, and all public spaces are human in scale. All the community amenities such as shops, entertainment facilities, cafes, and clubs are concentrated in one or two areas. It has thus been possible at Shay Gap to achieve an urban environment in the middle of a vast, inhospitable desert.

Although Shay Gap is a vast improvement over the earlier mining towns in Australia, the settlement, Nhulunbuy, combines the best in planning and management concepts; Nhulunbuy is located some 2000 miles north of Sydney.[14] The development timetable was two and a half years for the creation of an artificial community housing 5000–6000 people, a town with the appropriate living and associated community facilties. In order to avoid any hardships during the early period of its growth, the planners built a compact town centre, 'a little city', with sufficient allowances for its further growth (see **12.7**). A most successful solution to the planning problems faced in an arid region, Nhulunbuy established civic and commercial precincts designed for pedestrians: all the buildings, shops, coffee bars, eating, banking, medical, and commercial facilties, postal facilities, and cultural centres are linked by covered ways and centred around a town square.

The overall town scheme is, as **12.8** reveals, based on three neighbourhoods, each consisting of approximately 600 families and centred on a school. The first neighbourhood contains the central business area upon which the two other neighbourhoods depend for public, social, and recreational facilities. Essentially, the town is zoned for low-density housing (single cottages), as depicted in **12.9**, and medium-density housing (three-storey dwellings for small families and bachelors). Other areas include the town centre mentioned above; health, educational, and recreational facilities; light service industries; and open spaces including parks, gardens, and natural and green belt reserves.

The urban planners recognized very early the need to provide a permanent town where people could feel secure. There is such a feeling in Nhulunbuy because the town is solid and not made up of materials which will deteriorate with time and use. Nhulunbuy has been planned to last over 50 years (in line with the area's mineral potential); but with the possible future growth of other industries, it could well have an indefinite life span.

The close examination of external factors thus suggests that microclimatology, landscape design, and site planning are allied in several ways. The basic objective of all, of course, is to bring about good living close to nature without going to extremes that could nullify the benefits to be gained. To achieve domestic and social security by creating an urban atmosphere in sympathy with nature is a matter of making the right choices and planning delicately.

SOCIAL CONSIDERATIONS

From the discussion thus far, it is obvious that enough technical knowledge is available to guide a professional in planning and designing ways which reduce the adverse effects of the hot, arid environment on human comfort. However, evidence suggests that building new settlements also generates enormously complex problems of a psychological and social nature, problems which have had little or no attention paid to them.

In Australia we are dealing with people of two widely separated cultural backgrounds: people of European origin who have been attracted by the prospects of life in arid pastoral, mining, or administrative settlements; and aborigines and part-aborigines who have an extraordinarily complex tribal background and who traditionally play little or no part in the life of these new settlements. The Australian mining industry, for example, has reached a fairly sophisticated level of operation; being essentially capital-intensive, it requires a limited but highly skilled labour force. Because of their training and ability to perform the required tasks, Australians of European backgrounds dominate the labour force. Among them a high proportion are first generation migrants, particularly from such non-Anglo-Saxon countries as Italy and Greece.

The problems of adjustment faced by these people are not only physical acclimatization to what may best be termed an extremely harsh, hot and arid environment, but also mental adjustment to a whole new way of life. In addition, since single men constitute a large section of this labour force, there are further problems of instability; and the settlement must try to ensure conditions which will reduce the high rate of labour turnover. Likewise, many of these settlements had to be located in the heart of—or close to—what is essentially aboriginal country. The close proximity of the European population to the aboriginal community tends to create social tensions and undesirable interactions, such as the prostitution and exploitation of the native group leading to the disintegration of tribal cultures.

As mentioned earlier, it is now clearly possible to provide air conditioned housing and first class amenities such as night clubs, shopping centres, and swimming pools. But does this help the growth of the 'community'? If not, and evidence suggests that this is the case, then what constitutes a 'community'? This question leads one to ask a number of subsidiary questions such as, 'What is happiness?' 'What is "quality of life" and how does it relate to community morale, health, education, and what is often called a development of community spirit?' All these, of course, lead to questions about management. Management by whom—by the employer, by the company or the corporation, or by representatives of the people who are living there? What policies and practices are necessary to assist in the development of a stable population, happy and willing to accept the new place if they can feel territorially secure there?

In Australia, during the last decade, a number of social scientists have been concerned with the above problems. Their experience, though incomplete, points to some lessons which may be relevant to similar situations in many other parts of the world.

Remote Communities Environment Unit

Some years ago, as a result of frequent complaints about poor living conditions from residents of old and new towns in Australia's remote areas, CSIRO established a Remote Communities Environment Unit. Its main purpose was 'to demonstrate how conditions can be created in remote communities in tropical Australia such that (a) people will be attracted to these towns, and (b) residents will enjoy a quality of life as good as or better than that which would be their expectation in developed centres anywhere in Australia'.[15]

This unit has already conducted a series of exploratory attitudinal surveys in five selected settlements, Mt Newman, Dampier, Kununurra, Mt Isa, and Katherine, where residents were given the opportunity to comment on aspects of their living patterns ranging from relationships with neighbours to garbage disposal. The population surveyed was drawn from a fairly representative selection and included were long-term residents and many relatively new community members. The questions were mainly directed to family members rather than to single workers, even though in many cases, the latter formed a significant section of the town's population. This decision was based on the assumption that the stability of a town tends to improve with the proportional increase of families.

The surveys offered equal chances to all the varying age and sex groups within a family to participate. However, in order not to limit the study to specific questions formulated by social scientists, the participants were given what are known as 'incomplete statements questionnaires', such as 'Living here would be   if it was not for   ', in order to elicit information otherwise difficult to obtain. The results, particularly the answers to the 'incomplete statements questionnaires' were appropriately categorized into favourable and unfavourable comments on the existing community environment. Generally, in all the towns surveyed, the unfavourable aspects noted exceeded the sum of the favourable ones.

The detailed analysis of the results have enabled the social scientists to evaluate the public acceptability of existing community environment. They have been able to identify what improvements are needed and where greater economy is or could have been possible. It has also enabled them to compare the degrees of success of different towns, planned and developed according to different sets of planning concepts. With this information the social scientists make fairly clear and meaningful recommendations to planners, designers, and policy makers so that they are able to take appropriate action which may lead to the development of a better environment for people.

12.10 *Older mining town of Mt Isa. Housing located near mine site, making the human environment adverse*

Kambalda Study

Another significant study has dealt with human relations in the mining industry and their implications for planning new towns in isolated environments. This study is the work of the Human Relations Unit (HRU) which was set up by the Western Mining Corporation early in 1970.[16] Professor O. A. Oeser headed the unit which made some preliminary studies of the overall structure of management and then directed its efforts to a more detailed study in the township of Kambalda. Its aim was to report on attitudes such as satisfaction and dissatisfaction with the planning, housing, and living conditions generally. The only topics specially excluded were attitudes toward work and the subject of industrial relations.

The unit developed four main sets of questions which they entitled, respectively, 'Living in Kambalda', 'Teenagers of Kambalda', 'Education in Kambalda', and 'Voluntary Organizations'. The results, which are far too detailed and complex to be enumerated here, have been summarized in a series of 16 reports which have been submitted to the corporation management. However, according to Oeser, these reports point to the need to evaluate 'how people perceive their environment and themselves and how they manage the dynamics of social interaction that determine "the quality of living"'. 'This is the reason modern planning should be, and increasingly is, done by teams of experts in the engineering, the natural and the social sciences'.[17] Oeser is aware, however, of the obstacles blocking the establishment of such teams and their successful function. He therefore made important recommendations about the ways members of such teams can cooperate and thereby develop into an effective working group.

ENVIRONMENTAL CONSIDERATIONS

In this chapter the discussion has been restricted to the assessment of the environmental impact on housing and community development. It is, however, equally important to acknowledge the reverse situation, ie, the effect of building and mining activites on the natural environment. In a paper delivered at a UNESCO seminar on 'New Towns in Isolated Settings', Perry and McAlpine have stated that consideration of this effect is particularly important in a fragile desert environment where water, land, and people have to coexist in a finely balanced relationship.[18]

The older towns in Central Australia had a considerable adverse impact on their surroundings. Since mining was labour-intensive and since transportation, housing, and shops were largely uncoordinated, workers had to walk to the mine. Thus, the towns were located, as **12.10** shows, very close to the mine site. As a result, one central area was generally developed, and the environmental destruction varied according to the size of the town and the type of mining operations. Solid waste was often deposited on the extensive land which was previously used for grazing purposes. Almost all the trees for many miles around were cut and used for mine timber and domestic and industrial fuel, and the landscape was left totally unprotected from wind erosion and dust hazards.

It is now possible to avoid such mistakes in most of the new towns where modern techniques can be used and where modern fuel and construction materials can be obtained more easily and economically from distant centres than previously. Although these new towns will have built-up areas, these can be isolated from the mine sites with the impact of both on the surroundings therefore considerably reduced. Solid waste collection can be organized in such a way that it is dumped in small areas which can later be regenerated with fresh vegetation. The cutting of trees is also no longer necessary because there is increasing reliance on imported alternative fuel.

Perry and McAlpine have also suggested that it is necessary to control the activities of the townspeople within the surrounding countryside, and they have stated that:

> indiscriminate recreational use of vehicles and trail bikes, hunting and shooting, and collection or defacing of historical relics cause the main impacts. Mechanical damage to the landscape by vehicles will be greatest in the immediate outskirts of the town. The population of the town can be shielded from the resultant unsightly landscape and dust by a fenced reserve at least half a mile wide surrounding the town from an early stage. Beyond this the real extent of landscape damage by such activities can be reduced by concentrating them into specially selected sites. The impacts of such activities and others such as hunting, cutting down trees, collecting artefacts, defacing historical sites, etc. can be greatly reduced by education campaigns aimed at creating an environmental awareness in the population of the town.[19]

The earlier Australian experiences clearly suggest that the dangers of such environmental damage should be recognized right from the exploratory and building phases and that appropriate steps must be taken to avoid environmental destruction by following sound planning principles and practices.

CONCLUSION

Hot, arid regions such as Australia which are vast and underpopulated offer considerable room for expansion in an overcrowded world. The Australian example clearly shows that, despite a harsh environment, newly discovered resources in these areas are beginning to generate extraordinary developmental activity. The Australian case also indicates that planning of new settlements in arid zones must take into account the general principles established by traditional desert communities, suitably modified to meet the special needs—physical, psychological, and sociological—of people whose historic and cultural roots lie elsewhere. By using ingenuity, common sense, and appropriate technology, it is possible to find a happy compromise. What is most important is to ensure that the living environment of new towns is continuously evaluated and these evaluations are fed back to planners and decision makers.

NOTES

1 G. Blainey, *The Tyranny of Distance* (Melbourne: Sun Books, 1966); T. Brennan, 'Urban Communities', in *Australian Society*, ed. A. F. Davies and S. Encel (Melbourne: Cheshire, 1965), p. 298.

2 Of these, the best known is the telegraph station, currently being restored by the Northern Territory Reserve Board.

3 E. T. Weston, *Ventilation and Cooling of a House with an Attic Exhaust Fan*, special report no. 9 (Sydney: CEBS, n.d.); W. H. Badgett, *The Installation and Use of Attic Fans*, Bulletin no. 52 (College Station: Agricultural and Mechanical College of Texas, Texas Engineering Experiment Station, July 1940; 4th Series, vol. 11, no. 7).

4 J. Paynter, 'Evaporative Cooling—the Principles Involved and Their Application to Building in Western New South Wales', *Architectural Science Review*, 7, no. 3 (September 1964), 110.

5 R. N. Morse, 'A Rockbed Regenerative Building Cooling System', paper presented at the Conference of the Institution of Engineers on Thermodynamics and Fluid Mechanics, Brisbane, 7–8 December 1967.

6 R. V. Dunkle, 'Re-generative [sic] Evaporative Cooling System', *Australian Refrigeration, Air Conditioning and Heating*, 20, no. 1 (January 1966), 13–23.

7 N. R. Sheridan, 'Prospects of Solar Air Conditioning in Australia', in *Proceedings of United Nations Conference on New Sources of Energy*, 6, 87–97; 'Solar Air Conditioning', *Journal of the Institute of Engineering, Australia*, 33 (1961), 47–52; R. Andrew, *Domestic Use of Solar Energy*, technical report no. OR. 7 (Mt Isa: Department of Industrial Engineering, 1962); N. R. Sheridan and P. A. Juler, 'Engineering of the "Solair" Experimental House', *Architectural Science Review*, 5 (1962), 31–38; W. H. Carr, 'Solar Air Conditioned House', *Architecture in Australia*, 54, no. 1 (March 1965), 123–125; N. R. Sheridan and W. H. Carr, *A Solar Air Conditioned House in Brisbane*, Solar Research Notes no. 2 (Brisbane: Research Committee on Solar Energy and Tropical Living, University of Queensland, 1967).

[3] R. B. Deering, *Technology of the Cooling Effects of Trees and Shrubs*, Building Research Advisory Board, Conference Report no. 5 (Washington, DC: National Research Council, 1953).
K. L. Datta, 'Contribution of Landscape to Building Comfort', *Indian Construction News*, (1961–2), 1–14.
R. Geiger, *The Climate Near the Ground*, 2nd ed. (Cambridge, Mass.: Harvard University Press, 1950).
B. S. Saini, *Building Environment—An Illustrated Analysis of Problems in Hot Dry Lands* (London: Angus and Robertson, 1973), pp. 50–53.
P. Oliver, ed. *Shelter and Society*, London: Barrie and Rockliff. The Cresset Press, 1969), pp. 91–102.
L. H. Howroyd, 'Shay Gap', *Architecture in Australia, Journal of Royal Australian Institute of Architects*, 63, no. 3 (June 1974), 52–53.
J. M. Agius, 'Nhulunbuy—The Nabalco Gova Project', *Architecture in Australia, Journal of Royal Australian Institute of Architects*, 63, no. 3 (June 1974), 61–64.
T. B. Brealey, *Living in Remote Communities in Tropical Australia 1. – Exploratory Study*, Division of Building Research report TB 27.1 (Melbourne: CSIRO, 1972), p. 48.
O. A. Oeser, P. J. R. Redding, and J. K. Austin, 'Kambalda Case Studies', in *New Towns in Isolated Settings*, Australian National Commission report to UNESCO seminar on Man and the Environment, Kambalda, Western Australia, 18–23 August 1973 (Canberra: Australian Government Publishing Service, 1976), pp. 109–22, 201–226.

[17] Ibid, p. 110.

[18] R. A. Perry and J. R. McAlpine, 'Impact of Isolated Towns on the Environment', in *New Towns in Isolated Settings*, pp. 7–13.

[19] Ibid, p. 11.

SUGGESTED READING

Australian National Commission for UNESCO. *New Towns in Isolated Settings*. Proceedings of Seminar on Man and the Environment, Kambalda, Western Australia, 18–23 August 1973. Canberra: Australian Government Publishing Service, 1976.

Bryson, B., and F. Thompson. *An Australian New Town: Life and Leadership in a New Housing Suburb*. London: Penguin, 1972.

Rose, A. J. *Patterns of Cities*. Melbourne: Thomas Nelson, 1967.

Saini, B. S. *Building Environment—An Illustrated Analysis of Problems in Hot Dry Lands*. London: Angus and Robertson, 1973.

Siemens, L. G. *Interdisciplinary Research on Resource Frontier Communities*. Winnipeg: University of Manitoba, Centre for Settlement Studies, 1970.

—— *Planning Communities for the North—Some Social and Psychological Influences*. Winnipeg: University of Manitoba, Centre for Settlement Studies, 1970.

13 The influence of arid-zone climate on house design in Sudan

W. J. KIDD

The term Sudan is frequently loosely described as a broad belt of Africa situated between the Sahara Desert proper and the scrub country which lies to the south of it. Sudan stretches from Ethiopia in the east, through the region of Lake Chad to Senegal in the west. In this region, we often see cattle-owning nomads. This discussion, however, is concerned only with Sudan, the country which bestrides the eastern end of this belt. Formerly an Anglo-Egyptian condominium and since 1956 an independent republic, Sudan shares borders with Egypt, Ethiopia, Kenya, Uganda, Zaire, the Central African Republic, Chad, and Libya. It covers an area of nearly 1 million square miles and supports a population of some 16 million persons. Administration is centred in the capital city of Khartoum in the urban grouping of Khartoum, Khartoum North, and Omdurman, the 'Three Towns' situated at the confluence of the Blue and White Niles.

CLIMATIC CONDITIONS

The northern part of the country is virtually without rain. This means that buildings can exist for long periods. For instance, in the early 1960s in Nubia (the region which borders Egypt), remarkably well-preserved ancient monuments were found. Since then they have been transferred to a safe new location to prevent loss by flooding by the new lake, a consequence of the high dam at Aswan. In Khartoum in the centre of the country, the total annual rainfall is 7½in. This normally occurs from June to September and then only as comparatively isolated but heavy showers. The climate is dry and hot, and September brings a maximum temperature of 118°F in association with a low relative humidity of 20 per cent. Rainfall becomes progressively heavier and vegetation consequently denser toward the south of Sudan. Only the northern part of the country, therefore, can be considered an arid zone. The area which we will consider lies between the Egyptian border and an east-west line drawn through Kosti on the White Nile (see **13.1**).

Northern Sudan forms part of the Sahara Desert and consists of apparently limitless and fairly flat expanses of sand and stones with dried-up watercourses, *wadis*, and occasional small outcrop hills, *jebels*. Around Khartoum and farther south, there is more semidesert country with sparse dried grass and low scrub vegetation. The area is divided by the north-south line of the Nile along which the majority of the population is settled. The true desert people are nomadic and are sheltered by tents—light frameworks of poles held by ropes and covered with woven cloths. These tents can be readily erected, dismantled, and loaded onto camels for the next stage of the perpetual journey. Yet the settled population has permanent dwellings which complement the lives of their inhabitants.

The climate is responsible for the topography, and the topography, in turn, is responsible for the naturally available building materials. In the absence of timber of any reasonable size or quality, there remains virtually nothing but the earth itself.

BUILDING MATERIALS AND TECHNIQUES

The local earth is simply dug out, mixed with water to a stiff consistency, and worked by hand. This traditional method of building walls is known as *jaloos*. (See glossary of Sudanese Arabic terms at end of this chapter.) Depending upon the district and the individual builder, variations of mixture include different types of sand, gritty aggregates, or chopped straw. After the planned outline of the walls is scored on the surface of the ground with the lines about 1ft apart, a layer of the mixture about 1ft high is placed between the lines, beaten down, compacted, and moulded up with the hands.[1] When the layer, *suf*, has dried out, a second *suf* is laid and so on until the required height is reached. Each *suf* is formed with a slight inward batter and with a slightly rounded top surface to provide a better seating for the next one, which tends to slightly overhang. The lines between each layer, therefore, remain clearly visible like lift marks in concrete work. Door and window openings are cut through before the *jaloos* is finally dry.

Because each *suf* takes two to three days to dry out, it is obviously more economical if a builder can be involved with several nearby houses at the same time. Each house owner arranges directly for earth to be delivered to the site, but water for mixing is traditionally the responsibility of the builder. If none is immediately available, he and his men have to carry it there. The builder, *osta*, prefers to lay the *suf* for two houses at once in one day. While this is drying out, four other houses are simultaneously under construction. The walling team consists ideally of four men: two mix and carry the material, one throws it up, and the *osta* actually places it. At a normal, steady speed, the walls of a house can be completed in about eight or nine days. More practically, we can say that six houses can be completed in about 50 days.

The method of roofing is virtually the same everywhere. Assuming a standard room, which is approximately square, averaging 12 × 12ft, a central beam, *merig*, is placed to divide the span and is built into the top layer of *jaloos*. The *merig*, which has a cross section of about 4 × 4in, can be of almost any wood, though the most highly esteemed (and now the most expensive) is the *dom* palm. Wooden railroad ties and the steel rails and joists of steel have been used in the past and may still occasionally be seen. *Rassas*, rafters of poorer local timber about 3in in diameter and sometimes roughly adzed, are then laid across the two spans. Split bamboos, *ganna*, are placed on the *rassas* and tied down. If bamboo is not available, the interior palm frond stems, *jereed*, are used as a substitute.

Next comes reed or grass matting of some sort (*shokkab* or *shargania*, depending on the actual type) to act as a support for a layer of dry grass or straw, usually requiring four camel loads. This is known as *homra* or *nal* (again depending upon the type) and acts as insulation for the roof. It is covered with a layer of the *jaloos* mixture about 6in thick, which is laid to falls and hand moulded to form projecting rainwater spouts, *sabaloka*. Nowadays, these are usually made of metal and frequently from beaten out petrol tins. Finally, both walls and roof are covered with *zibala*, a waterproof rendering of earth, chopped straw, and animal dung which is mixed with water and left to ferment before application. Four men can build a roof over one room in about a day and a half if all the materials are available; this does not include *zibala* work, which is carried out by a different tradesman.

Entrance doors may be wooden and very simple or metal and highly ornate with applied decoration and colour. In the absence of any other embellishment on houses, the door is the only medium by which the owner can express his personality or wealth.[2] Internal doors are as simple as possible. Because of the scarcity of good timber, windows have a low priority. As has already been mentioned, window openings are cut in the still wet *jaloos*. Often the marked out rectangle will not be removed until the owner can afford the luxury of a window. The rectangular piece is then pushed out and the window (with a simple timber frame with wooden shutters) inserted. *Jaloos*, then, is the basic earth construction material of northern Sudan.

A more sophisticated method of building walls is by using individual bricks and mortar. Sun-dried mud bricks are set in mud mortar and rendered with what is virtually the same mixture. The whole wall remains essentially one material; but, by virtue of its construction, it is not as truly homogenous as *jaloos*. Brick work, however, has the advantages of straighter lines and sharper corners. It can also be built continuously without the need of waiting for layers to dry. Even more durable walls may be built with red bricks properly baked in a kiln and often laid in cement mortar. A fairly commonly used compromise construction technique is *gishra*, mud brick backing with a red brick facing. These more durable walls normally have proper footings of red brick. The difference between mud bricks and red bricks is not necessarily as great as the nomenclature might imply. It has been usual to reject up to 50 per cent of red bricks delivered to sites. Traditional domestic construction is therefore a direct outcome of the arid climate. But what of the forms of the houses? What are the plans, circulation patterns, and climatic indicators of the lives of their occupants?

THE TRADITIONAL HOUSE

As everywhere throughout the Muslim world, domestic privacy is of paramount importance to the Sudanese way of life. For females, this amounts to absolute seclusion. This necessitates high surrounding walls to compounds within which individual rooms are located. Frequently when a plot is obtained, the boundary wall is immediately erected to mark the territorial establishment. Then nothing further may be done for some time. Indeed, if finances are low, all available money is spent on the wall; further work waits until the financial situation improves. The most basic arrangement of living accommodations within the compound, *hosh*, consists of two adjacent rooms with long east-west axes and with opposite window openings in the northern and southern walls to allow optimum cross ventilation (see **13.2**). The rooms may both be the same size (about 12 × 12ft), or one may be smaller. There could be an internal interconnecting door and some sort of lean-to veranda. Even with these few simple

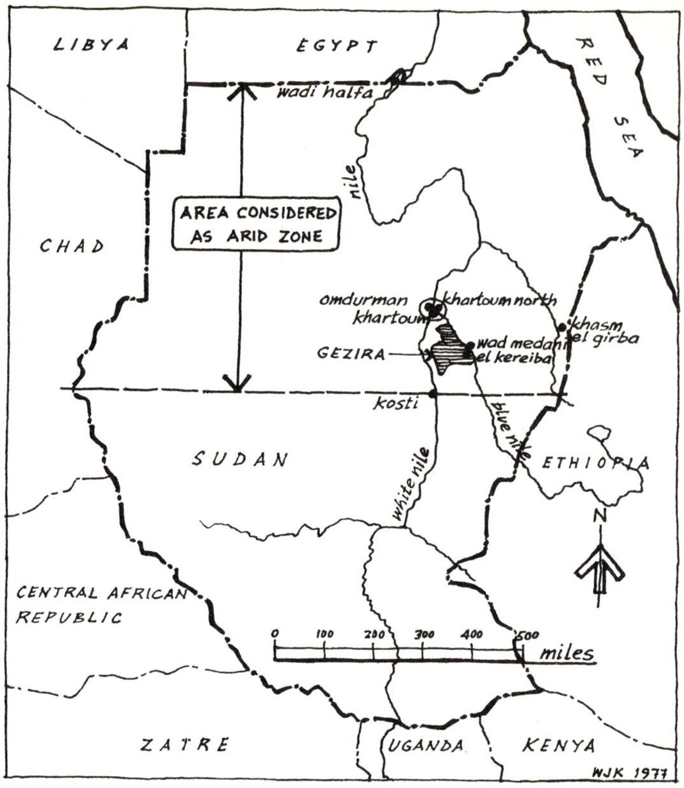

13.1 *Map of Sudan*

elements, there is room for all manner of variations and additions.

Compared to the main room's construction, the lean-to veranda, *kashasha*, usually has a roughly made framework of poor quality wooden poles thatched with millet, *dura*, straw or *shargania* matting. Less frequently, a completely freestanding structure, *rakuba*, is similarly built. Although such verandas shade the walls against which they are built and thus perform the function of overhanging roof eaves, their main purpose is to provide shade for domestic animals. Often the *kashasha* projects no more than 3ft so that the thin strip of shade is of little use for human activities but is acceptable for huddled sheep and goats. Often a fence of *shargania* matting or thorn branches marks off part of the *hosh* as a *zereiba*, animal compound.

Figure **13.3** illustrates a slightly more elaborate but typical rural house in El Kereiba village—the cotton growing area of the Gezira, the triangular land between the two Niles to the south of Khartoum. The village itself is three miles west of Wad Medani on the Blue Nile , 120 miles southeast of Khartoum. In this typical house, the line of rooms is interrupted by a space roofed as a *kashasha* to form a covered entrance hall. The *hosh* is divided into two parts. One serves the easternmost room. When this is used as a guest room, the visitor has his

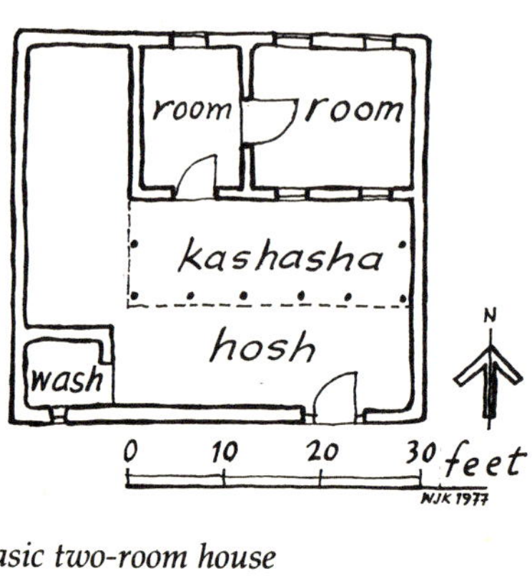

13.2 *Typical basic two-room house*

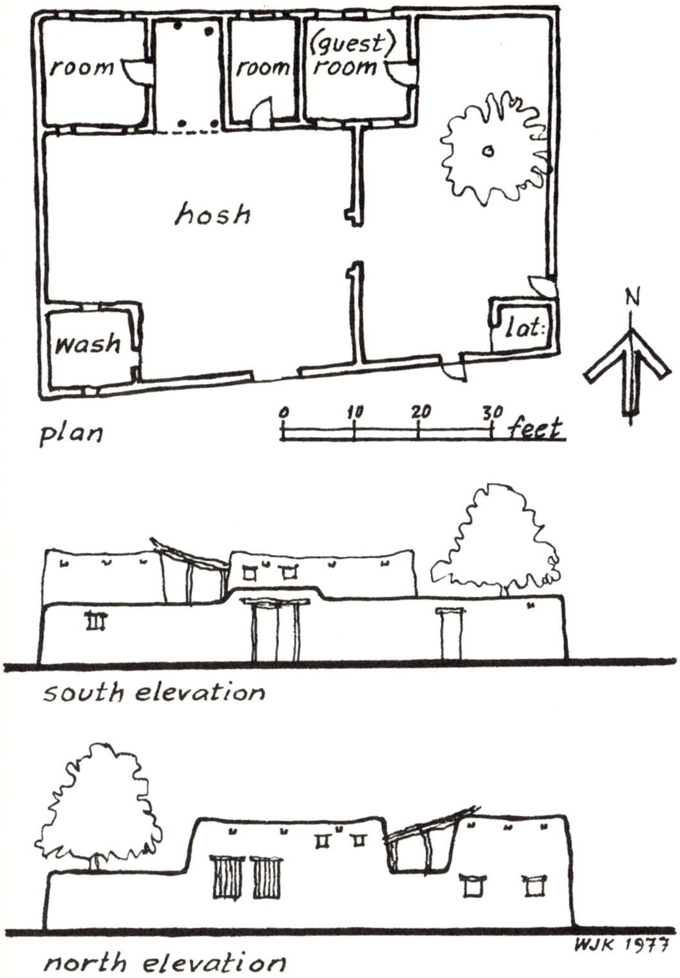

13.3 *House in El Kereiba village*

BASED ON A 1962 SURVEY IN THE DEPARTMENT OF ARCHITECTURE, UNIVERSITY OF KHARTOUM

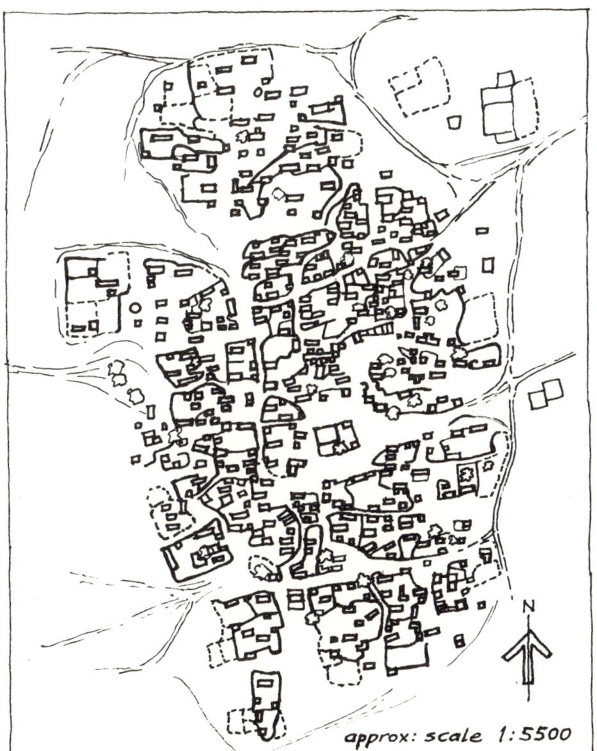

13.4 *Plan of El Kereiba village*

BASED ON A 1962 STUDENT SURVEY IN THE DEPARTMENT OF ARCHITECTURE, UNIVERSITY OF KHARTOUM

13.5 *Plan of part of Omdurman*

BASED ON A PHOTOGRAPH BY THE AUTHOR

own private area. The accommodation and entertainment of guests have always been of primary importance throughout Sudan, as indeed throughout all Islam. Even in the basic two-room house (see **13.2**), one room is always at the disposal of visitors. The separation of the sexes determines that the minimal domestic accommodations must consist of two rooms. The washroom and latrine are placed in corners of the *hosh*, partly to be as far as possible from the habitable rooms and partly to economically use the existing boundary walls.

A plan of the whole village is shown in **13.4**. Although the arrangement of buildings may appear haphazard at first, we can see that almost all of them obey the east-west axis rule and that all have a reasonable amount of open space around them in order to make the most of air movement. Such a plan affords ample opportunity for flexible and organic development. New rooms can be built in corners, and the building programme is adaptable to gradual change. As families grow, more rooms can be incorporated. Conversely, unwanted rooms can be left to gently collapse. Exchanges between adjoining properties can be effected with minimum disruption simply by breaking through new openings and filling up existing ones.

In my discussion, I have referred to all living units as rooms without specific designation. Inside each and every room may be found almost any combination of furniture and personal possessions, tables, chairs, beds, storage utensils, and so on. Each, then, is a general purpose living cell which cannot be referred to specifically as a traditional bedroom or as a living room. A family room is perhaps the most fitting description. However, for the greater part of the year, people sleep, sit, and talk outside. Furthermore, although one room may seem to be more like a kitchen than anything else, cooking is almost invariably done outside because of the smoke and smell. The *hosh*, therefore, is much more than just a courtyard or a garden; it is an interconnecting series of unroofed rooms. The whole area within the boundary wall is the whole house with a cellular arrangement of spaces. Some are covered; some, serving a multitude of purposes, are open to the stars at night. This honeycomb of cells may be open or closed and may melt into ever-changing variations of basic shapes to suit the occupants' lives.

An aerial view of part of Omdurman (see **13.5**) shows this organic quality clearly. In comparison with the rural layout of El Kereiba, there is a closer grouping which reduces open spaces between rooms and so minimizes air movement. However, a reduction in comfort conditions in an urban environment is a frequently paid price for the debatable privilege of living near the bright lights and the amenities offered by a large town. If an aerial photograph of such an area could be taken annually and the series projected cinematically, the result would appear as a slowly moving, constantly changing, living organism. It also seems that this would be a suitable environment for slow-paced, flexible people. It is an environment which has exercised the minds and technological resources of well-meaning architects for some time past.

CLIMATE AND THE IDEAL HOUSE

It has already been established that the Sudanese climate is hot and dry. Meteorological records for Khartoum show that temperatures of 100°F and over occur in every month of the year, that the mean maximum temperature exceeds 100°F for six months of every year, and that the remaining six months have a mean maximum average of 93.8°F. The average, annual relative humidity is only 33 per cent.[3] Proceeding northwards toward Nubia, conditions become hotter and drier. Relieved only slightly during the winter months and aggravated rather than relieved by the brief rainy season with its increased humidity and discomfort, this is one of the most arid climates in the world. The dust storm, *haboob,* is particularly prevalent during May and June, though blown sand is everywhere at all times. Because of clear skies and light-coloured sand with little or no vegetation, glare is intense.

The ideal house for such conditions would presumably be the textbook example of the tropical house for a hot, dry climate. It would be sited with its major axis running east-west so that the longest walls facing north and south would be shaded either by the overhang of the main roof, by a lean-to veranda roof, or by applied shading devices. Window openings, also shaded, would be small enough to give adequate light and ventilation without excessive glare and would be capable of being firmly shuttered against dust storms and pilferers. The roof would have a reflective top surface to minimize heat absorption and a construction incorporating some sort of insulation, most likely with intervening air spaces. The surroundings would be laid out with as much surface vegetation and tree planting as feasible in order to lessen glare, encourage air movement, and break up the ground surface to reduce reradiated heat.

How then, does the traditional Sudanese house measure up to these ideal standards? Certainly most houses are sited with their main axes east-west, and the north and south facing windows are usually placed more or less opposite each other for cross ventilation. It is fortunate that the prevailing winds are either from the north or south, depending upon the season of the year. The windows are small, seldom more than about 3 × 2ft. They essentially are wooden shutters which certainly provide security (though it is doubtful whether they ever fit sufficiently closely to effectively exclude blown sand; actually, only a completely hermetically sealed building could ever be fully dustproof). Windows are also small because of the high cost of timber for their manufacture and because of the lintels which largespan openings necessitate. Often there is no lintel, and the head of the window frame is supposed to hold up the wall above it.

Again because of timber scarcity, the traditional roof has evolved without projections to help shade the walls. Lean-to verandas offer only minimal relief; it already has been noted that the *kashasha,* whose construction can only be regarded as temporary, is primarily for the use of animals. Because of its finish, the roof provides neither a good reflective surface nor any air spaces, though the layers of matting and grass between the *ganna* and the *jaloos* obviously give some measure of insulation.

Earth walls are necessarily thick and dense and are thus slow to absorb and radiate heat Consequently, the internal temperature of an earth building is modified, and the peaks and troughs of the graph indicating diurnal temperature variation are less extreme. They will also lag behind the times of the external temperature maxima and minima and will tend towards a more constant temperature. On the other hand, a lightweight building made of a framework covered with metal sheeting will both heat up and cool down more quickly; its internal temperature will correspond much more closely with that outside.[4]

It has been proposed elsewhere that the ideal house for such a climate should consist of two parts: one of heavyweight construction such as *jaloos,* which would act thermally and would be for daytime use, and the other of lightweight construction.[5] The latter, say a wooden frame with pivotable panels in place of walls and a corrugated metal roof, would be unbearably hot during the day but would cool rapidly as soon as the sun went down. Then it could be completely opened up. Such a room would be used for evening living and for sleeping. The traditional Sudanese house actually conforms to this ideal. The heavyweight *jaloos* rooms, dimly shuttered against heat and glare, are for the day. After dark, the open areas of the *hosh* become the equivalent of the lightweight night room. A roof, as will be testified by anyone who has experienced the singular pleasure of sleeping under the Sudanese sky, is unnecessary.

The ideal of surrounding vegetation is sadly lacking. Growing plants demand water; but this is husbanded for more precious needs, and it would be unreasonable to expect any Sudanese to cultivate a garden for its own sake. Only in the 'Three Towns' close to the rivers is domestic planting common. Yet even here, many Sudanese appear to have an irrational dislike of greenery; it is not uncommon to see large trees cut down and mature hedges replaced with untidy fences of matting or corrugated metal sheets. People do this in the interest of maximum privacy and probably do not realize that the environment consequently suffers visually as well as climatically. It seems reasonable to think that every single shade-giving leaf would be nurtured with gratitude in such a harsh climate. The inexplicable and almost perverse antipathy towards vegetation is something which is hard to understand.

Apart from the lack of shading for its exterior and its lack of greenery, the traditional Sudanese house manages to measure up to most of the desired climatic criteria; since the inhabitants do this within their limited basic resources, the result is remarkable. By subtle manipulation of the house's elements, these people also provide an acceptable and flexible living environment. Its construction allows easy demolition and rebuilding, though this is not necessarily a good thing. In fact, the greatest disadvantage of this type of building is its impermanence.

INNOVATION AND EXPERIMENT

Earth walls crack easily, particularly since inadequate foundations (or no foundations at all in *jaloos* construction) can hardly prevent differential settlement resulting from uneven soil movement due to moisture. The common black, so-called cotton soil is extremely susceptible to such movement in the rainy season. Cracking leads rapidly to partial, then complete collapse. Although a well-rendered surface is reasonably waterproof, rain will penetrate cracks and cause further deterioration within. Bases of walls are regularly eroded by splash-up from the surrounding ground. While some parts of the world provide a rubble plinth, this cannot be done in Sudan; the lower parts of walls suffer accordingly. Because tops of walls are unprotected by overhanging eaves or any sort of coping and lack adequate guttering for roof drainage, they also suffer. The mud of the roof itself, even with its *zibala* covering, is especially susceptible to cracking. Directly exposed to solar radiation, it is also subject to penetration of water and potential deterioration.

All of this, however, only occurs during the seasonal and brief rainy season. If there is reasonable annual maintenance, there is little likelihood of further deterioration during the dry period the rest of the year. Since labour is cheap and exercise is generally regarded as a communal activity, such maintenance as filling cracks and eroded cavities and repairing the external surfaces of walls and roof with *zibala* is inexpensive and is tradionally accepted as a normal part of life. The occupant of a traditional Sudanese mud dwelling would therefore be unwilling to pay for more sophisticated and consequently more costly building materials. Nevertheless, progress cannot be ignored.

For many years there has been a steady stream of experiments in Sudan to try to improve the performance of traditional building materials. Work has been carried out in the past by government departments and local authorities, and more recently by international organizations, the University of Khartoum, and the National Building Research Station. The two principal subjects for investigation have been the stabilization of earth to produce a more durable wall unit and the composition of a superior waterproof rendering for external surfaces. Traditionally, local builders had mixed other ingredients with the basic earth in order to improve its performance. Apart from chopped grass and straw, animal dung was the most commonly used material, though juices and dyes have been known to be used elsewhere.[6]

Cement and bituminous products are the normal stabilizers in recent experimentation. The amount of cement added to an earth mixture is, of course, critical: too little or too much can be equally useless. As in all concrete work, quality control is essential. This explains why some experimental stabilized earth blocks that seemed adequate in the laboratory failed after having been made under somewhat different conditions on site. It has been proved that the most successful blocks can only be made on site under intelligent supervision. Pressure is applied to the cement by a blockmaking machine; later, these blocks must have a long, controlled period of drying out.

A successful experiment in low-cost housing was initiated in 1956 by the National Housing Authority.[7] Using cement-stabilized earth blocks of an interlocking design formed in a simple hand operated block-making machine, builders constructed several schemes in different parts of Sudan. In 1976, I revisited some of these housing areas; apart from some irregularity in the joints (though stability was maintained by the interlocking shape of each individual block), they had endured with little or no maintenance during the intervening 20 years.

In the Managil extension of the Gezira cotton growing area, the Gezira Board built some experimental houses in the 1960s. The builders employed a form of construction which was basically traditional but which incorporated a minimum of upgrading.[8] Foundation trenches were dug, and red brick footings were laid and continued about 9in above ground level. They were to act as a hard plinth for the rest of the wall which was of mud brick backing with red brick facing topped with a red-on-edge parapet. The mud mortar joints were raked and pointed with cement mortar. A damp-proof membrane of sheet polythene was laid within the roof's thickness, and similar sheeting was placed as an underground waterproofing apron around the outside. There was a general upgrading in the quality and finish of timber, and proper lintels were provided over openings with three small-section beams side by side. Some rooms had timber boarding in the roof in place of the split bamboos or palm fronds; some had their normal beaten earth floors paved with red bricks.

Other recent housing in Khartoum designed by the Ministry of Housing in 1975 makes use of semicircular vaulted roofs in an attempt to upgrade construction and provide a ventilated space. Yet another recent experiment involved bituminous material and mud bricks stabilized with waste oil. A prototype house was built to the east of Khartoum North on a site designated for police housing in 1975. I examined this house after a rainy season and found it in better condition than nearby houses of similar design constructed of ordinary mud bricks. Corners were sharper, and surfaces were less eroded. Nevertheless, several more years must be allowed to elapse before any attempt at a realistic performance evaluation is made.[9]

Surface renderings incorporating both cement and bituminous additives have been tried, but none has yet been generally successful. Cement mixes tend to crack and flake off because of the problems of adhesion to the weaker material behind. The fixing of metal lathing or even chicken wire to provide a better key introduces another element which is economically undesirable. Bituminous mixes also have difficulties of adhesion, and oils darken the colour which hardly improves heat absorption. Further painting of surfaces introduces not just another cost element but also another item for recurrent maintenance. Silicone treatment has been used successfully but only on an experimental level; initial cost and the need for regular application make it unacceptable practically. Experiments like these designed to improve standards of house construction have been continuing for many years. It is natural to want to make improvements. Tradition is not enough. The balance between tradition and progress is always delicate and is particularly so in developing countries where progress inevitably involves imported ideas and materials and hence, to a greater or lesser degree, the superimposing of aspects of an alien way of life.

TRADITION AND PROGRESS

In 1959 the National Housing Authority issued a booklet for the guidance of those wishing to build their own houses under a self-help scheme.[10] After giving information on tools and materials and instructions on actual methods of construction, this organization provided specimen plans. One of these is shown in **13.6**. The rectangular walled plot measures 45 × 60ft. Two adjoining rooms, each 12 × 12ft, are abutted on two sides by a veranda. This whole unit divides the *hosh* into two approximately equal parts. A third room, suggested for the use of guests, is in one corner; and the washroom and latrine occupy another. If further rooms are eventually needed, there are two more vacant corners. The climatic considerations of orientation and cross ventilation are standard. The construction for any village house, except for straight lines, is basically the same.

The plot has been drawn with tee-squares and set-squares and is intended to adjoin other similar plots in long straight rows. The walls are not of *jaloos* but are of brickwork. The lintel over the door opening on the section is shown as reinforced concrete. The veranda is no rural *kashasha* but a regularized structure of machined timber. The traditional form remains; but because of the upgrading of materials, the appearance has changed. Not only the individual house, but more significantly, groups of houses constituting the urban environment are different. This touches, once again, upon the controversial relationship between tradition and progress and the degree of balance which must be maintained between them.

Clearly, traditional construction has its limitations due to the weakness of earth walls and the sort of timber available for spanning roofs. This explains why the basic unit of the house is standardized for a room some 12 × 12ft and why its appearance is also standardized. It is the same colour as and has a similar texture to the surrounding earth of which it is made. The lack of straight lines, the rounded corners and tops resulting from hand moulding, and the irregular relationship of all the parts combine to produce an organic architecture capable of natural growth and decay and economically acceptable and visually pleasing.

Yet, maintenance is required, if not annually then certainly after two rainy seasons. The average life of a *jaloos* wall is about 12 years; however, with lack of settlement and good maintenance, such walls can last for about 25 years. Nevertheless, buildings demanding a greater degree of permanence, even in remote rural areas, are generally built of mud brick or sometimes of red brick and are roofed with corrugated sheeting with overhanging eaves. These offer the walls both shading and protection from rain even though internal comfort conditions are worsened during the day because of the quick buildup of heat by the thin roof material. Schools, medical centres, mosques, and similar nondomestic buildings come into this category of superior construction; but probably the house of the head man and perhaps one or two other houses whose owners can afford to display their affluence also receive special treatment. The desire to prosper demonstrably, to keep up with one's neighbours, or to try to surpass them, is universal. It is an attitude which cannot be derided, though it is probably true that today the visual environment suffers rather than gains from such manifestations of prosperity. Fortunately in the rural parts of Sudan, the impact of change is still slight; the aesthetic of traditional house building is still existent. Hopefully, the community will not significantly change, though the radiating influence of Khartoum cannot be ignored.

Khartoum is a capital city and consequently has long had an international population. It needs the large and varied buildings of a capital city and now has its own architects and associated professions to construct them. The main cultural influences naturally tend toward the Middle East, so it is not surprising that recent hotels and office blocks reflect the anonymous concrete-framed lumps of Athens, Cairo, and Beirut. Actually, the character of central Khartoum has changed decidedly for the worse during the past 20 years as a result of these misapplied and misunderstood alien importations. Yet, according to many of the inhabitants, they must be good because they are new and fashionable as opposed to the former buildings which must have been bad, even in some way shameful, because they were old.[11]

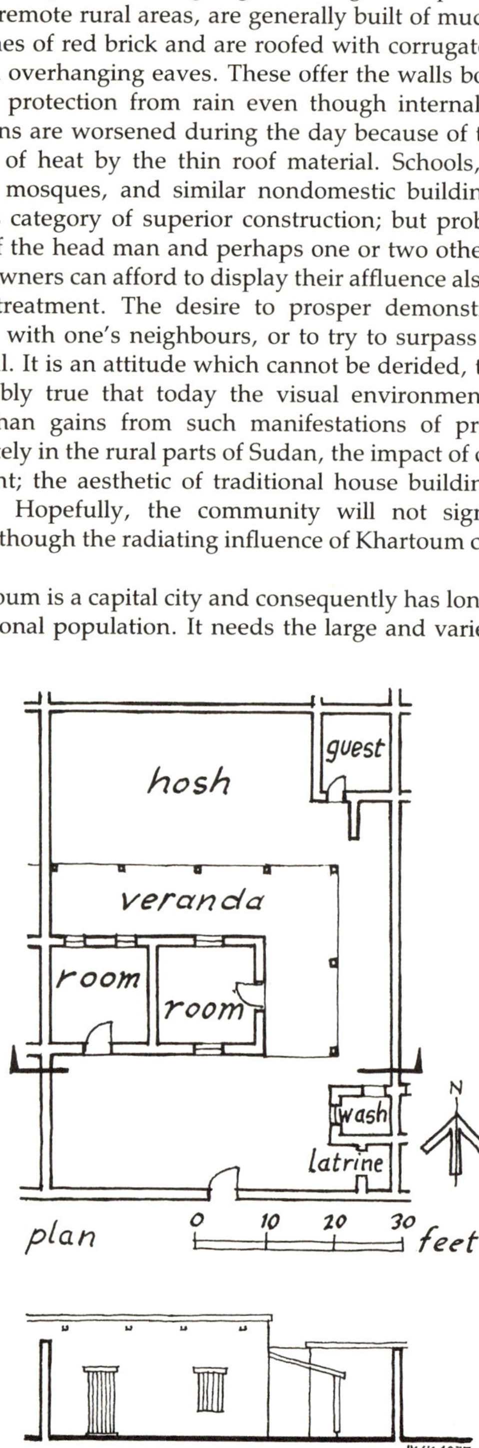

13.6 *Specimen house drawing for do-it-yourself builders*
FROM *Build Your Own House* (KHARTOUM: MINISTRY OF LOCAL GOVERNMENT, 1959), p. 37

Apart from being the capital, Khartoum (or to be more exact, the 'Three Towns') is the only major conurbation in the country. It is also geographically in the centre. It therefore acts as a magnet to attract people, especially young people, from other parts of the country. Those who do eventually return to their villages like to show where they have been. Yet, many who come to the big city stay, and the population grows disproportionately. All the alien influences are there: cinemas and clubs, the Hilton and the international airport, trade and money. Consequently, vast numbers of new houses are appearing on the outskirts of Khartoum, particularly to the south, and are spreading around the airport to the river at Gereif. To a lesser extent, they can also be seen around the edges of Omdurman and Khartoum North. Probably one in every 50 of these houses is architecturally acceptable by international standards. When we look at the design and construction of the others, we can see that they all contribute to an appalling visual environment. There is no need to elaborate upon the self-evident problems of uncontrolled sprawl, services, transportation, and the financial viability of such development in relation to the present economic state of the country.

Here is a paradox. On the one hand, there are careful and sensitive attempts to improve traditional standards, largely in rural areas and largely by dedicated persons in international agencies, government departments, and universities. On the other hand in urban centres, there is a sweeping aside of traditional values in the face of brash modernism. Can the former survive against the latter? Was the holding of a balance between tradition and progress ever more difficult to achieve? Office buildings and cinemas patently cannot be constructed of mud and poles, and no sensible person would expect them to be. Nor would he expect them to ape traditional forms like monstrous pavilions in an international exhibition. Nor would he be so unrealistic as to search for a Sudanese style in present day architecture.

But he would surely look for good, basic construction, for an intelligent use of local materials, for a sympathetic scale and detailing, and for an appreciation of climatic conditions. Above all, he would surely expect to see buildings reflecting the love of their designer for the place and for the people. This paradox is not peculiar to Sudan, nor even to other developing countries. In spite of recent propaganda publicizing conservation and the preservation of heritage, Western Europe and America can hardly own to a blameless record when it comes to senseless destruction of old buildings and their replacement by new ones which nobody wants. Only since the early 1960s has Sudan been training its own architects, and only during the past half-dozen years have they been qualifying in significant numbers.[12] Before it is too late, we must hope that they are aware of the mistakes made elsewhere and that they are in a position to perpetuate the best aspects of their visual heritage. The future remains uncertain. However, it is wise at this stage to look at a specific example from the present, or at least from the fairly recent past.

PLANNING AND PEOPLE

The most dangerous aspect of planning in Sudan in recent years has been the ability to draw straight lines. They presumably are not drawn in any conscious effort to go against tradition; it is just easier for a planner sitting at a drawing board to use a tee-square and set-square than to doodle imaginatively. A few straight lines produce a series of regular rectangles. They become rows of plots for houses. The occasional space is left open; roads run like racetracks between plots, and from this pattern a town is born.[13]

At the end of the 1950s, Egypt started the construction of the high dam at Aswan. The dam would eventually hold back the Nile to form a new lake which would flood many riverside villages in Nubia as well as the important town of Wadi Halfa on the second cataract. This was the train-to-boat transfer point and customs post on the Sudanese-Egyptian border and was also the centre of an area well-known for the cultivation of dates by a specialized traditional system suitable for the virtually rainless climate. The local population had to be moved elsewhere. An obvious solution might have been to move them as short a distance as possible to the line of the new lake shore. This would have let the people retain their accepted way of life with little interruption. If this plan failed, the people could have moved to nearby established centres within the same ethnic region. Instead, governmental decree ruled that the population of Wadi Halfa should be moved to the newly irrigated area round Khasm el Girba. This is some 700 miles away in the southeastern part of the country near the Ethiopian border. Further, it is a district with a different indigenous population, a different climate, and a potential for a different agricultural economy.

> Even more drastic was the change in the architectural environment. Wadi Halfa had developed in the organic way common to the mud constructed settlements of the Middle East and North Africa. It was untidy, interesting, plastic and, above all, human. Khasm el Girba on the other hand was a new settlement based on grid-iron plans which when translated into reality became a drab collection of rows of blockwork houses—tidy, dull, rigid and quite inhuman.[14]

Some 30 villages and a central administrative centre were planned. Each had standard houses with precast concrete panel walls round the *hosh*. The rooms had concrete block walls and corrugated asbestos sheet roofs with no insulation underneath. Apart from the mean appearance of the houses, the most terrifying impression came from the physical implementation of the planners' straight lines. Apparent were long wide streets, open spaces which would never be used for their designated purposes (whatever they might have been), and the feeling of unnecessary distance between everything. It is not surprising that the newly arrived inhabitants reacted against this alien environment, and many soon left either for Khartoum or to return to Nubia. Now, nearly two decades later, those who stayed are well-settled; thanks to human resilience and adaptability, they have helped to make the new area a workable proposition.

However, it is regrettable to think that people who have a hard enough struggle against the climate and the hardships of daily existence also have to contend with extra conditions imposed by planners and other doubtless well-meaning authorities. It is also unfortunate that authorities can be so blind to human needs and aspirations. Again, this indifference is something universal; but in Sudan, it is prevalent. The Gezira irrigated area for cotton growing is centred administratively on Wad Medani and, as mentioned earlier, lies across the triangular land between the two Niles south of Khartoum. Its controlling body, the Gezira Board, has pursued an enlightened policy towards the people who live and work within its area ever since the first stages of irrigation were planned in 1907. Villages have been allowed to develop naturally within an overall distribution network. New settlements have been built by their inhabitants in traditional ways. Existing villages like El Kereiba, now very much part of the Gezira organization, have existed for nearly 400 years and have continued to develop in the old organic way. It is a much more pleasant and human experience to visit a Gezira village like El Kereiba than any of those around Khasm el Girba.

CONCLUSION

The northern part of Sudan has an extremely hot, dry climate and consequently is a desert. The only natural building material is the earth itself. This is used in conjunction with the absolute minimum of timber and grasses to build houses which are, within their limitations, remarkably successful in thermal performance when measured against ideal standards. The basic unit is the square room; this, planned in free combination with similar units and the open spaces left between, makes houses suitable for the life-style of the inhabitants. The major disadvantage is impermanence, and there has been considerable experimentation to try and upgrade materials and techniques. All progress, for financial reasons, remains at a low technological level. No mention has therefore been made of high-grade imported materials, mechanical ventilation, or air conditioning.

Innovations, of course, are potentially dangerous, particularly when alien influences are used thoughtlessly and when planning proposals ignore tradition and human beings. Innovation need not be the enemy of tradition, but it must be introduced sensibly and sensitively. By respecting tradition, the Gezira Board has maintained one of the most successful hierarchical systems of village organization.

Close examination must be made of the way in which people live and work both inside and outside their houses. It is better to accept the donkey than to assume its immediate replacement with a Land-rover. It is better to provide shaded narrow alleys for donkey traffic than glaring, long, wide, straight roads. It is better to keep daily travel distances short so that motor transportation is needed only for less frequent, longer journeys. It is better to think of the environment in terms of people, first, and their relationship to one another and to work, schools, medical centres, other villages, and towns, rather than of an overall environment where people will somehow fit.[15] An equilibrium between the forces of tradition and progress must be sensibly maintained if respected human values are to survive. There is not better example as a starting point than the traditional Sudanese house.

NOTES

1 The unit of measurement is the *diraa* which varies from 22–24in. *Jaloos* walls are ½ *diraa* thick and each *suf* is ½ *diraa* high. Room walls are normally 7 *diraa* high (12ft 10in–14ft); and *hosh* walls, 4 *diraa* high (7ft 4in–8ft).

2 Nubia was the only part of the Sudan with a tradition for decorating houses externally. This was done by setting plates and saucers in the walls. Today even this custom is waning. Compare Northern Nigeria, a region with a fairly similar climate, where highly decorated mud houses are usual.

3 K. L. S. Murdoch, 'Air Conditioning', *The Sudan Engineering Society Journal*, no. 8 (1963), 58.

4 I once lived in a house in Khartoum which was partly roofed with a concrete

slab covered with black asphalt and partly with corrugated metal sheeting. The difference in internal comfort conditions at different times was very marked.

[5] At the University of Khartoum, I proposed such a house with a heavyweight *jaloos* ground floor and a lightweight first floor which would fully shade the ground floor and would be more susceptible to air movement. It was soon realized, however, that the lightweight part was not viable economically, and in any case, it was unnecessary.

[6] Waste material from the local indigo dyeing industry is used in Kano in Northern Nigeria.

[7] Republic of the Sudan, Ministry of Local Government, *Report on Research and Experiments on Low-Cost Housing in Interlocking Self-Aligning Blocks of Sand/Soil and Cement made by Carlestein Block Making Machine* (Khartoum: National Housing Authority, 1956).

[8] P. Crooke and B. Kimmins, *Report on Technical Work for Low-Cost Housing Trials in El Huda Village* (Khartoum: University of Khartoum, 1963).

[9] The earth mixture for the bricks is stabilized by an asphalt emulsion which is a by-product of the petroleum refinery at Port Sudan. The material has been named Asphadobe.

[10] *Build Your Own House* (Khartoum: Ministry of Local Government, 1959) (in Arabic).

[11] This unfortunate but very natural attitude doesn't apply only to buildings. This year's model of virtually anything must be better than last year's because it is newer and its possession produces a feeling of superiority in the owner.

[12] When the Department of Architecture at the University of Khartoum started in 1957, there were four students. The 1978 intake was about 55.

[13] Without defending this attitude to planning, it must be remembered that the country is very flat and that irrigation canals are very straight.

[14] W. J. Kidd, 'Comment', *Urbanization*, 7 (1968), 8.

[15] Much work in this field has been done in the Department of Architecture, University of Khartoum. Most of the information on El Kereiba comes from a survey of the village made in 1962 by students under the direction of P. Crooke and S. D. Russell. Further investigations into hierarchical planning have been made by M. Kadic, El Bedri Omer Elias, and others. A detailed study of planning and servicing in the Gezira is currently under preparation by Abdel Basit Ahmed Awad El Kerim at the Queen's University of Belfast.

GLOSSARY OF SUDANESE ARABIC WORDS

Diraa	Unit of measurement, about about 2ft though often slightly less.
Dom	Type of palm tree with strong fibrous trunk.
Dura	Millet, local staple grain.
Ganna	Split bamboos used in roof construction.
Gishra	Brickwork consisting of red (ie, kiln baked) brick footings and facing with mud brick backing.
Haboob	Dust storm.
Homra	Insulating layer of grass in roof construction.
Hosh	Compound or series of open spaces within the perimeter wall of a house; used for most aspects of family living, especially after sunset.
Jaloos	Earth mixed with water and moulded as a material to make walls and roofs.
Jebel	Hill, usually an inselberg type of rocky outcrop.
Jereed	Spines of palm leaves after the rest of the leaf has been stripped; inferior to *ganna*. An even more inferior material of thin irregular tree branches known as *fallakab* is used for poorest quality work, eg, storerooms.
Kashasha	Lean-to veranda.
Merig	Main beam in roof construction.
Nal	Straw, local variant of *homra*.
Osta	Builder, usually the leader of a building team.
Rakuba	Similar erection to *kashasha*, but freestanding.
Rassa	Wooden rafter.
Sabaloka	Rainwater spout or gargoyle.
Shargania	Matting woven from *nal* straw.
Shokkab	Local variant of *shargania*. Another rather more durable type is *birsh* matting.
Suf	Layer of *jaloos* walling.
Wadi	Dried-up watercourse.
Zereiba	Compound for animals, usually separated from the rest of the *hosh* by rough fencing.
Zibala	Waterproof rendering of earth, chopped straw, and animal dung mixed with water and fermented. It is smeared over walls and roofs to a thickness of about ½in.

14 Settlements and housing in an arid-zone developing country: case study in Hausaland, Northern Nigeria

J. C. MOUGHTIN

Unprecedented advances in building technology in the Western world over the past century are rooted in economic expansion; but these advances and equally spectacular progress in industry, education, and science have not produced solutions to mass housing problems. As Charles Abrams notes, 'the simple refuge affording privacy and protection against the elements is still beyond the reach of most members of the human race'.[1]

HOUSING PROBLEMS IN DEVELOPING COUNTRIES

In developing countries within the arid zone, some of the causes of housing problems are rapid expansion of population, inadequate financial resources, weak institutional framework for the organization of an infantile building industry, and totally inadequate urban infrastructure. Furthermore, the process of modernization and change from traditional cultures to a complicated urban society has found both the traditional methods and the still nascent 'Western' methods of housing incapable of satisfying the needs of a rapidly increasing urban population. The result of this process has been provision of housing for the wealthy and the middle classes by the modern sector of the building industry, while the urban poor crowd into existing dwellings or build squatter settlements having few if any of the essential urban services. However, it is heartening to read of those occasional brave attempts which are made by African governments to house low-income groups in low-cost housing schemes.[2]

Despite recent wealth created by the oil industry, housing conditions in Nigeria are similar to those found in other developing countries and are summarized in Nigeria's Third National Development Plan as follows:

> Although housing conditions are generally inadequate in the rural areas, Nigeria's housing problem, like those of other developing countries, is essentially an urban problem. Rapid urban growth associated with accelerated tempo of socio-economic development hàs seriously aggravated the shortage of dwelling units in Nigeria resulting in overcrowding, high rent, slum and squatter settlements which are visible features of the urban scene throughout the country.[3]

G. A. Onibokun in his article 'Housing Finance in Nigeria' clearly demonstrates that the financial arrangements for the modern sector of the building industry in Nigeria do not meet the needs of the low-income groups:

> The two main sectors that can undertake the responsibility of solving the housing and the associated environmental problems of Ibadan Region are faced with inadequate financial resources. The citizens are bogged down in poverty, the three levels of government are beset with meagre resources, and the public agencies are operating under marginal financial capability, while at the same time, the population is rapidly increasing.[4]

It may seem as if the answer to the problem is the channelling of more resources into housing; but even in Nigeria with its new found wealth, this is not possible. Clearly, it is out of the question for many less fortunate developing countries.[5] This chapter, therefore, sets out to explore an alternative approach to satisfying housing need by questioning the definition of the problem and by examining the ways in which traditional methods of building can be used in Hausaland, a cultural region of Nigeria.

LOCATION AND ENVIRONMENT OF HAUSALAND

Hausaland, *Kasar Hausa*, is in the northwestern part of the Federal Republic of Nigeria and lies between the confluence of the Niger and Benue rivers.[6] It extends approximately from latitude 3.5° east to latitude 11.0° east—or 525 miles (844 km)—and from longitude 10.5° north to longitude 14.0° north—or 250 miles (400 km) (see **14.1**). Hausaland consists of mature or old-age plains upon which the forces of erosion operate slowly. Weathering has taken place to a great depth and to such an extent that the rocks have been decomposed, and it is not possible to recognize their original type. Although there is a wide variety of underlying rock structure, large areas of Hausaland are covered by a common end product, laterite, with limited exposures of fresh rock.[7] From the viewpoint of traditional building operations in the area, laterite is the single most important material in use.

As R. J. H. Church points out, 'Nigeria has a greater variety of climate than any other West African country; the Cameroon, Equatorial, Semi-seasonal Equatorial, Seasonal Equatorial, Southern Savannah, Jos Plateau, Savannah and Sahel types of climate being found'.[8] Hausaland has a climate with pronounced wet and dry seasons. The rainfall drops from the south to the north of Hausaland and varies from a savannah type to a sahel climate in the extreme northern districts. Hausaland's climate is influenced by two principal wind systems: a hot dry northeasterly wind blowing from the Sahara and a southwesterly, warm, moist air current from the southern Atlantic Ocean. The *harmattan* or northeasterly wind brings dust with it from the Sahara; when it blows from November to April, day temperatures are high in the afternoon and low at night and in the morning.[9] During this time of year, humidity is low; it is frequently hazy with visibility

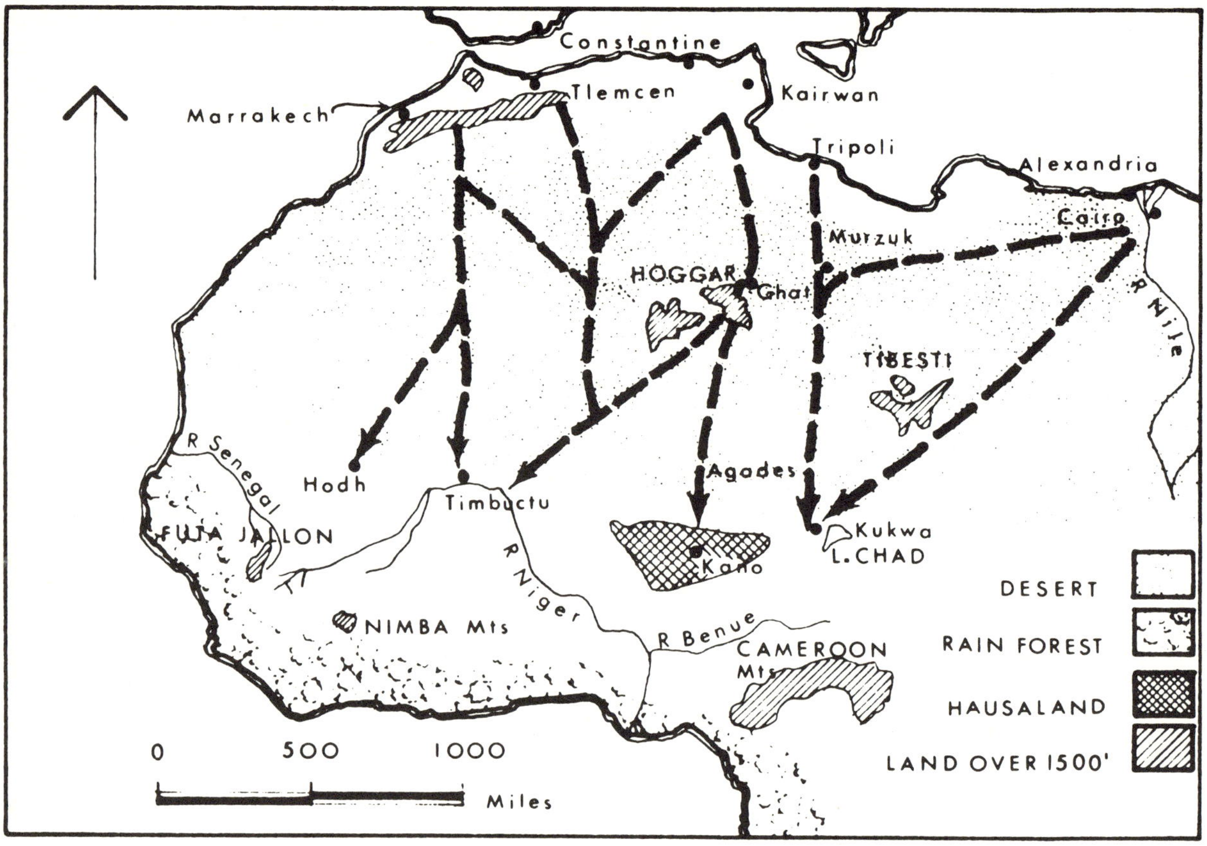

14.1 *Location of Hausaland, Nigeria*

SOURCE: J. C. MOUGHTIN, 'THE TRADITIONAL SETTLEMENTS OF THE HAUSA PEOPLE', *The Town Planning Review*, 35, no. 1 (1964), 22

14.2 *Topographic map of Hausa, showing location of the seven city states*

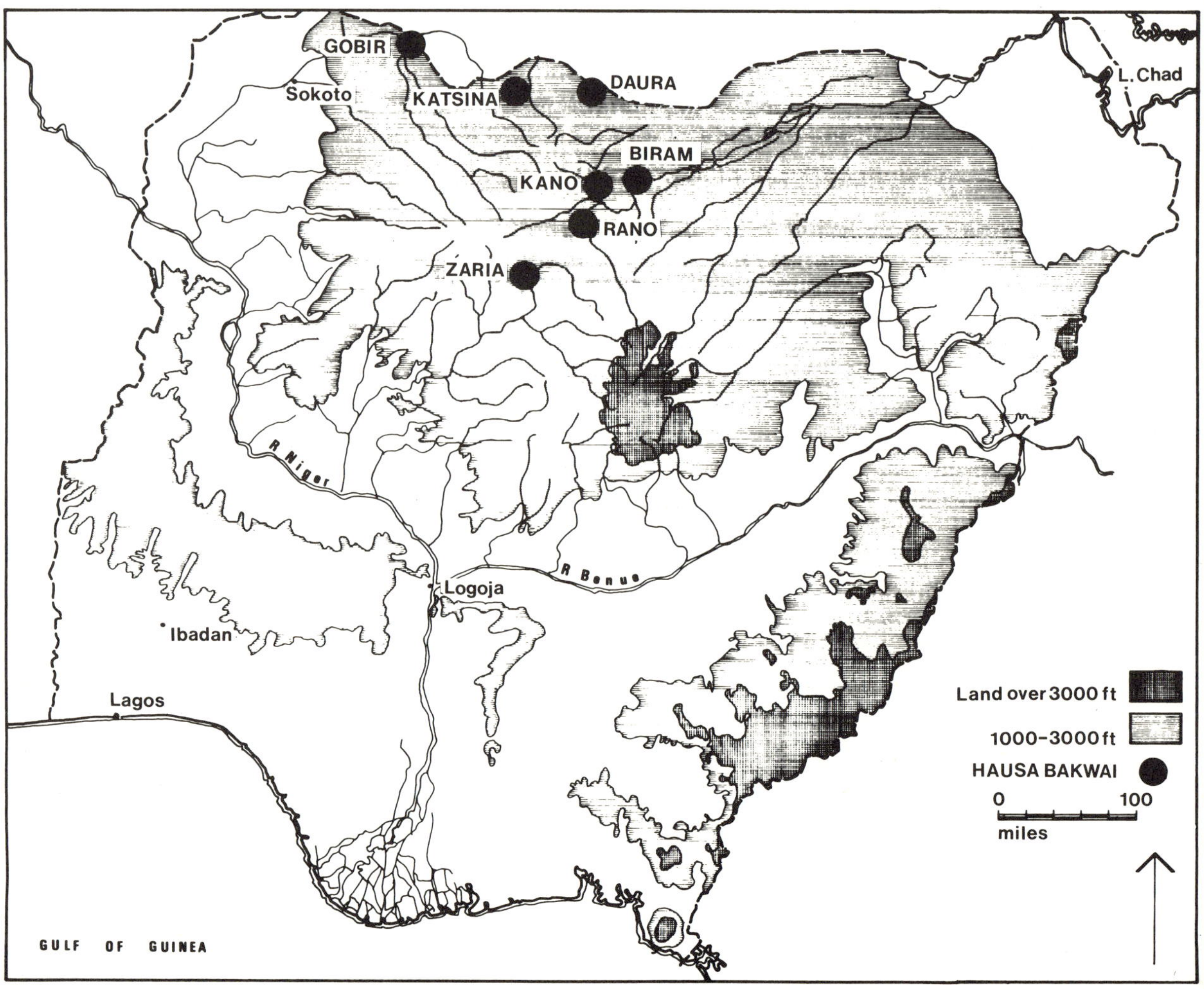

reduced—sometimes to less than half a mile. During the remainder of the year the southwesterly monsoon winds bring moist, warm air, most of the rainfall, and high humidity. Surface winds tend to be generally light; the only exception is the line squalls which are at the beginning and the end of the rainy season and can gust up to 90 mph.

Climatic conditions in Hausaland affect building operations in two important ways. Shelter has to be designed for two distinctive climatic conditions: a warm, humid climate where cross ventilation is important and for a hot, dry climate where it is important to keep the interior cool and enclosed during daylight hours. In simplistic terms, the first condition requires light walls with many openings screened by the roof, and the second requires thick walls with few and small openings. Unfortunately, laterite, the chief building material in Hausaland, is highly unstable in rainy conditions; this imposes a whole series of problems for the builder to solve. The measures adopted to solve these problems give Hausa architecture some of its distinctive features.

There are two main vegetation zones in Hausaland: the Northern Guinea zone and the Sudan zone. A zone of transition between the vegetation types runs broadly east-west through Northern Zaria province. The climax vegetation of the Northern Guinea zone is a western extension of the great *miombo* woodlands, the *Isoberlinia-Brachystegia* of east, central, and southern tropical Africa. It has been suggested that the climax vegetation of the Sudan zone was probably similar to the *mutemwa* vegetation of Zambia which consists of large emergent trees with open canopy, an understorey of deciduous shrubs, and sparse or no grass.[10] In both zones the natural vegetation has been completely modified as a result of human occupation because the bush has been cleared and burnt for cultivation, hunting, and cattle grazing.

In the areas near large settlements such as Kano and Katsina, natural bush vegetation is almost completely absent. Trees such as *Acacia albida, Tammarrind indica* and *Butyro spermum parkii* have been planted in settlements for shade and fruit, but timber itself is rarely used in building operations. An important exception is the dumb palm, *Hyphanene thebaica*, which occurs in the Sudan zone either singly or in dense groves. *Azara*, the wood from this tree, is a fibrous material which is free from attack by white ants. Consequently, it makes excellent roofing joists and may also be used as reinforcement for mud buildings. In addition to *azara*, other vegetable material used in building includes small branches for light framing, grass thatching, grass matting of various types, and vegetable additives to mud for waterproofing purposes.

THE DEVELOPMENT OF HAUSA SETTLEMENTS

The built forms evolved by different cultural groups to solve similar environmental problems are many and diverse and are dependent almost entirely upon the perception of those problems.[11] Such perceptions can vary from almost totally ignoring environmental problems and placing greater emphasis on socioeconomic or symbolic considerations at one extreme, to an attitude which places a premium on design for optimum internal environmental conditions in which built form takes on the form of a climatic control mechanism. For this reason it is important to understand the cultural history of the Hausa in order to appreciate the attitude of these people toward the development of the architectural programme and to understand the values attached by the Hausa to the design factors governing traditional built forms.

Historical aspects

Until the coming of the Europeans, Hausaland was cut off from the coast by an impenetrable barrier of tropical rain forest; all contact with the outside world was by way of the Sahara. For many centuries the great bullock trains, and later the camel trains, which crossed the Sahara were the commercial lifeline that connected Hausaland to North Africa. Along these routes travelled the wealth of west Africa, gold and slaves in return for salt and Mediterranean produce.[12] Up to the first millenium AD, the areas just south of the Sahara were undergoing a change in the structure of their vegetation; in part this was due to worldwide climatic changes, but later man's agricultural activity brought about great changes to this

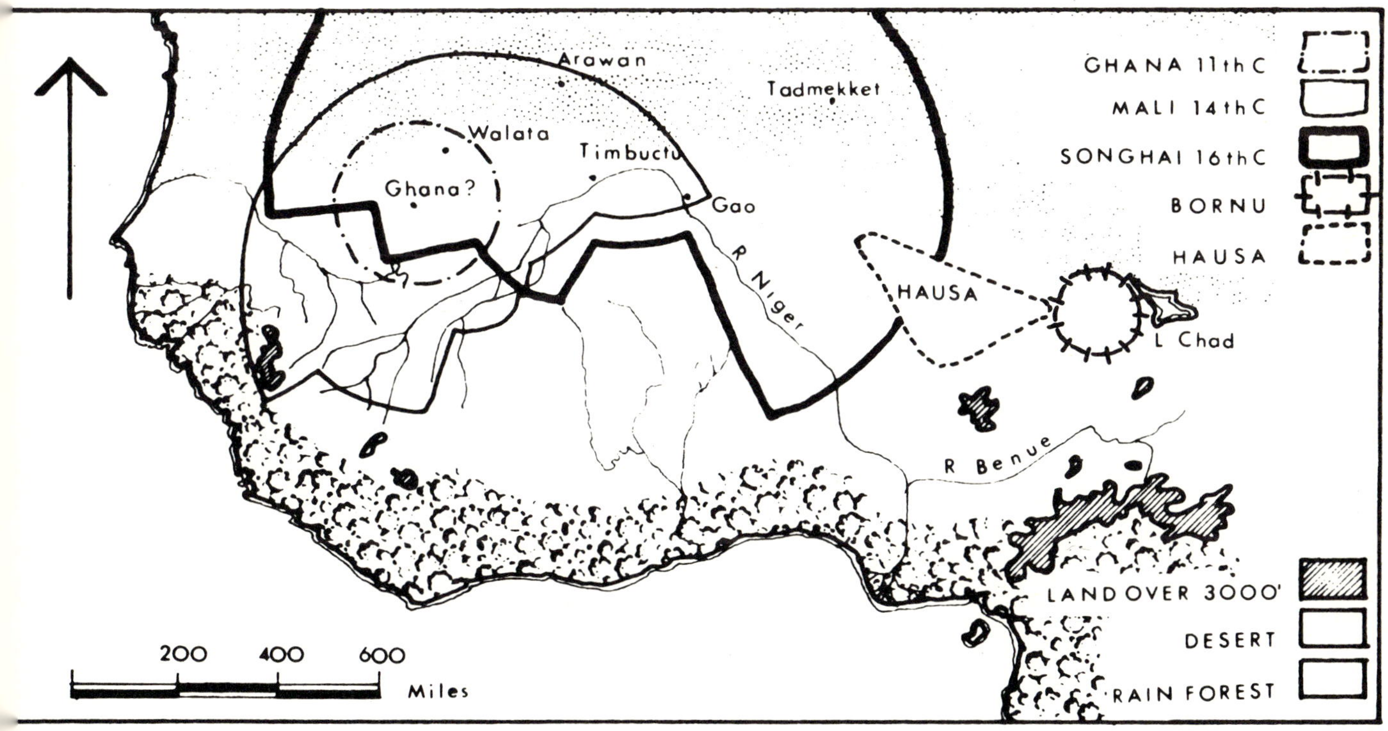

14.3 *Location of Hausaland with respect to her neighbours–a historical perspective*

SOURCE: J. D. FAGE, *An Introduction to the History of West Africa* (CAMBRIDGE: CAMBRIDGE UNIVERSITY PRESS, 1955), pp. 19, 25, 29

region. There was, therefore, in what is now the sahel vegetation zone, a consequent increase of population pressure which produced a southward movement of people in Hausaland and other areas of west Africa.[13] This great southward movement of peoples is recorded by the Hausa and other Nigerian peoples in their legends which generally place their origin in the north.[14]

In the savannah lands between the great northern bend of the Niger and Lake Chad was founded a group of seven small related city states: Daura, Kano, Zaria, Gobir, Katsina, Rano, and Biram, called the *Hausa Bakwai* or true Hausa. The origin of the Hausa states is obscured by legend, but it is probable that they owe their origins to a series of related invasions of peoples from the western shores of Lake Chad in the first millenium AD (see **14.2**).

Beginning in the ninth century AD, Hausaland was polarized between the powerful states of Ghana, Mali, and Songhai to the west and Kanem-Bornu to the east. Hausaland occupied the no-man's-land between these two power blocks and acted as a buffer state (see **14.3**), alternating under the domination of either west or east. Although in constant subjection to foreign rulers and despite having their farmlands laid waste and their trade dislocated by the occasional attacks of their more powerful neighbours, the Hausa states managed to retain their identity and their local autonomy which enabled them to develop an elaborate social organization.

Because the Hausa states were located at the terminus of the central Saharan trade route, they were able to develop as a series of high density urban centres with highly structured political and administrative systems of government and with a prosperous economic base. Continual tribal warfare before the British occupation (1900–6) of northern Nigeria added impetus to the establishment of a pattern of densely occupied, walled settlements in areas of unoccupied savannah land. In addition to the major administrative cities such as Kano and Zaria, there are also many smaller walled settlements ringed by bush hamlets.[15]

Cultural elements

Islam has had a considerable effect on many aspects of Hausa culture. Building techniques and architectural style are no exception. Imported ideas of what constitutes a good building have changed the indigenous architecture from a savannah style to one that is more suited to an arid Saharan climate. The early contact of the Hausa states with Islam were by way of the western Sudan and the Maghrib; tradition says that Islam was brought to Hausaland by merchants from Mali in the 14th century. For many centuries Timbuctu remained a renowned centre of Muslim scholarship, and it was from this source that the Hausa received their introduction to Islam. In the 16th century the Hausa city of Katsina became an important centre of Islamic learning with its own special quarter in the town for students.[16]

Long cultural contact between Hausaland and the African Mediterranean coast affected building traditions in two main ways. First, the acceptance of Islam, and more particularly, the practice of segregating the sexes (*auren kulle*) has brought about changes in the detailed arrangement of the various components of the house plan; second, the style of Hausa architecture has evolved from a savannah type to one based on the forms usually associated with the Islamic architecture of the Mediterranean world. In more recent times, contact with the West has introduced other ideas and concepts into Hausa culture, particularly the changes brought about in the new suburbs and developments outside the traditional settlements. Traditional building is now being affected mainly in terms of detail or accommodation to imported fittings and fixtures including the motorcar.

ZARIA

City structure

It is difficult to generalize about the layout of the major Emirate cities. Most of them, however, have well-defined quarters separated from the old city. The layout pattern of Zaria, in particular, emphasizes the separation of the various quarters. The barriers between the quarters are the railway track, the river, or the city wall. The settlement resembles a cluster of independent urban centres rather than a single unit, and the pattern it forms reflects religious, racial, and cultural differences overlaid by the recent need to accommodate the motorcar, modern educational establishments, or administrative and commercial institutions (see **14.4**).[17]

The one-time 'European' or residential quarter is laid out with large avenues of *madaci* and *neem* trees. At its centre is the Zaria club and racecourse, and in this area are found administrative offices, law courts, and the post office. The *Sabon Gari* (new town) is occupied chiefly by southern Nigerians who work in the industries, commercial firms, railway, and administrative organizations. The street pattern in the *Sabon Gari* is set out in the form of a rigid grid. The buildings are mainly concrete block walls with tin roofs totally unsuited to the environmental conditions. At its centre is its market, and everywhere there are small commercial firms, bars and night clubs. In fact, it is a shanty town, full of brash character, noise and colour, but tremendously overcrowded with polluted open drains and a primitive night soil collection system.

Tudan Wada, another distinct quarter just outside the walls of the old city, is inhabited chiefly by Muslim outsiders. This area, like the *Sabon Gari*, is laid out in a grid street pattern conforming with health and bylaw regulations introduced by the British earlier in the century. Here, however, many of the buildings are regularized versions of traditional Hausa buildings, although the ubiquitous tin roof is fast replacing the mud roof, and concrete block construction is also in evidence.

A large residential area in *Samaru* has been established in the last 15 years to serve the Ahamdu Bello University, which

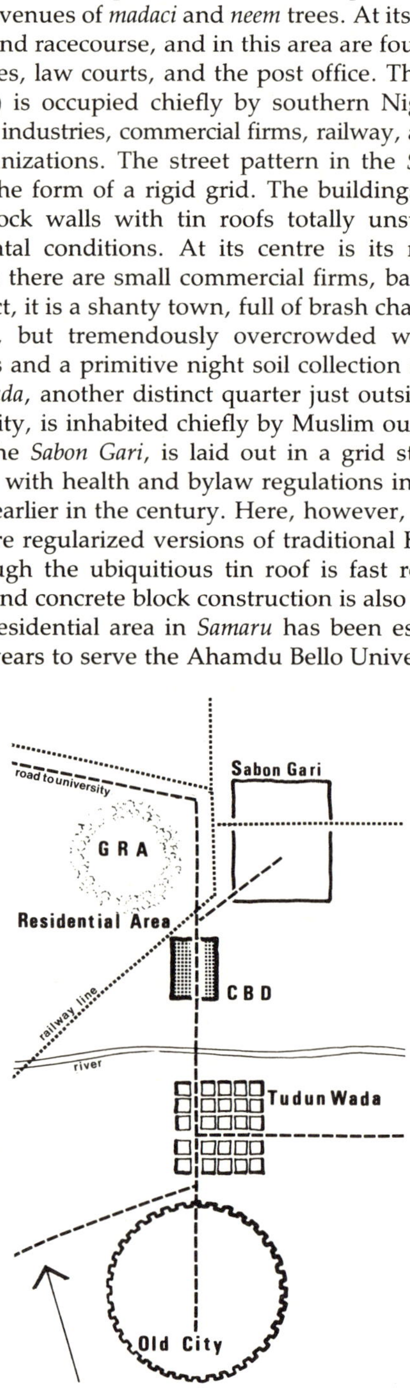

14.4 *Layout of Zaria, Hausaland*

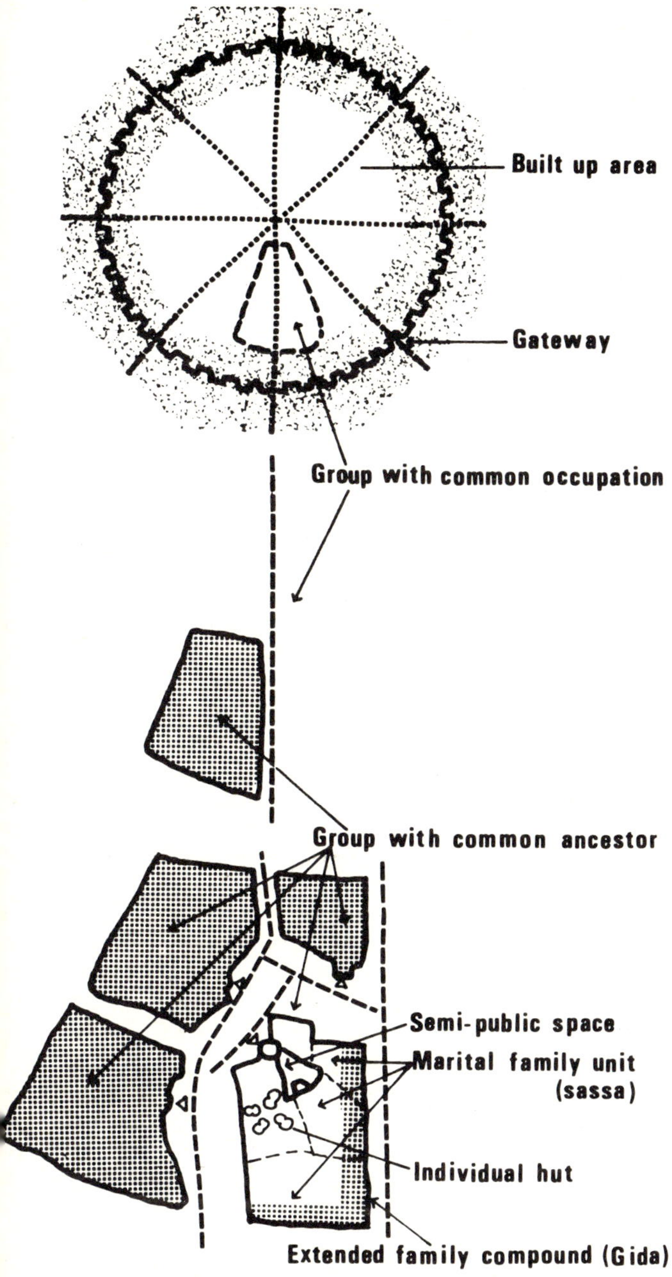

14.5 *Physical framework reflecting social structure of traditional city of Zaria*

is some five miles from the city. This modern area of *Samaru* resembles the *Sabon Gari* in form, density, and social structure and exhibits all the environmental problems that can be found there. Between the university and the older parts of Zaria, there is ribbon residential development consisting of modern suburban areas for senior administrative, commercial, or university staff. There are also traditional expanded Hausa villages following closely patterns of development in the ancient city and low-cost housing schemes built by government agencies.

Traditional city of Zaria

Within the apparently formless complex of mud dwellings that exist inside the walls of the ancient city, there is a basic physical framework which reflects its social structure. The residential area in the old city is subdivided into zones usually associated with one of the main gateways. Each zone is occupied by groups having certain similarities, common occupations (dyeing, weaving, potting, etc) or membership in one of the royal families. These zones are further subdivided into

14.6 *Daura, an outdoor praying area*

areas of a few street blocks occupied by families having a recent common ancestor (see **14.5**). The basic element within this complex pattern, however, is the street block, irregular in shape and enclosed by a high mud wall. The street block contains the main social and economic unit of the extended family. The family compounds are linked by pedestrian routes which connect the market place, the Friday Mosque, and the gateways. These routes also connect outdoor praying places, shaded sitting and teaching areas, and the large formless borrow pits and take the form of a sequence of spaces, some of which are narrow passages barely wide enough for a fully laden donkey (see **14.6** and **14.7**).

FAMILY STRUCTURE AND HOUSE FORM

Hausa families are of three types: individual families; a married man, his married sons and their dependants; or a group of collateral agnates and their dependants. These varieties of Hausa co-residential kin groups are not formal alternatives but manifestations of the same rhythmic and dynamic cycles (see **14.8**). [18] The effect of this household cycle is most evident in the organic nature of the settlement pattern. New family units are constantly being formed, maturing and breaking up. During the dry season each year, new homes are built; and during the rains, the unused parts of decaying compounds are

14.7 *Street scene, Zaria, showing modern geometrical patterning*

reduced, first to rubble, then to simple mounds of laterite. This process of growth and decay is assisted by the impermanent nature of the building materials. Once vacated, a building soon disappears, either naturally or by being demolished; its materials are then reused for building on another site.

A typical house plan follows the traditional African pattern with rooms arranged within or around a courtyard. The compound within a wall is also an important feature of traditional African housing and contains the main economic unit of the extended family which works the same fields, shares the same grain store and eats from the same pot. The extended family is broken down within the compound into marital family units or *sassa*, each of which occupies a separate part of the house delineated by low walls of mud or matting made from guinea corn stalks.[19] Figure **14.9** shows the plan of a typical house in Zaria which provides privacy for the womenfolk. Entrance to the house (*gida*) is through the entrance hut (*zaure*). Circulation from the entrance hut to the family part of the house (*cikin gida*) is through one or more courtyards. In these courtyards are huts for unmarried youths and male guests, and around them are further huts which are screened entrances to the various quarters of the individual marital groups. Within each *sassa* each wife has one or two huts which she decorates with her dowry and other belongings and where she sleeps with her children. In addition, there may also be a hut for the husband and huts for his other dependent relatives.[20]

Changes in the family structure result in either the subdivision of the compound between the new compound heads, that is, the male inheritors of the estate, or the group may break up completely, and one or more sections build new cells on a new site. If the compound is subdivided, the divisions between *sassa* become harder, and each new compound has its own perimeter wall and external gateway. Field investigations indicate that the man who sets up a new compound on a virgin site builds first his perimeter wall and then an entrance hut and huts for his wife and himself. As his family grows, he adds huts where and when they become necessary; later still, walls are built to subdivide the compound into *sassa*. When the family shrinks, through death or the loss of a breakaway *sassa*, the land soon returns to agricultural use. Like the settlement, the house is not just an accumulation of accommodation which is static during a man's lifetime; it is constantly adapted to suit his changing needs with the unit of growth and decay being the hut.[21]

In traditional Hausa society, land and the buildings erected upon it have no market value, the ownership of land being usufructuary only.[22] The fabric of the house acts mainly as a shelter and is not regarded as a long-term investment. Until recently Hausa builders were not preoccupied with the durability of building materials. However, even in the rural areas of Hausaland and over large areas of the Emirate cities, there is evidence of experimentation with new forms of material and with new systems of construction. For example, concrete block walls are used with mud walls, and corrugated iron roofs are used with either wall system. Mono-pitched iron roofs are quite common in Kano with gable walls taken above the roof line. The traditional outlook, however, of the Hausa builder has been a concern to prolong the life of the building to suit the needs only of the current occupants.

The great ingenuity of the Hausa builders has produced many and varied techniques of weatherproofing for the critical surfaces of the building. Figure **14.10** shows Hausa men at work repairing a well-weathered structure. The result of the system of construction which has been developed is a perfect match between the changing needs during the life-cycle of the family and the organic nature of the houseform. Experiments,

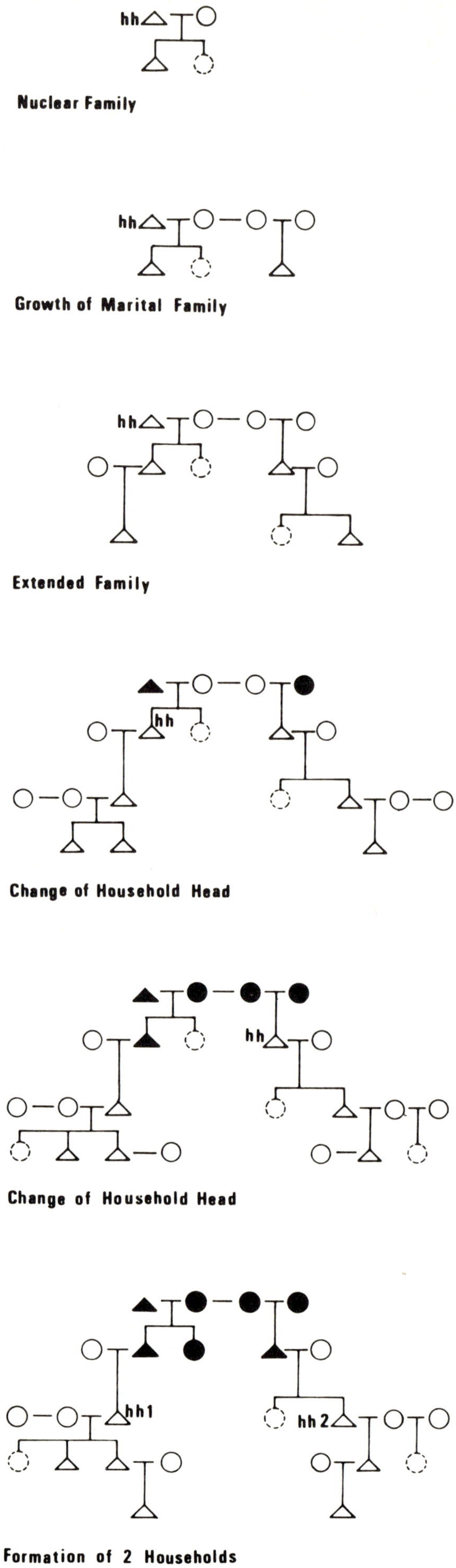

14.8 *Chart indicating Hausa coresidential kin groups*

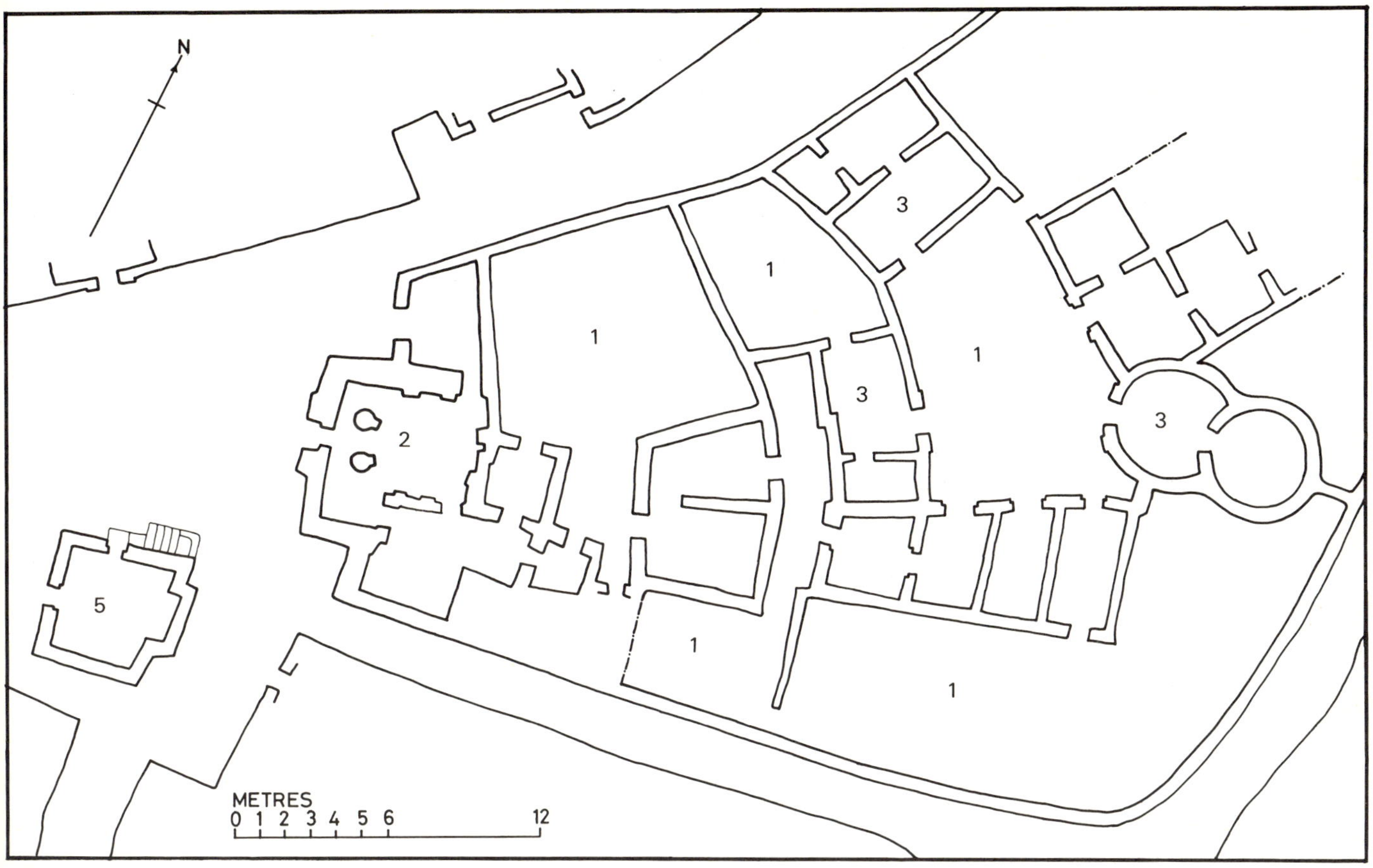

14.9 *Layout of house of Babban Gwani, Zaria: 1 courtyard; 2 entrance hut* (zaure), *3 wives' huts; 5 mosque*

SOURCE: J. C. MOUGHTIN, 'THE FRIDAY MOSQUE, ZARIA CITY', *Savanna*, 1, no. 2 (1972), 156

14.10 *Repair of well-weathered perimeter wall of large family home, Zaria*

SOURCE: MOUGHTIN, 'TRADITIONAL SETTLEMENTS OF THE HAUSA PEOPLE'

therefore, with methods of extending the life of buildings either with new construction methods or new additives to the existing materials are not entirely appropriate unless confined to such areas of the home as the *zaure* (entrance hut), which often is expected to outlast the occupants' life-span.

14.11 Tubali *construction techniques in Hausaland: A four layers left to dry after construction (foreground); a completed wall with mud plaster on the outside (background); B seven or eight layers of* tubali *laid to reach the level of a door lintel*

SOURCE: ILLUSTRATION A, MOUGHTIN, *'The Traditional Settlements of the Hausa People'*, OPP. p. 29

CLIMATIC INFLUENCES ON HOUSE STRUCTURE AND LAYOUT

As indicated earlier it is difficult to design for comfort in both types of climate in Hausaland when traditional building techniques are used. To find a solution to this problem in the form of a single building type is expensive and can be achieved only by using a modern structure with a mechanical support system. The Hausa, therefore, have adopted a compromise using different structural types for different times of the day or year.

The house normally consists of structures with different properties: rooms built entirely from mud; simple structures consisting of a light frame and a large grass roof with shade trees and large walls providing additional shaded areas. The all-mud building (*soro*) has a heavy mud roof and thick walls

14.12 *Main hall, Friday Mosque, Zaria*
SOURCE: MOUGHTIN, 'THE FRIDAY MOSQUE, ZARIA CITY'

14.13 *Cross sections, Friday Mosque, Zaria*
SOURCE: MOUGHTIN, 'THE FRIDAY MOSQUE, ZARIA CITY'

with few openings which give it good properties of thermal insulation. Within a building of this type, it is possible to trap the cool night air and prevent the interior from being warmed by the sun or external air; conversely, during cold nights the warm air is retained inside the building. In this way, the extremes of external temperature associated with the dry season are reduced, and a more even internal climate is achieved. In contrast, the areas beneath the shade trees and under the grass-roofed open structures are ideal in the humid rainy season when air movement through the building is important to keep skin temperatures cool.

In discussing the adaptation of built form to climatic conditions in the case of the Hausa, it is important to consider the complete range of buildings and spaces to be found in the family compound. It becomes clear that the Hausa system of house planning provides comfortable living conditions at all times of the year. Climate control cannot be as effectively achieved with cheap modern materials, particularly on a tight urban site. Therefore, there should be no departing from the traditional system without good reason.

CONSTRUCTION TECHNIQUES

The construction process for a mud building is a long one, and careful preparation of the materials is important. The mud walls are made up of regular courses of unbaked bricks (*tubali*) laid in rough courses and set in mud mortar. The *tubali* are

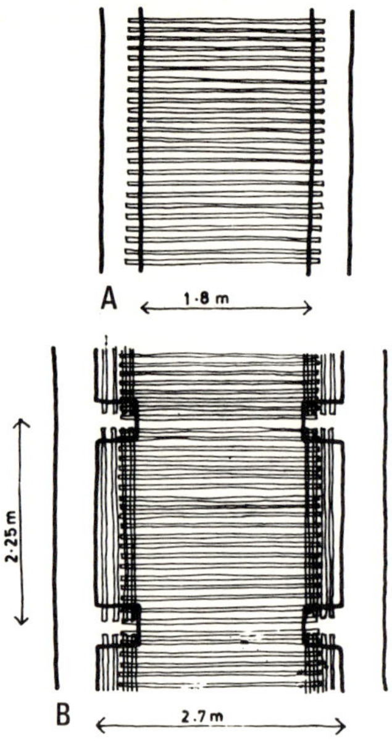

14.14 *Mud roof construction: A basic span 1.8 m wide between supporting mud walls; B increased span allowing room width of 2.7 m by means of corbelling and a beam*
SOURCE: MOUGHTIN, 'THE FRIDAY MOSQUE, ZARIA CITY', 146

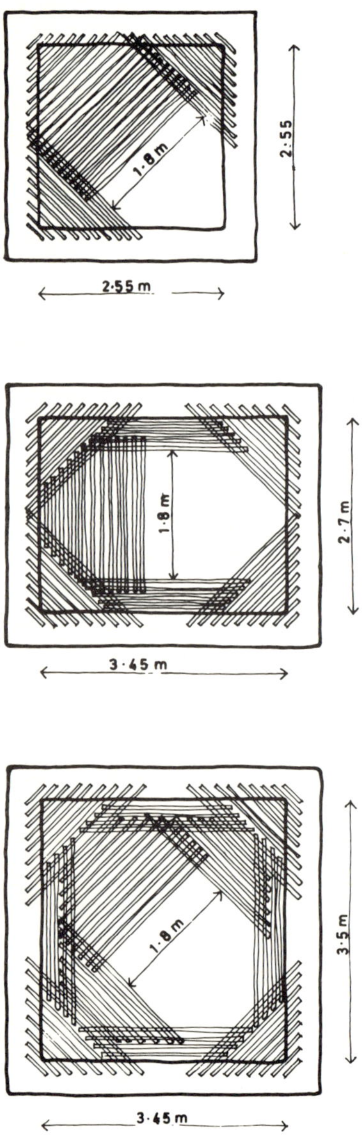

14.15 *Method of constructing mud roof for room 3.45 × 3.45 m*
SOURCE: MOUGHTIN, 'THE FRIDAY MOSQUE, ZARIA CITY', 147

coneshaped, and in Zaria they are usually about 15.2 cm in diameter.[23] It is considered good practice to lay only two or three courses of mud bricks a day; when the wall reaches the height of the door lintel, the work is suspended for 24 hours so that the walls are thoroughly dry before finishing the top courses (see **14.11**).

The normal method of building is for the builder to sit astride the top of the unfinished wall. Mud bricks and pats of mortar are thrown up to him; as he finishes the day's courses which are within his reach, he moves away backwards over that part of the wall which was built the previous day. Scaffolding and ladders are then needed only for the construction of the larger buildings, such as the Friday Mosque in Zaria (see **14.12** and **14.13**). Heavy roofs require the construction of thick walls; for example, those of the Friday Mosque in Zaria measure 1.2 m at the base. All such walls are further strengthened with *azara* laid transversely across the width of the wall, on top of which are placed layers running longitudinally, at about 0.6–1 m above the ground and again just above the height of the door. Piers and columns are also reinforced by using groups of bound *azara* bonded together with mud mortar and surrounded by a thick coating of mud.

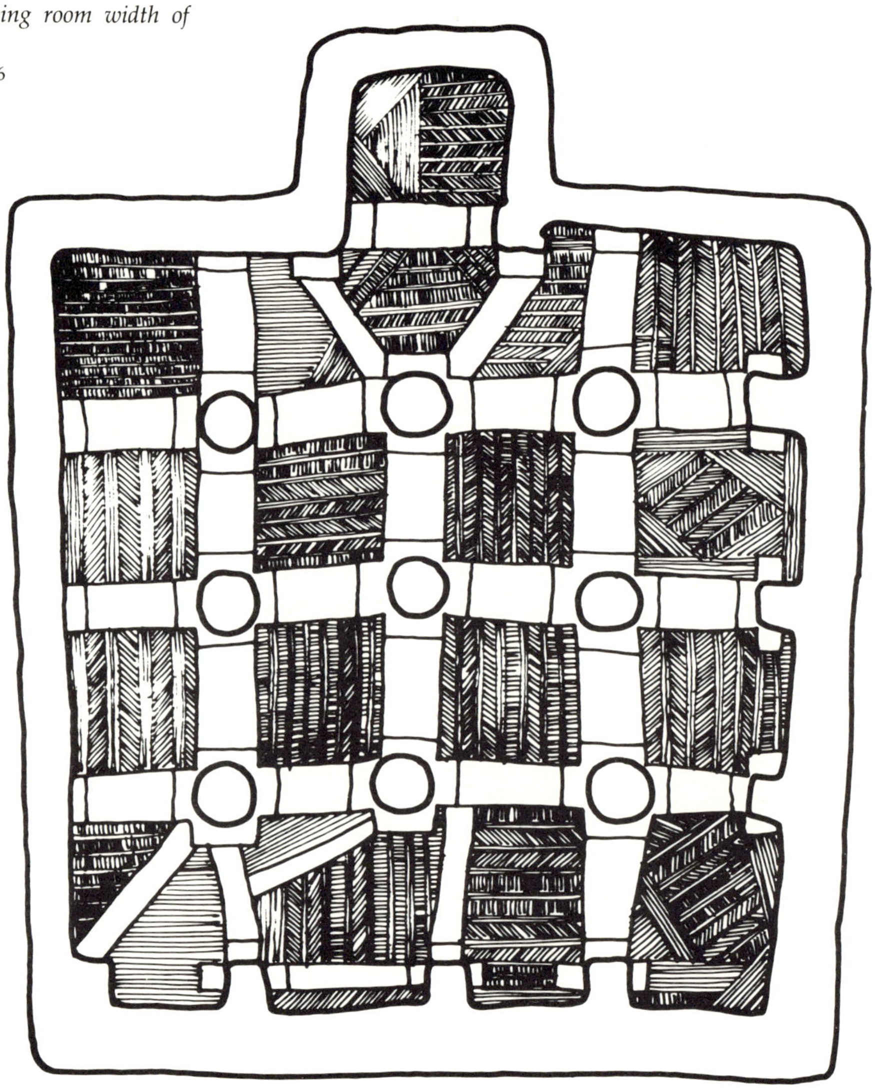

14.16 *Mud roof plan, Friday Mosque, Kazaure, using columns and beams in addition to the corbelling and diagonal* azara *used for smaller rooms*
SOURCE: MOUGHTIN, 'THE FRIDAY MOSQUE, ZARIA CITY', 148

The most basic type of mud roof is formed simply by spanning a space 1.8 m wide between supporting mud walls (see **14.14A**). Spaces which are larger than this economic span for *azara* can be made by forming a series of mud corbels at the tops of the walls. These mud corbels are reinforced with several layers of *azara* and project 45 cm from the face of the wall. They are usually at about 2.25 cm on centre, and the space between them is spanned by a beam made up of several layers of *azara*. From this beam, *azara* joists span across to the other wall in the usual way. Using this construction system, the room width can be increased to 2.7 m (see **14.14B**). To increase the room size to about 3.45 × 3.45 m, *azara* must be placed diagonally across the corners of the room (see **14.15**). For the construction of very large rooms, the roof may be supported on a series of columns connected by beams. The mud column is usually surmounted by a simple capital, consisting of two or four *azara* corbels which are used in order to increase the spacing between the columns to about 2.4 or 2.7 m. The spaces between columns are spanned by mud beams reinforced in the usual way with *azara*; the roof joists then span between the beams. Figure **14.16** is a drawing of the roof plan of the Mosque at Kazaure which illustrates the roof form of a building with a simple trabeated construction and with some use of corbelling techniques.

Spanning spaces larger than 3.5 × 3.5 m (as, for example, the Friday Mosque depicted in **14.12**) requires the use of the mud arch; and with this structural system, it is possible to construct rooms 8.0 × 8.0 m. The mud arch is essentially a series of reinforced mud corbels one on top of the other. Figure **14.17** is a section through a mud arch showing a typical arrangement of reinforcement. In better class construction, the layers of *azara* reinforcement should not project more than about 70 cm, nor should the change in angle between succeeding *azara* be too great.

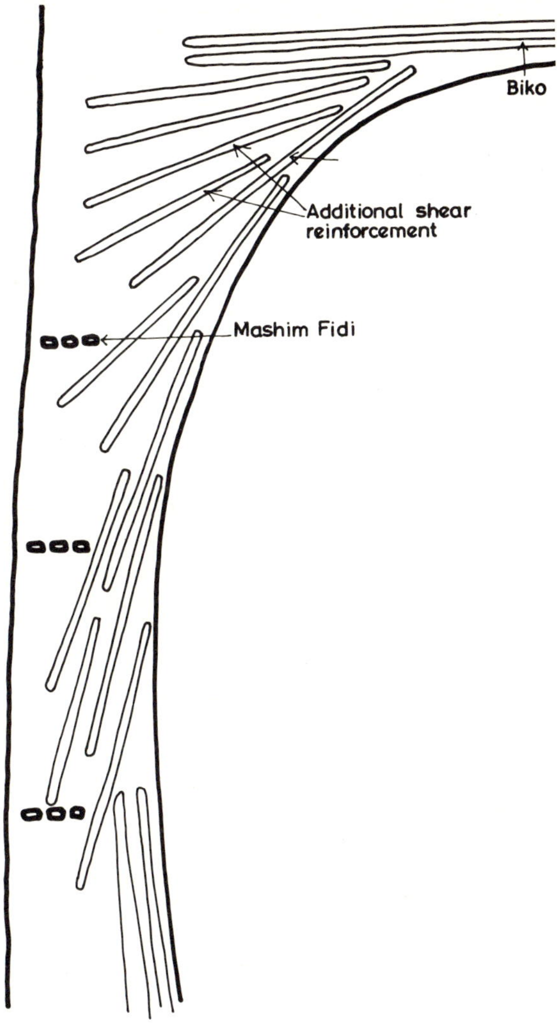

14.17 *Schematic diagram construction of mud arch*
SOURCE: J. C. MOUGHTIN, 'THE FRIDAY MOSQUE, ZARIA CITY'

14.18 *Schematic representation, mud roof construction for room without internal piers and more than 4.5 m on its shortest side*
SOURCE: MOUGHTIN, *'The Friday Mosque, Zaria City'*, 149

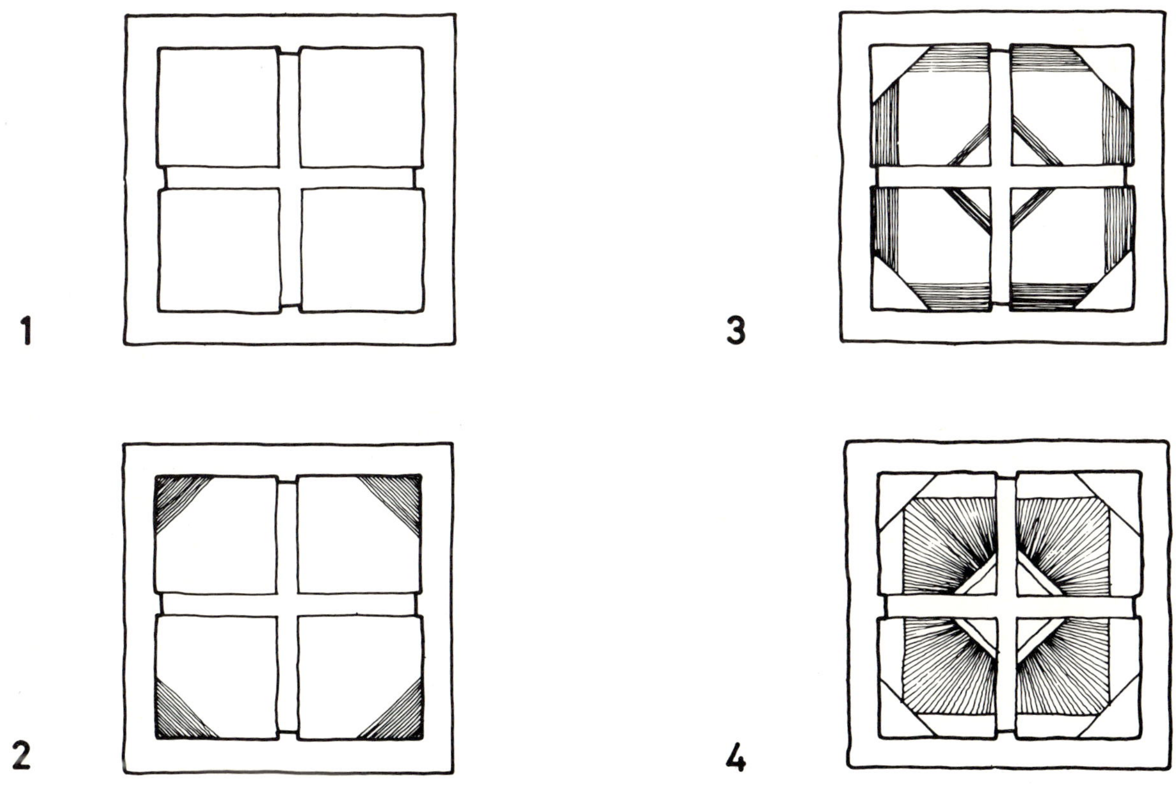

For these reasons the mud arches should start quite low down, and they are normally constructed in the following manner. Each layer of *azara* reinforcement (*kafi*) is tied back to the preceding one; when the gap between the two halves of the arch is small enough, horizontal *azara* (*biko*) are used to complete the arch. Each corbel is allowed to dry overnight before the next one is constructed, and in this way the arch can be built without the use of centring or scaffolding. Additional lengths of *azara* are used for reinforcement and are placed at right angles to the wall and project into the body of the arch. Lengths of *azara* (*mashim fidi*) are also placed in the wall at the back of the arch and at right angles to it so that cracking is prevented by distributing any thrusts through a large area of the wall.

In a rectangular room where the shortest side is less than about 4.5 m, the room is usually divided into bays of 2.1 m; and simple arches span across the room parallel to the shortest side. The area between arches, which is usually about 1.8 m, is covered in the usual way with *azara* rafters. When the shortest side of a room without internal piers exceeds 4.5 m, it is usually spanned by two or more arches built at right angles to each other. When this system is used and the arches have been built, diagonal lengths of *azara* are laid across the corners of the room; from the triangular platforms so formed, *azara* beams are carried over the backs of the arches. The arches and these beams form a rigid framework on which the *azara* rafters run from the apex of the arch over to the wall beams, presumably in order to throw most of the load directly on to the wall and reduce the loading on the arches (see **14.18**).[24]

Mud is an extremely unstable building material when used in a climate such as that of northern Nigeria where rainfall is high and frequent. The Hausa, however, have developed ingenious methods of adapting mud construction to the climate. Rainwater is directed away from the roof in one of two ways, either by long rainwater spouts projecting just over 50 cm from the face of the wall, or down deeply incised vertical channels cut into the face of the wall and lined with a waterproof finish. The base of the wall is usually protected by a plinth projecting about 50 cm. The Hausa, too, have developed many types of waterproof finishes which cannot be described in detail here, but which prolong the life of the building.[25] Each finish has a recommended use and a recognized life-span. For example, *laso* which is made from *katsi* is a by-product of the dyeing trade. It is used for the most exposed surfaces, such as roofs, pinnacles (*zankwaye*), and rainwater channels; it has a life expectancy of five to six years when used for walls. The best and most expensive finish for walls is *cafe* which is reputed to last for many years without maintenance. It usually consists of black earth mixed with the solution of the pounded seeds of the *bagaruwa* tree.

Hausa building techniques are highly developed and extremely complex; it is difficult to conceive of improvements to the system without destroying its integrity. The extension or continuation of its use will require that building codes will allow its use along with other modes of construction so that it can be confined to those areas and for those problems for which it is best suited. In such circumstances, it will simply involve the management of the basic and necessary resources for the construction process, ensuring that palm plantations are renewed, that the building craft remains attractive to young men, and that proposed developments are properly related to mud borrow pits.[26]

LIMITATIONS OF THE HOUSE SYSTEM

Most technological systems have strict limitations on their use, and the Hausa system for providing shelter is no exception. For example, the Hausa system is thought to contribute to the spread of epidemics such as meningitis. For part of the dry season, the *harmattan* blows from the desert, bringing with it fine particles of suffocating dust. The only method the Hausa know to reduce the effect of the *harmattan* is to retire within sealed rooms; it is at this time that dust-borne epidemics occur. Doctors believe that these epidemics are caused by the confinement of many people in overcrowded and stuffy rooms where the chances of contact with affected cases increase. This failure of the system in terms of the control of the spread of epidemics may be due in part to the growth in population and its concentration in large urban complexes such as Kano; as such, it may represent a fundamental limitation of the system's ability to cope with the problems associated with rapid urbanization. It must be pointed out, however, that epidemics are just as prevalent in areas such as *Sabon Gari* and *Samaru* where the physical designs are quite different; therefore, eradication of the disease may require more fundamental treatment than a simple change of built form.

Yet another problem associated with scale and density of development relates to sewage disposal and water supply. It is the Hausa custom to provide both a water supply from a well and also a pit latrine within each compound, and it is not known under what conditions this aspect of the system will constitute a health hazard. Therefore, rather than concentrating on research, experimenting with climatic control durability, etc, it may be wiser to investigate the limits of traditional systems such as this one in order to determine the critical factors and to suggest methods for extending the boundaries of the system.

LOW-COST HOUSING SCHEMES: KANO

The official answer to the problems of rural-urban migration, overcrowding and poor housing conditions has been the implementation of low-cost housing schemes. Figure **14.19** illustrates such a housing project on the Zaria Road in Kano which was designed and constructed by the Metropolitan Kano Planning and Development Board. While it represents a brave attempt to solve the housing problem by using a design based loosely on traditional house forms, it fails in many respects. The estate is laid out in a grid pattern with road access to all dwellings and with large open spaces between the blocks. There are one- and two-storey patio houses and a few three-storey blocks of apartments. The buildings are solidly built with all the necessary basic services including kitchen, bathroom, and toilet.

Table 14.1 Housing costs in the Kundila Housing Estate on the Zaria Road, Kano (1977)

2-room type	N 6082.80*
3-room type	7658.30
4-room type	10918.90
dwelling in block of flats	11987.50

*N, Nigerian Niara; 1.18 N = £1 sterling
Source: The Metropolitan Kano Planning and Development Board.

Table **14.1** shows the costs of the dwelling units in this estate; they are in excess of costs for dwellings in a similar estate in Ibadan which Onibokun analyzed and found to be well beyond the means of the population most in need of such housing:

> Again, the housing experiment, *per se*, could not be regarded yet as salvation for really low income groups in terms of their housing needs. Compared with past building costs, the prices of the experimental houses are very low, but in relation to the average income of a substantial proportion of the people in Nigeria, the prices are still too high. Many low income people cannot afford even the

low cost homes at the prices indicated. As succinctly put by a messenger, 'the buildings are very good, but can a person like me, a messenger, earning N 900 per annum build or buy such?'[27]

In terms of the detailed design of the project, the space about buildings is excessive when compared with the provisions of private spaces within each house. In the traditional Hausa compound, large areas are devoted to private space, while external space outside the compound wall is limited to useful areas such as agricultural land, borrow pits, or small praying areas. This project sharply contrasts with traditional built forms, providing only a 17 × 24ft courtyard for each house while lavishly providing for the motor car and public garden space.

With so much land devoted to external open space, it is difficult to understand the rationale for introducing such an alien form as the three-storey block of apartments into the project. This is particularly strange since the cost of each unit is far more than for the one- or two-storey houses. It must be stressed, too, that the form of the block of apartments is such that many rooms are badly oriented from a climatic viewpoint with unprotected windows facing the morning and evening sun and with no orientation to prevailing winds.

Unlike the scheme in Ibadan analyzed by Onibokun, the Zaria road project in Kano has not been the subject of an experiment using local building materials. In the case of the three-storey apartments designed with a reinforced concrete frame, the design is a positive waste of imported resources. Onibokun calls for additional experiments with local building materials in order to reduce costs; he also suggests the implementation of government loans and the organization of finance to enable people with low incomes to finance and build their own homes.[28] These are necessary steps which must be taken if low-cost housing is to become a realistic proposition, but low-cost housing alone will not solve the housing problem.

CONCLUSION

There is no single solution to Nigeria's housing problem—no panacea for instant success. The solution, if indeed there is a solution, is the following of an integrated series of policies which treat housing as one aspect of regional planning and

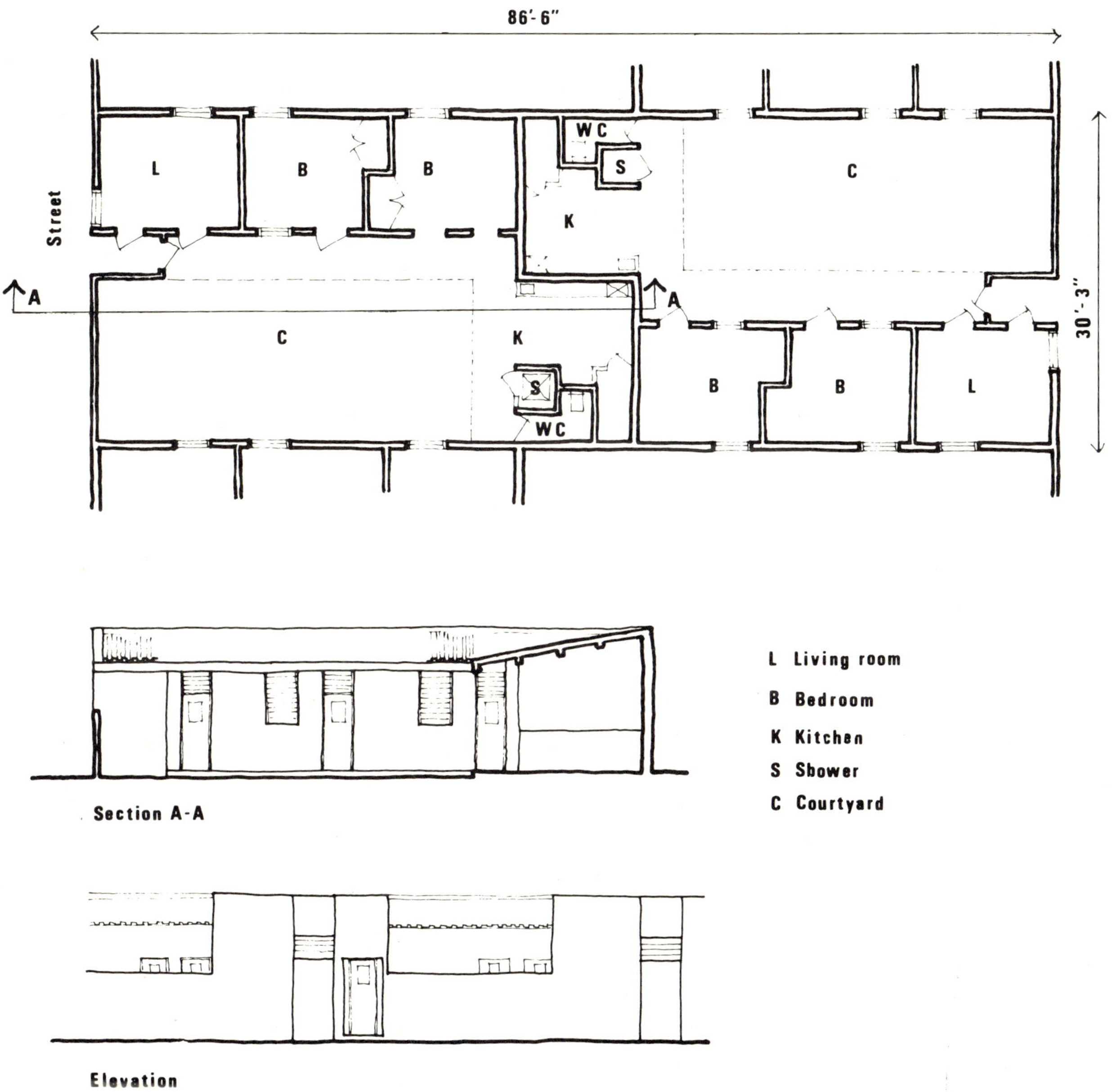

14.19 *Typical low-cost house built on Zaria road, Kano*

BASED ON DRAWINGS PREPARED BY THE METROPOLITAN KANO DEVELOPMENT AND PLANNING BOARD which take a radical view of the building industry as a continuum of possible developmental inputs ranging from the traditional informal sector through a series of sectors variously influenced by modern construction techniques or financial arrangements to the completely formal sector. Each element of the building industry which can be distinguished should then be assigned its appropriate role in the developmental process.

Housing itself, of course, is just one element in a vast web of interconnected problems which can be discussed effectively only as part of a general view of the problem of urbanization facing Nigeria. Inadequate urban services and the formation of a large and growing class of urban poor are among the results of rapid, uncontrolled urban growth with inadequate levels of industrialization. The larger cities in Nigeria thus exhibit many of the evils associated with cities in the West and few of the advantages of the Western metropolises. The 'urban solution', that is, urban growth based upon industrialization, has been the policy for regional and national development in Nigeria as in other developing countries in past years. In the mad rush for urban growth, traditional cultures such as that of the Hausa people are in danger of destruction before truly urban life patterns are established.

A natural consequence of urbanization is the development of a building industry. This is an important employment outlet, but the growth of the building industry often results in the neglect of local building crafts and leads to their ultimate decline. The standard of traditional Hausa building, for example, even in its strongest centre, Zaria, has declined noticeably in the past 15 years. Western building codes and standards of construction and the durability and thermal properties of materials are considered to be important, and building units are designed to fulfill preconceived occupancy levels. As we have seen, the achievement of these standards is an expensive business; when the costs are compared with the ability to pay, standards are inevitably lowered. When the full dimension of the problems is perceived, governments react by providing site and service schemes. The system results in individuals fending for themselves in the urban market in order to procure inferior materials from which they build their homes. This then, is the new realism in housing.

The problem of shelter in a developing country can take on an added dimension provided attitudes to urbanization itself are changed. The self-fulfilling prophecy of urban growth must be challenged and balanced by a goal for achieving rural development. In Nigeria an important aim, therefore, should be to retain the population in rural areas wherever possible. This, of course, would mean urbanization of rural areas by developing the rural infrastructure to an extent that the need and desire to move to overcrowded urban areas would be checked. Having established such a policy, the method of building shelter in a Third World country takes on a new meaning; and once more, traditional craft has an important role to play. The expansion of indigenous settlement patterns in ways which conform to the tradition of a particular community gives a new perspective to the whole set of problems associated with housing. Western standards are no longer the goal by which success is measured; instead, the existing system is examined in detail, and the points which can, and should, be improved are noted.

It is safe to assume that the growth of urban areas will continue despite any effective rural development programme. In order to utilize Hausa building techniques and to provide for Hausa life-styles in an urban situation, the first requirement is the designation of sufficient land for housing purposes. Most of this land should be allocated as private space within the control of individual householders; any additional space available on the periphery of all new residential areas should be reserved for the development of urban farms. It is also important to maintain the traditional attitude toward user rights to land in order to keep housing costs within the reach of all classes of people.

The traditional system of water supply and sewage disposal is largely based upon pit latrines and wells, often sited within short distances of each other. This dual system of water supply and sewage disposal is probably an important factor in the cause of many diseases in northern Nigeria and should be the subject of a concentrated research and development project. For the Muslim Hausa practising wife seclusion, the provision of communal standpipes for water would be totally inadequate and would probably lead to the digging of private wells in the compound. A solution to this problem, therefore, may be the provision of a piped water supply to each compound. If water supply to individual compounds is economically feasible, and if urban farms are a recognized part of each residential layout, then the pit latrine could continue its traditional function of providing an essential source of manure for agricultural purposes.

The single feature which has had the greatest effect on Hausa settlements in the most recent past has been the increasing use of the motorcar. The effect of the car in the future will probably continue to change Hausa settlement form; any plan which does not take the automobile into account is hardly likely to provide for the basic needs of the future. It seems unlikely that there will be funds available to provide surfaced roads to each small group of houses in every urban area in Hausaland; a sensible alternative would be the planning of large areas (by Western standards) of traffic-free housing where the traditional mode of movement would prevail: foot, bicycle, or mule. If such a system is found practical, the resulting pattern would be a series of wholly pedestrianized housing areas separated from each other by large allotments and strung like beads along the sides of a few well-made roads.[29]

For this proposed system of housing for the Hausa to be successful, the role of the planners and administrators must be kept in low profile to permit an ancient culture to adapt itself through the experimental response of its citizens to the problems of providing shelter for their families.

NOTES

1 Charles Abrams, *Housing in the Modern World: Man's Struggle for Shelter in an Urbanizing World* (London: Faber, 1966), p. 1.

2 United Nations Economic Commission for Africa, *Economic Housing in Africa* (New York: United Nations, 1976).

3 Federal Republic of Nigeria, *Third National Development Plan, 1975–80* (Lagos: The Central Planning Office, Federal Ministry of Economic Development, 1975), I, 307.

4 G. A. Onibokun, 'Housing Finance in Nigeria, a Critical Survey of Private and Public Sources', *Town Planning Review*, 42, no. 3 (July 1971), 290.

5 Onibokun and Adenyi argue for a more comprehensive approach to the problems of housing in Nigeria and suggest that Nigerian resources would be sufficient if effectively used. See G. A. Onibokun and Oloruntobi Adenyi, 'Housing and Inflation', unpublished paper, Nigerian Institute of Social and Economic Research, Ibadan, 1975.

6 A better translation of *Kasar Hausa* is 'the land where Hausa is spoken'.

7 K. M. Buchanan and D. Pugh, *Land and People in Nigeria* (London: University of London Press, 1955), p. 2

8 R. J. H. Church, *West Africa*, 6th ed. (London: Longmans, 1957), pp. 448–449.

9 R. K. Udo, *Geographical Regions of Nigeria* (London: Heinemann, 1970), p. 2.

10 R. W. J. Keay, *An Outline of Nigerian Vegetation* (Lagos: Federal Government Printer, 1959), pp. 22–25.

11 Amos Rapoport, *House Form and Culture* (London: Prentice Hall International, Inc., 1969), p. 47.

12 J. D. Fage, *An Introduction to the History of West Africa* (Cambridge: Cambridge University Press, 1962), p. 9. Also see S. J. Hogben and A. H. M. Kirk-Greene, *The Emirates of Northern Nigeria* (London: Oxford University Press, 1966), pp. 30, 31.

13 Abdullah Smith, 'The Early States of the Central Sudan', in *History of West Africa*, J. F. A. Ajayi and Michael Crowder ed. (London: Longman, 1978), I, pp. 158–201.

14 E. J. Arnett, trans., 'A Hausa Chronicle', *Journal of the African Society*, 9, no. 34 (1910), 161–67.

15 J. C. Moughtin, 'The Traditional Settlements of the Hausa People', *The Town Planning Review*, 35, no. 1, (April 1964), 21–34.

16 Hogben and Kirk-Greene, pp. 163–166.

17 Moughtin, 'The Traditional Settlements of the Hausa People', p. 23.

18 M. G. Smith, introduction to Mary F. Smith, *Baba of Kano: a Woman of the Moslem Hausa* (London: Faber and Faber Ltd, 1954), pp. 21–22.

19 J. C. Moughtin, 'New Homes for Old Societies', in *Proceedings of the Town and Country Planning Summer School*, University of Manchester, 1968, pp. 74–75.

20 Moughtin, 'The Traditional Settlements of the Hausa People', p. 26.

21 Moughtin, 'New Homes for old Societies', pp. 74–75.

22 In some districts of the older cities, houses are bought and sold; but in general, this procedure does not conform to the traditional system of land tenure.

23 A. F. Daldy, *Temporary Buildings in Northern Nigeria*, technical paper no. 10, (Lagos: Public Works Department, 1945).

24 J. C. Moughtin, 'The Friday Mosque, Zaria City', *Savanna*, 1, no. 2 (1972), 143–163.

25 F. W. Taylor and A. G. G. Webb, trans., *Labarun Al'Adun Hausawa Da Zantatukansu* [Accounts and conversations describing certain customs of the Hausas] (London: Oxford University Press, 1932), pp. 168–191.

26 J. C. Moughtin, 'Building Materials in the Third World: a Case Study in Northern Nigeria', in *Proceedings of the Town and Country Planning Summer School*, University of Nottingham, 1976, p. 75.

27 Adepoju Onibokun, 'Low Cost Housing: An Appraisal of an Experiment in Nigeria', *Journal of Administration Overseas*, 16, no. 2 (April 1977), 118–119.

28 Ibid, p. 119.

29 This suggestion was first made in 1968. See Moughtin, 'New Homes for Old Societies', p. 75.

Index